Axonal Conduction Time and Human Cerebral Laterality

Axonal Conduction Time and Human Cerebral Laterality

A Psychobiological Theory

by

Robert Miller

Department of Anatomy and Structural Biology
University of Otago Medical School
Dunedin
New Zealand

harwood academic publishers

Australia • Canada • China • France • Germany • India • Japan • Luxembourg
Malaysia • The Netherlands • Russia • Singapore • Switzerland • Thailand
United Kingdom

Emmaplein 5
1075 AW Amsterdam
The Netherlands

British Library Cataloguing in Publication Data
Miller, Robert
Axonal conduction time and human cerebral laterality: a psychobiological theory
1. Axonal transport 2. Cerebral dominance 3. Laterality
I. Title
612.8'25

ISBN 3718658658

CONTENTS

ACKNOWLEDGEMENTS

The most important influence which led to the production of this book was a set of experiments I did in the early 1970s. I was fortunate to be able to work as a post-doctoral researcher in the laboratory of George Gordon, of the University Laboratory of Physiology, at Oxford. I would like to thank him especially for the opportunity he provided me with at this time. During that two-year period I recorded antidromic responses in cortical neurones of the cat, projecting to various other brain structures. I saw with my own eyes that a proportion of cortico-cortical connections have latencies approaching 50 milliseconds, an interval which encroaches on the time frame for many psychological experiments, especially those on perception. Twenty years later there are still only a handful of people who have personally made this observation. Over the years since then my work has shifted compeletely from experimental to theoretical neuroscience, and I have been able to use this experience in a number of aspects of the theory of forebrain function. Another important influence came from my contacts with the Max Planck Institute for Biological Cybernetics, under the leadership of Valentino Braitenberg. He also has for many years been advocating the idea that axonal conduction delays have a cybernetic role, particularly in the cerebellum. I now cannot remember whether my own interest in this topic came from reading his work, or arose independently. In any case, I thank him very much for continued support and intellectual stimulation. In 1987 I published a short paper in *Psychobiology* which was a forerunner of the present work. From this developed the immediate event which impelled me to start work on the present book: About five years ago my short paper became the focus for a discussion I had with Friedemann Pulvermueller about the biological basis of cerebral lateralization. This discussion, mainly by electronic mail, ended in confusion, because the concepts I tried to communicate were far more complex than I realized at the start of our exchange of ideas. Thus, the present work is really my attempt to unravel the confusion created in that discussion. I thank Friedemann for this provocation and hope he gets a better idea from this book of what I was trying to say five years ago. Apart from these major influences, I would like to mention three books, by authors I have not met, and some other sources of inspiration, which have provided some of the additional premises I have used in the present work. In E.R. John's book, *Mechanisms of Memory* I was introduced for the first time to Hebb's idea of rule-dependent modification of synapses, and to some of the earliest experimental work on this concept. The Tubingen neurotheoreticians, Gunter Palm as well as Valentino Braitenberg, introduced me to a closely related concept, Hebb's idea of cell assemblies. In A.L. Blumenthal's book *The Process of Cognition* I learnt of the importance of temporal structure in psychological processes. Moshe Abeles' book *Local Cortical Circuits* provided an important insight into the temporal structure of neural assemblies, which has been developed extensively here. My contacts in the University of Otago with researchers studying motor behaviour and motor physiology, have kept in my mind the distinction between open- and closed-loop control, which becomes important in the later parts of this book. I particularly want to thank Greg Anson in this context, who directed my attention to papers on the temporal structure of motor programs. I also thank Rolf Koetter for drawing my attention to early morphological studies in the German literature. Apart from this, I am of course indebted to a very large number of researchers in many countries who, over many years, have provided such a wealth of biological and psychological evidence on the subject of cerebral laterality. I have met hardly any of them, but would like to thank them all. I would also like to thank Gareth Jones for making it possible for me to work on large undertakings like this. I am indebted to Robbie McPhee, for all the diagrams in this work, which, I believe, have been very successful in illustrating complex concepts in graphic form. I thank Ross Marshall for expert help in final preparation of the manuscript. I thank the staff of Otago University medical library, especially Ralph-Peter Hendriks for his unfailing efficiency in handling inter-library loans. Finally I would like to thank the Health Research Council of New Zealand. This work has been carried out entirely during the period that I have received support from the HRC as Senior Research Fellow, and would not have been possible otherwise.

R. Miller

LIST OF FIGURES

LIST OF TABLES

1 Introduction

The linguistic signals we receive through our ears, and those we emit with our own vocal apparatus are, as physical events, both complex and distinctive. However, just because they are the main vehicle for the seemingly-miraculous faculty of communication between people, does not make them metaphysically different from other physical events. The incoming messages which constitute language can still be regarded as physical stimuli comparable to other stimuli with which our sense organs deal. Similarly, the vocal gestures we emit when we speak can still be regarded as fragments of motor performance comparable to other portions of our complex stream of motor activity. In either case the physical attributes of these speech signals can be analysed scientifically, using the same measurements of physical variables as would be used for other stimuli or motor responses. Thus, the sensory mechanisms by which we perceive linguistic signals can be regarded as an example of sensory processing generally. Perhaps these mechanisms are specialised in characteristic ways compared with non-linguistic sensory processing, but they do not differ from them in a categorical way. Similarly the motor mechanisms by which we emit linguistic signals may be specialised in comparison with other aspects of motor performance, but are not of a fundamentally different nature from them, just because they are linguistic. Assertions to the contrary might be seen as undermining the possibility of a truly scientific study of the brain mechanisms of language.

A central biological fact about language, based on a massive amount of evidence, is that in 90% of people, the representation of most aspects of language is localised to the left hemisphere. One can then ask the following questions: What are the physical characteristics which distinguish speech sounds from other sensory stimuli? What are the physical features which distinguish the linguistic behaviour we emit from other behaviour? If one can answer these questions, the next, and more difficult question is: In what way(s) does the left hemisphere differ from the right, so that it becomes specialised for the sensory analysis of speech, and for the motor production of speech?

In this book the answer given to the first two questions is that for both the sensory analysis and the production of speech, complex temporal patterns must be represented, with a very detailed and precisely-measured temporal structure. Thus the basic elements from which spoken language is built may have a duration of the order of 100 msec or less. All longer strings of spoken language can therefore be analysed as a temporal chain of events, whose essential characteristics cannot be grasped unless temporal resolution at this fine time scale is possible.

The *central hypothesis* of the book is that the physical basis of the left hemisphere specialisation lies in the properties of the axons connecting together different regions of the cortex in that hemisphere. Specifically, it is suggested that in the left hemisphere there is a greater proportion of the fine calibre, slowly conducting axons than in the right. This means that, as far as direct monosynaptic cortico-cortical connections are concerned, there can be a greater temporal dispersion of any signal in the left than in the right hemisphere. This in turn, it is argued, gives the left hemisphere a better capacity than the right for linking the representation of events which are separated in time (especially for short intervals of the order of 100 msec). This then allows the left hemisphere to represent short temporal patterns in greater detail and with greater accuracy than can the right hemisphere.

Two important *caveats* to this thesis should immediately be pointed out. First, this central hypothesis is not about left hemisphere specialisation for *language* as such. As will be seen in later chapters, the superiority of the left hemisphere for representing temporal patterns in sensory input, and in planning temporally structured output, is not by any means limited to language, though this may one of be the most obvious functions for which the left hemisphere is specialised. Secondly, the left hemisphere should in no way be regarded as the "dominant hemisphere", whether for language or for other lateralized functions. The right hemisphere also has its own distinctive manner of specialisation. According to the central hypothesis this depends on there being a richer repertoire of cortico-cortical connections, with rapid (and, effectively, instantaneous) transmission in the right than in the left hemisphere. This leads the right hemisphere to have a superiority over the left in representing patterns which are complex "instantaneous" conjunctions of signals, rather than complex temporal patterns. The obvious example of such "instantaneous" patterns are visual spatial patterns, and the term "spatial" will often suffice as a general label for the functions for which the right hemisphere has an advantage. However, we should not limit the description of right hemisphere superiority just to the "spatial" category, as used in a strict and literal sense. Moreover, in addition to such "instantaneous" patterns, the right hemisphere comes to have a superiority in slowly fluctuating temporal sequences which lack the fine temporal structure of the words of spoken language. These slower temporal patterns, dealt with better by the right hemisphere, may play a significant part in language as a whole, for instance in the prosody which accompanies spoken sentences. Thus it is incorrect to designate language as the unitary function for which the left hemisphere is dominant. Expressed in terms of basic physical variables language is not a single function, and so it need not be lateralized monolithically to the left hemisphere.

These, then, are the main themes of this book. The starting point for the argument is a discussion of the physical nature of linguistic signals, emphasising their temporal structure (Chapter 2). Following this, in Chapter 3, the manner in which such temporal structure might be represented in the activity of neural tissue is discussed. It is concluded that axonal conduction delay is the physiological property of nervous tissue most likely to be involved in representation of precise temporal structure on the time scale of elementary speech sounds. From this, it is an easy step to the central hypothesis that the axonal interconnections within the hemisphere which is dominant for many (but not all) aspects of language have quantitatively different delay properties from those of the other hemisphere.

Having set up this hypothesis, predictions can be made to test it, in many areas of empirical research. These predictions can be divided broadly into "biological" and "psychological" ones. The "biological" predictions are those about morphology and physiology of nervous tissue in the two hemisphere. Chapter 4 discusses these matters. The most obvious implications of the central hypothesis are morphological, but in fact empirical study of these in a direct way is scarcely possible, because of a combination of ethical and technical problems. However, there are a number of indirect morphological implications, which are discussed in detail. A little relevant data is also available from electrophysiological comparisons of the two hemispheres, also dealt with in this chapter.

"Psychological" predictions are those made about the behaviour of the whole organism, that is the human experimental subject, rather than about tissue in different parts of his or her brain. Discussion of this evidence occupies the largest part of this book. The central hypothesis about axonal conduction delays arises most obviously from considering the *sensory* processing of language. Therefore lateralization of perceptual functions is considered first. In Chapter 5, a series of predictions is made in the field of perceptual psychophysics, and the evidence on each of these predictions is discussed in Chapter 6. As a development from these perceptual differences between the hemispheres, the content of memory also differs between hemispheres. This subject is dealt with in Chapter 7, at least in so far as memory can be probed with words as stimuli.

Cerebral lateralization certainly applies to *motor* functions, such as linguistic expression, as well as to perceptual ones. The central hypothesis requires considerable extension if it is to explain the lateralization of these motor functions. The theoretical arguments on this topic are developed in Chapter 8, leading again to a set of predictions. These concern lateralized motor functions in general, not just speech production. Chapter 9 surveys the empirical evidence relating to these predictions.

Chapter 10 deals with a variety of other lateralized functions, at a higher level than immediate perception or programs for immediate motor performance. Because of the complexity of these functions, this chapter does not set up explicit predictions in advance of the evidence, but is rather an attempt to discuss and interpret the published evidence in the light of the central hypothesis. The topics covered include short term memory, general alertness (arousal), selective attention, the perception of the emotional value of stimuli, and the expression of emotion.

Chapter 11 is a short one, pulling together evidence from all the previous chapters about the intercorrelation of various measures of laterality, and their correlation with gender. Chapter 12 includes a summary of the whole argument, set out as a numbered set of main points. The reader may find it valuable to read this section before starting the main part of the text. This final chapter also highlights the predictions made at various stages in the book, where adequate evidence is so far unavailable. This may serve as a guide for future experimental work.

It is necessary to state explicitly that the perspective from which this book is written is one of neurocybernetics. Thus, its primary aim is to explore the relations between a simple but global premise of neuronal interconnectivity (differing between hemispheres) and a wide range of implications it has for strategies of information processing in each hemisphere. Since this is the perspective, there are some areas of literature, relevant to a comprehensive coverage of cerebral laterality, which are scarcely touched upon. One of these is the developmental and endocrine processes which lead to differences between the hemispheres in either gross morphology, or in cytology. Another is the evolutionary trends amongst the mammals and (other groups of animals), which have been forerunners of human cerebral laterality. Finally, it needs to be stated that although human language is the starting point for consideration of laterality, it is not the main topic of this book. Therefore, detailed structure of language, especially of the high level structure (syntactic and greater) is not attempted. There may be important issues here which can be understood in the context of cerebral laterality. However, they must be dealt with in another book, and probably by another author.

2 The Physical Nature of Linguistic Signals

The most fundamental type of human language is that which is spoken and heard. Therefore, the best clues we can get to help us understand the left hemispheric specialisation for language probably come from considering this type of language, rather than language which is written and read, the sign language used by deaf people, or Braille used as written language by blind people.

One of the most basic characteristics of spoken and heard language is that it is a complex temporally-organised pattern of sounds. The time scale of speech patterns varies greatly according to which linguistic feature is under consideration. Elementary speech sounds (vowels and consonants) are temporal patterns whose components may last 50–200 msec. Single syllables or monosyllabic words are patterns with durations up to about 0.5 sec. Sentences (with their distinctive pattern of rise and fall of the voice, called prosody) are organised on a much longer time scale.

Of course, all streams of incoming sensory information or outgoing behaviour (not just those in the linguistic domain) can be defined in terms of temporal structure. For many aspects of language processing the time scale for description is no different from that for other aspects of sensory processing or motor control. For the shorter speech signals (elementary speech sounds and syllables) it may be asked whether their essential temporal structure is sharply different from that of most non-linguistic sequences of sensory stimuli or motor output. Specifically, when the cerebral cortex deals with non-linguistic sensory processing and motor control, can it deal with such complex temporal structure (on the fine time scale of the order of 1/10 sec) as it does when processing language? In answer to this suggestion, it can easily be pointed out that musical sequences, musical performance and many other distinctively human motor functions can also have such a fine temporal structure. However, these motor functions are also lateralized, which may be a reflection of some more fundamental difference between hemispheres in humans which leads to lateralization of language as well as other functions. Moreover, non-linguistic sensory stimuli or motor tasks can be arranged in an experiment to require sensory analysis or motor programming with temporal structure as fine as that of the sounds of speech, and not just in the auditory modality. Lateralization of the brain's ability to deal with such non-linguistic information is described in subsequent chapters. Nevertheless, in normal everyday circumstances it is probably spoken language which puts the most severe demands on the capacity of our brains to analyse temporal structure in sensory input, or to organise temporal structure in the motor output. Whether the representation of such fine temporal structure in sensory perception and in motor planning is a distinctively human attribute is an open question, since very little psychophysical work has been done to determine the temporal resolution of these processes in experimental animals. However, since the principle of fine temporal structure is most obvious for humans, let us consider the principle in more detail as it applies to humans, dealing in turn with the sensory stimuli which have to be analysed as speech *perception*, and for the motor mechanisms underlying speech *production*.

2.1 SPOKEN LANGUAGE AS SENSORY STIMULI

The microscopic structure of speech is a code of small segments, each of defined frequency, and of defined duration within the range 50 to 200 msec. Consonants have a finer temporal structure than vowels (Tallal and Schwartz, 1980), and a single stop consonant may be resolvable into a few (e.g. two or three) frequency components, in sequence, each lasting only a few tens of milliseconds. This fine temporal structure can be discovered in acoustic analyses, but is too fine to be consciously recognised as a sequence by the unaided ear. Klatt (1976) reviews evidence that the *duration* of acoustic segments is important for making many of the distinctions amongst speech sounds on which normal language perception depends. Amongst other details, he mentions that stressed vowels have an average duration of 130 msec (though there can be an eight-fold variation in this duration, according to context). Unstressed vowels, and consonants have an average duration of 70 msec. In consonant clusters, individual consonants are usually shortened. The sound /s/ varies between 50 and 100 msec in duration, according to context. The "just-noticeable-difference" for a single phonetic segment, estimated by various authors, lies between 10 and 25 msec (close to the psychophysical limits for discrimination of duration of stimuli in the range of the phonetic components of speech). In other words, if a speech sound is longer or shorter than usual by more than this amount, the listener will perceive it as unusual, or different from normal. Klatt lists a series of distinctions made on the basis of duration of (or time relation between) phonetic components. Cutting (1974) constructed synthetic speech sounds. In this somewhat artificial situation, the part of the spectrogram needed for subjects to identify an initial stop consonant lasted 120 msec, that for liquids ('l' and 'r') lasted up to 225 msec, while for vowels it had a duration up to 400 msec. Tuller and Kelso (1990) have produced evidence to show that identification of speech sounds does not depend on absolutely invariant temporal patterns in the sound. A particular speech sound is identified for stimuli which vary over a defined range of timing. However, when the relative timing of different components of a speech sound varies beyond the defined range, perceptual identification of the sound may shift quite abruptly to a different phoneme. Thus timing of the components of the sound can vary continuously, but at the level of perceptual identification there is sharp categorisation of the sounds.

2.2 SPOKEN LANGUAGE AS MOTOR BEHAVIOUR

In *production* of speech sounds, timing is just as important. Gracco and Abbs (1986) studied the patterns of movement of upper lip, lower lip and jaw while subjects pronounced the word “sapapple”. While some aspects of performance showed substantial variability in timing, in other aspects timing was highly consistent within a subject (and to a lesser extent between subjects). Quite small timing differences were critical. For instance, averaged over four normal subjects, the time between onsets of upper and lower lip movements was 23 msec, while that between lower lip movement and jaw movement was only 12 msec. The authors conclude that muscles are activated in such a manner that the timing of certain critical patterns of movements of the speech apparatus occur in a relatively invariant fashion. Such movement patterns appear to be “designed” to produce certain specific acoustic consequences. The relation between the patterns of movement and the acoustic consequences are quite complex and certainly not derived by any sort of one-to-one correspondence between movements and sounds.

To produce the complex movement patterns needed to articulate speech sounds, a large number of small muscles need to be activated, again with timing as of critical importance. Technical difficulties in the electromyographic recording of these patterns has meant that data on this aspect of speech production is as yet very far from complete. The data that are available suggest that timing is accurately controlled on the scale of a few tens of milliseconds. For instance Leandersen *et al.* (1970) find that the EMG activity in facial muscles associated with pronunciation of the consonant /p/ shows systematic differences in timing of 10–15 msec according which vowel is adjacent to it. Lofqvist *et al.* (1984) show that peaks in the activity of the cricothyroid muscle are timed to an accuracy of less than 50 msec in relation to voice onset. The relation between activation of different muscles involved in articulation and phonation, and the necessary movements they produce is complex, and not yet well defined. Gracco and Abbs (1986) write that “agonist muscles underlying oral closing ... do not appear to be activated simultaneously or in the same sequence as their movement onsets or velocity peaks.” Differences in inertia of the different structures for articulation and phonation mean that there can be no one-to-one correspondence between muscle activations and movements.

On the fine time scale, the timed sequence of events which produces each elementary speech sound appears to correspond to a “central program” without moment-by-moment sensory guidance. This conclusion comes from the work of Gracco and Abbs (1988) who showed that perturbation of lower lip movement during pronunciation of the word “sapapple” had relatively little effect on movement duration and sequencing of peak velocities in the three chosen articulators (although significant compensations of *magnitude* did take place). However, although there appears to be a central program governing the timing of the several motor events for production of each segment, the segment as a whole *is* subject to sensory guidance. This can be seen very clearly from the potent disruptive effect produced when auditory feedback of speech is delayed (an effect which is most potent for delays of about 180 msec.), as discussed in Chapters 8 and 9.

Just as there is no one-to-one correspondence between muscle activations and articulator movements, or between movements and speech sounds, there can also be no one-to-one correspondence between segments of neural activity in this central program and the muscle activations it produces. This statement can be made because the different effectors involved in speech production are connected to the forebrain regions which presumably generate the central program by pathways of rather different latency. In particular, the innervation of the larynx via the recurrent laryngeal nerve follows a much longer route than the innervation of tongue, lips and jaw, and this nerve actually has a significantly longer course on the left than the right side. Thus, the "commands" for the several parts of an articulatory and phonatory pattern need not be initiated in the motor speech regions of the brain in a sequence congruent with their performance by the various effectors.

To summarise, in speech production, several complex temporal patterns have to be coordinated. These are: the sequence of commands in the central program for each speech sound, the sequence of muscle activations, the sequence of movement of different effectors, and the sequence of sounds finally emitted. Clearly speech production is acquired during infancy, and, during that acquisition, many precise details of emitted speech come to correspond closely to the speech which the infant actually hears from his or her parents. For instance, our speech uses not only the same language, but also the same accent as we heard spoken by those from whom we acquired our speech. A major problem comes in conceiving how the temporally detailed central programs which underlie our speech can be acquired by listening to others speaking.

2.3 COMPARISON OF SPEECH SIGNALS WITH OTHER SENSORY STIMULI OR MOTOR OUTPUT PATTERNS IN HUMANS

It was suggested above that, in everyday life, linguistic signals may demand a more detailed temporal analysis, and speech production may demand a more precise temporal organisation, than do most other commonly-used human sensory and motor functions. However, it is incorrect to infer from this that there is a fundamental distinction between linguistic and non-linguistic information processing. Similar capacities for sensory analysis and motor production with high temporal resolution may apply to non-linguistic sensory signals or motor patterns. In fact, in this book representation of temporal structure in *non-linguistic* signals comes to be of particular importance for making comparisons between the hemispheres. This is because, in principle, cerebral processing of linguistic signals could show lateralization because of a variety of characteristics of language other than the degree of detail in their temporal structure. Studies of lateralization using linguistic material therefore cannot distinguish whether the basis of lateralization is the degree of fine temporal patterning of the signals used, as opposed to other aspects peculiar to linguistic signals. However, by using non-linguistic signals one can compare the basic performance capacities of the two hemispheres in processing fine temporal sequences, with less of a confounding influence coming from these other features of language.

3 The Representation of Temporal Structure in the Dynamics of Neural Networks

3.1 INTRODUCTION: ASSOCIATION, HEBBIAN SYNAPSES AND CELL ASSEMBLIES

Both speech perception and speech production involve the neural representation of temporal patterns which are highly structured at the scale of 50–200 msec, and also highly stable, within a defined range (given some variation according to the context in which each sound appears). Since the representation of speech sounds is acquired rather than innate, it is necessary that there exist mechanisms by which *specific associations* can form between neural representations of events which occur at different times.

This requirement resolves itself into three more specific questions:

(i) At the cellular level, what is the process of association itself?
(ii) What is the general relation between single nerve cells, the nerve networks they form, and the meaningful information which nerve tissue as a whole can represent?
(iii) Given answers to the first two of these, how can associations between non-synchronous neural signals be formed?

Dealing with just the first of these processes, neural tissue is certainly capable of *association*, and this appears to be a fundamental property of many individual nerve cells. Ultimately association relies on synaptic strengthening, according to Hebb's postulate (Hebb, 1949): Coincident activation of the pre- and postsynaptic sides of an excitatory synaptic junction are considered to be necessary for synaptic strengthening to occur. Since the post-synaptic neurone is more likely to be activated above the relevant threshold if several afferent synapses activate it together than if just one does so, synapses upon the soma and dendrites of a neurone will tend to be strengthened in groups, the same groups which tend to be activated together. This mechanism can in principle account for both simple forms of associative learning, such as Pavlovian conditioning, and more complex forms, such a pattern recognition.

In the last twenty years, Hebb's postulate has become the focus of a great deal of experimental investigation, both in the intact brain and in various brain slice preparations (see reviews by Voronin (1983); Brown *et al.* (1990); Tsumoto (1992)). Excitatory synapses with Hebbian properties have been found in several parts of the central nervous system, but especially in the cerebral cortex. In that structure the population of excitatory modifiable synapses is likely to be mainly those made between the cortico-cortical axonal arborization of one pyramidal cell and the dendrites of another. In particular synapses made upon dendritic spines are probably the most important variety of modifiable synapse (Wickens, 1988). Another aspect of these modifiable synapses is of special importance for the present work: This is the definition of the degree of synchrony of co-active afferents which will permit their co-operation to trigger the mechanism of synaptic modification. We may define as the *Hebb interval* the maximum temporal separation of the activation of two synapses, converging on a single neurone, at which they can still co-operate to produce synaptic change. Two papers have estimated this interval as 20 msec or less in the hippocampus (Levy and Steward, 1983; Gustafsson and Wigstrom, 1986), while a third, more recent paper gives an improved and somewhat lower estimate, of a little below 10 msec (Hashemzadeh-Gargari *et al.*, 1991).

Concerning the second of the above questions, it is unlikely that any single nerve cell in the cerebral cortex is the unique representation of any single item of meaning. The latter rather simple-minded doctrine, caricatured as the "grandmother cell" hypothesis, has been much debated. The consensus view of most students of the cerebral cortex is that meaning is represented as overlapping combinations of many nerve cells. Such combinations are envisaged to be more effectively connected amongst themselves than with other neighbouring neurones, which are not part of the said combinations. The combinations in this hypothetical scheme of representation are known as "cell assemblies" another of Hebb's important concepts. Originally the concept of the cell assembly was formulated in close conjunction with his rule for synaptic modification. Hebb (1949) considered that the operation of his rule for synaptic modification within a large quasi-random network of neurones would lead to the formation of such strongly-connected assemblies, a small selection from the cortical network as a whole. Given the development of this "structural" aspect of cell assemblies, their functional operation can be characterised as "ignition" of the strongly connected groups (the phrase used by Braitenberg 1977, 1978), when the neural traffic impinging on the assembly exceeds a certain critical level.

Although this scheme leaves many questions in search of an answer, the cell assembly concept is scientifically attractive in a variety of ways. Meaningful information would be *diffusely* represented by cell assemblies, in a manner compatible with the many studies of the loss of learned information after lesions of the cerebral cortex. Information would be stored more *economically* than in the "grandmother cell" scheme, because information storage would be more closely related to the number of synapses than to the number of neurones. Furthermore, information would be represented in a very *versatile* fashion, preserving all the complex interrelationships of information in the outer environment, rather than confining the representation so that all items of meaning are mutually exclusive categories. Several weaknesses of the original concept underlying cell assembly theory have been recognised and solutions have been proposed to them (Braitenberg, 1977, 1978; Palm, 1982; Miller, 1989, 1991; Miller and Wickens, 1991). Some of these will be mentioned in later chapters of this book. The whole body of cell assembly theory is the best answer we yet have to the question of how nervous tissue represents information, an essential bridge between neurocytology and psychology. Such theory is still developing.

3.2 TEMPORAL STRUCTURE IN THE ACTIVITY OF NERVOUS TISSUE

One notable extension of Hebb's concept of cell assemblies was made by Abeles and co-workers (Abeles, 1982a; Vaadia and Abeles, 1987; Villa and Abeles, 1990; reviewed by Abeles, 1982b). It is of great relevance to the theoretical side of the present work. These researchers conducted elegant electrophysiological experiments, involving simultaneous recording from several neurones, using either the same or a different electrode, in the auditory cortex of cats and monkeys. They analysed the relation between impulse firing in the different neurones using the *cross-renewal density analysis* method. This method (also known as *cross correlation*) determines the likelihood of an action potential occurring in the record from one neurone at specific times before or after an action potential occurs in the record of another neurone. The later study by Villa and Abeles (1990) sought *"favoured" temporal patterns*, in single spike trains or shared between spike trains from different neurones.

With these methods precise temporal dependencies of the firing of one neurone upon one or more others could sometimes be found. When the dependency of the firing of one neurone on that of a *single* other one was plotted, they found either no evidence for direct interaction, or only short-delay interactions. However, when the dependence of a neurone's firing was analysed as a joint function of the firing of *two* other simultaneously-recorded neurones, a few very striking long delay correlations were seen. This allowed them to make statements such as "neuron 5 fires at a high rate within 25 ms after neuron 4 fired, but only when neuron 1 fired 375 ms earlier" (Abeles, 1982b). Even more dramatic demonstrations of temporal structure were found by Villa and Abeles (1990), in their search for "favoured temporal patterns", though it must be admitted that this method is computationally very complex, and when favoured patterns are found it is difficult to express the statistical likelihood of their having occurred by chance.

The intervals in such dependencies could be as long as several hundred milliseconds. While the amount of evidence of such dependencies produced by Abeles' group is not overwhelming, other researchers using similar techniques give some support. Studies of the EEG conducted earlier this century show that there may be precise temporal patterning over many tens of milliseconds (Morrell and Jasper, 1956; Yoshii and Hockaday, 1958). Other studies using single unit cross-correlation techniques demonstrate direct connections with mostly quite short transmission delays, of 1–2 msec at most. However, there are exceptions to this. Gassanov *et al.* (1985) used single unit records made simultaneously from visual and motor cortical areas, during conditioning tasks in cats. The authors mention a few cases (and illustrate one) where there were direct connections with delays up to 60 msec. Dayhoff and Gerstein (1983) found that spike trains from single cortical neurones may show regularly repeated temporal patterns lasting over 100 msec. Steriade and Amzica (1994) found the neurones from cortex displaying spike-and-wave epileptiform activity could show correlated firing with lags of up to 50 msec. A study with similar objectives to that of Villa and Abeles (1990), but using a different computational algorithm (Frostig *et al.,* 1990) has detected recurring temporal patterns of firing in the frontal cortex of unanaethetised cats. Patterns of up to ten spikes were identified, distributed amongst spike trains of up to four neurones, with pattern durations up to 100 msec, and timed to a precision of one millisecond or less. In this study the exact probability of such patterns recurring by chance could be calculated. In many cases a relation to cardio-respiratory rhythms could be excluded, and it was also shown that the recurrences occurred in both waking state and during rapid eye-movement sleep, but much more rarely during slow-wave sleep.

From such experimental observations, Abeles (1982b) was led to the conclusion that networks of connected nerve cells in the cortex could have firing patterns which were very precisely structured in time. Abeles gave these time-structured chains of activity the name *"synfire chains.* In essence they are a version of the cell assembly concept which are defined by temporal as well as spatial structure.

3.3 AXONAL CONDUCTION DELAY AS THE MEASURE OF TIME IN TIME-STRUCTURED NEURAL ACTIVITY

An important set of questions is raised by Abeles' concept of synfire chains, which, as yet, have no agreed empirical answer: *How* can cortical neural networks come to have such precise temporal structure in their dynamic activity? What property of nerve cells or nerve networks gives rise to the temporal dependencies which have the precision actually observed in these experiments? This brings us back to the third issue raised above, about how associations between non-synchronous neural signals can be formed. This of course is of central relevance for this book, because of its close relation to the coding of temporal structure in language (such as the specific delays between different elements of a speech sound) and in other aspects of sensory, motor and cognitive information processing.

The post-synaptic co-operativity found in Hebbian synaptic modification depends on the summation of excitatory post-synaptic potentials (EPSPs). Hence one would expect that such association at the neuronal level would rely upon synchronisation of co-active afferents to an accuracy of about 10 msec, the approximate duration of an EPSP (and, as mentioned above, of the "Hebb interval"). Within the soma and dendrites of a single neurone there are no known built-in delay mechanisms by which selective linkage could occur between events occurring as much as 50 or 100 msec apart. Admittedly there are slow biophysical changes possible in some neurones (e.g. thalamic relay neurones), and neuronal membranes may have a number of different voltage thresholds for the different ionic conductances. However, the precise shape and size of the resulting potentials are determined by a wide variety of factors, and it is difficult to envisage that the exact *time* at which a slowly-increasing membrane potential crosses a particular voltage threshold can be set in a highly stable fashion. Thus, if an association between events is to be represented, it must ultimately depend on *synchronous* activation by the different events at single somata and their dendrites, to an accuracy of about 10 msec. Indeed the empirical evidence showing that this degree of synchrony is necessary for Hebbian synaptic change to occur (Hashemzadeh-Gargari *et al.*, 1991) supports this view.

At this point it can easily be said that something as complex as language cannot be reduced to the properties of single neurones. It might, for instance, be suggested that transmission delays within a complex network of nerve cells can easily be much longer than 10 msec, if only because many synaptic delays are involved. However the delay at an individual synapse is quite short (less than 1 msec), so one must envisage chaining of tens or even hundreds of such delays if patterns on the time scale of the "psychological moment" (such as those involved in speech sounds) are to be represented. This, in effect, is how Abeles (1982b) interprets the temporal dependencies in his synfire chains. One *might* envisage that such multisynaptic chains encode the time relationships between different parts of a speech sound: The later acoustic events would then be recognised because they match, in the temporal as well as the frequency domain, the representations activated by the initial parts. However, there is a severe difficulty in this scheme: It becomes impossible for the linkage between the element at the beginning and that at the end of such a chain to be a specific one. The reason for this is that as soon as there are intervening neurones between the initial and final neurones whose activation needs to be linked, massive divergence of the signal will be inevitable at every step of the chain. In such a temporally-structured network where there are very many synaptic links between initiation and termination, the network as a whole becomes very vulnerable to global changes of level of arousal, or metabolic or hormonal influences on synaptic transmission, which may disrupt the stability of expression of temporal structure represented in the network.

It may still be argued that other much more complex, but unspecified mechanisms mediate the representation of temporal patterns, such as speech. However, an explicitly-specified mechanism which can be examined in detail (and if necessary rejected) is always to be preferred to an unspecified mechanism which is not susceptible to rational scrutiny. Such a specific mechanism is postulated here:

There is one basic process in neural circuits which has a built-in delay which is both highly specific and highly stable. This is axonal conduction. In the central nervous system, this process may be studied by recording antidromic responses in single neurones. For a particular site of electrical stimulation the antidromic response recorded in a single neurone has a characteristic latency which normally varies by only a few percent.

Admittedly antidromic latency can show greater variation under some circumstances. For instance the second of two closely-spaced stimuli elicits an impulse which travels more slowly along the axon than normally (Swadlow and Waxman, 1976). However, it is unlikely that impulses in cerebral cortical axons occur so close together under physiological circumstances that axon conduction velocity is substantially affected by this factor. A second type of variation has been documented by Swadlow (1982, 1985) on the basis of experiments in which he recorded antidromic responses in chronic preparations. He has shown for a proportion of axons (30–50% of those recorded) that the latency may gradually change over many weeks (usually increasing, less commonly decreasing in latency). The reason for this is not clear, but since the "drift" of latency is usually more pronounced in the earlier weeks of the recording period in these experiments, the drift may well be a pathological phenomenon associated with the proximity of either the stimulating or recording electrode to the neurone under study. Thus, neither of these sources of latency variation significantly alters the conclusion that axonal conduction delay is a highly stable feature characteristic of each axon, or segment of an axonal tree.

How could such stable "delay-lines" become useful for representing temporal patterns? A neural network in which neurones are interconnected by such delay lines would allow one neurone ("A") to influence a neighbouring neurone ("B") after a precisely specified delay has elapsed. This fact could be used to form an association between two events slightly separated in time: Suppose that the first of two events to be associated activates neurone A and the later event activates neurone B, with the conduction delay in a connection from A to B corresponding approximately to the temporal separation of the two events. Under this circumstance, neurone B will receive nearly-simultaneous convergence of signals triggered by both events. Provided that the simultaneity is accurate to within about 10 msec, it will then be possible for summation of EPSPs to take place, so that the joint occurrence of both events, with the specific delay between them, is the necessary condition for the activation of neurone B above threshold. Similarly, if the EPSPs produced by either input to neurone B are subthreshold, but their coincident occurrence is suprathreshold, the Hebbian mechanism can strengthen the two synapses, so that neurone B comes to represent the temporal relation between the two events. This is represented in Figure 3.1.

Of course, for a *network* to use this simple principle in a manner conferring versatile representation of a variety of temporal patterns, one needs far more numerous delay lines than just the single one whose delay happens to fit a particular pair of events. A network that has a capacity to represent the full repertoire of the brief temporal patterns of events which may occur in (say) spoken language would need a correspondingly rich repertoire of delay lines. Between any two neurones there would need to be a number of alternative routes, giving a wide range of imposed delays. In other words, transmission of any signal would necessarily involve *temporal dispersion* using a large population of stable delay lines. The converse of such dispersion (but implicit in it) is that there could be a wide variety of patterns of temporal *convergence* upon single neurones.

Given such a network configuration the process of association between temporally separated events can easily be conceived as a joint product of axonal conduction delay, and

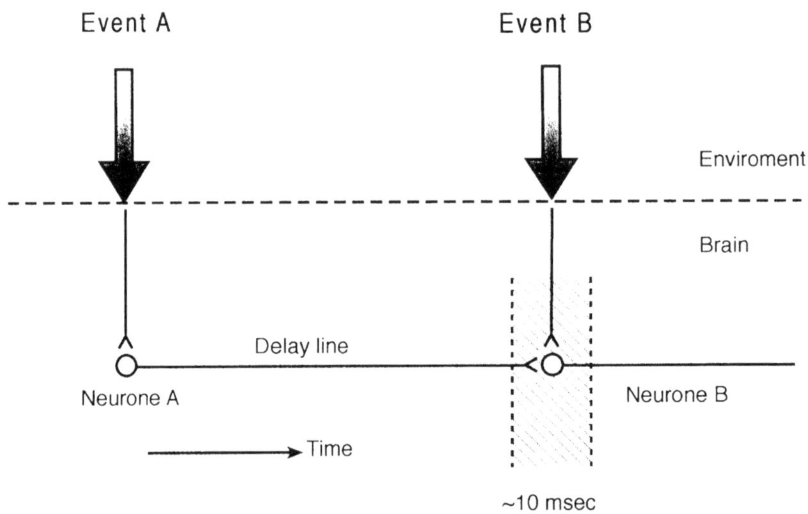

Figure 3.1 Use of Delay Lines, to Allow Association of Events Which Occur Non-Synchronously. See text. Vertical band (indicated by dashed lines and oblique lines) represents the neuronal integration time for summation of EPSPs or for activation of the Hebbian mechanism for synaptic modification. The same convention is used in later figures.

Hebbian synaptic modification. Initially one may assume that synaptic strength is uniform across synapses. With such assumptions, the influence of any event impinging on the neurone would be dispersed non-selectively over multisynaptic chains, involving many neurones and over an indefinite time after the event. Without a principle to *select* which of these pathways become active, the dynamics of the network would be quite disorganised, by such unlimited spread. Temporal convergence, necessary for summation of EPSPs to the point where they exceed threshold, could represent mere coincidence between events occurring a long way apart in time, and quite unrelated to each other, rather than closely correlated events occurring regularly in precisely-timed sequence.

Of course there is such a principle of selection, the premise of Hebbian modification of excitatory cortical synapses. This has already been referred to as producing strengthening of selected synapses when several of them activate a neurone synchronously or nearly so. In the case of a network in which axonal conduction delays achieve temporal dispersion of a signal's influence, the Hebbian principle will also strengthen those groups of afferent synapses which regularly activate a neurone synchronously, even when the time at which the soma contributing each afferent synapse was activated would be non-synchronous.

This process (in some ways the *mirror* of the temporal dispersion which the network produces when operating without synaptic modifiablity) has been referred to by Tank and Hopfield (1987) as "concentrating information in time". They suggest that, in order to recognise sequences, "information in the data stream between time 'T – t' and time 'T' must have been implicitly stored for availability at time 'T'." "If a subset of units in a network is to be strongly driven and to turn 'on' at time 't' to indicate that [the sequence] occurred, then this information must be concentrated on those units in space and in time." This is exactly the mechanism envisaged to underlie the encoding of temporal sequences such as elementary speech sounds. The "implicit storage" corresponds to the activation "in advance" of that set of axons whose action potentials will arrive at a single neurone in approximately synchronous fashion, while the "concentration in time" is the selection of these synchronously active afferents to the neurone by the Hebbian mechanism.

There is an important limit to how accurately temporal information can be stored in this way. This limit is set by the Hebb interval. It will be recalled that empirical evidence suggests this interval is about 10 msec. Envisage a fragment of a neural network, whose component neurones are connected spatially, and, via axonal conduction delays, temporally. One may also imagine that a certain threshold number, **g**, of synapses is needed to activate the Hebbian mechanism in the post-synaptic neurone. Suppose that this fragment of the network contains a neurone upon which **g** afferents converge, and their conduction delays from the preceding somata are all short, that is, less than 10 msec. If all afferents are activated synchronously (for instance if they are come from a small visual pattern in space), the neurone on which they converge will also be activated synchronously, to an accuracy of 10 msec. It is then likely that these **g** synapses, as a group, can become strengthened by the co-operative effect of the different parts of a synchronous pattern. If however, the same number of afferents converge upon another neurone, but their connections have a variety of different conduction delays, mostly much longer than 10 msec, that collection of synapses can become collectively strengthened only if the afferents are consistently activated in the correctly-timed sequence. The timing of each component of the sequence must be accurate to within 10 msec. Thus, *as concerns the representation of temporal patterns by Hebbian synaptic modification and axonal conduction delays, the temporal resolution*

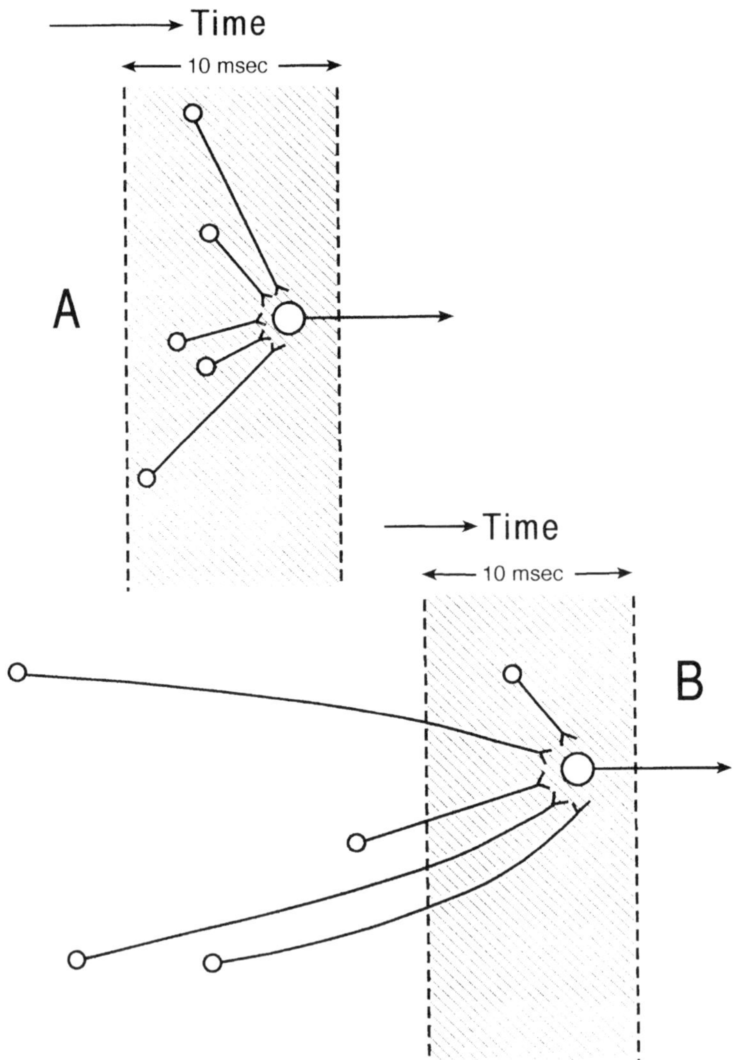

Figure 3.2 Role of Single Neurones in "Spatial" and Temporal Pattern Recognition, Both Based on Hebbian Synaptic Modification. In both *A* and *B*, the horizontal axis represents time, and five convergent active afferents are required to trigger the mechanism for Hebbian strengthening (i.e. **g** = 5; see text). The vertical band in each diagram represents the "Hebb interval" over which integration between inputs can occur. In *A* all the afferent connections are short latency (<10 msec), so the neurone on which they converge represents the parts of a more-or-less simultaneous pattern. In *B*, the afferent connections have a variety of different latencies, and the neuronal site of convergence thus represents a pattern extended in time. In both cases the active inputs need to converge on a single neurone with a simultaneity accurate to within 10 msec.

can be no better than the Hebb interval, that is, roughly 10 msec. The two styles of pattern representation described above are illustrated schematically in Figure 3.2. The temporal convergence illustrated in Figure 3.2(B) can be considered as depicting the process of acquisition of a speech sound with five frequency transients occurring in sequence.

3.4 QUANTITATIVE ASPECTS OF TEMPORAL REPRESENTATION IN NEURAL NETWORKS USING AXONAL CONDUCTION DELAY

With this principle of selection of synapses for representation of temporal patterns, another problem arises as to whether realistic neural networks have sufficient connections, and with sufficiently diverse conduction delays to achieve the required associations between temporally discontiguous events. It can be argued that, if a signal emanating from one or a small number of neurones is temporally dispersed along axonal links with a variety of time delays (including quite long delays) it is less likely that there be sufficient synchrony at any post-synaptic neurone to actually produce EPSPs which exceed threshold. In other words, the consequence of temporal dispersion of a signal is loss of amplification at any single postsynaptic neurone. This question requires some simple mathematical analysis, and computation, ideally based on quantitative data from the real brain.

Consider a block of cortical grey matter containing **N** neurones, each giving **n** connections randomly linked to other neurones in the block. Then, assuming **n** << **N**, the fraction of neurones receiving a single connection from any other specified neurone in the block is approximately **n/N**.

Assume, initially, that the conduction delays in this population of connections are randomly allocated between connections, and, overall, are described by a flat distribution between 0 and 100 msec. Since the temporal resolution can be no better than the Hebb interval, taken to be 10 msec, we may consider that each connection is allotted to one out of ten 10 msec time-delay "bins". The fraction of neurones receiving a single connection whose conduction delay falls within a particular "bin" will then be **n/10N**. In other words, as a result of temporal dispersion, the influence of any one neurone on those it is connected to is reduced ten-fold, on average. More generally, one can designate the number of time "bins" over which temporal dispersion occurs as **w**; and amplification is thereby reduced **w**-fold.

Consider now the chances of neurones in the block coming to represent a single speech sound. Assume that this sound is a stop consonant, consisting of three identifiable frequency components, 1, 2 and 3, packed, in measured sequence, within a duration of 100 msec. Let the number of neurones in the network being directly activated by each of these three sensory events be Q_1, Q_2 and Q_3. Assume also that **triple** convergence is required to activate the Hebbian mechanism (i.e. **g = 3**)*. Let **P** be the fraction of neurones receiving simultaneous triple convergence from the each of these components. We need to calculate **P**.

The fraction of neurones receiving a direct connection from neurones representing component 1, along lines with a specified delay, will be $Q_1 \times n/10N$. The fraction of neurones receiving an input from neurones representing both component 1 and

* *Triple* convergence may not, in itself be sufficient to trigger the Hebbian mechanism: The value g representing the number of co-active afferents needed to trigger Hebbian modification, mentioned in the previous section is probably rather larger than 3. However, it is assumed that under the circumstances of alert attention, when learning occurs optimally, membrane potential of the relevant cortical neurones is controlled so that it lies poised only a little below threshold for firing. Under these circumstances, the three synapses representing each component of the speech sound are assumed to be those upon which Hebbian modification critically depends, though they need not be the only ones active at that time.

component 2, each along lines with a specified delay will be $Q_1 \times Q_2 \times (n/10N)^2$. The fraction of neurones receiving input from all three components, each along lines of specified delays,

i.e. $$\mathbf{P = Q_1 \times Q_2 \times Q_3 \times (n/10N)^3.}$$

The **number** of neurones receiving such triple convergence so that signals initiated by all three events arrive at the neurone within a single 10 msec time bin, will be given by $\mathbf{P \times N}$.

Assuming that $Q_1 = Q_2 = Q_3$ (= Q), one can calculate the value Q, the number of neurones needing to be activated by each component of the speech sound in order to provide the required triple convergence in a given number of neurones ($\mathbf{P \times N}$).

Assume N = 1,000,000 (equivalent to a block of cortical tissue of dimensions about 10 mm^3 in the mouse cortex, or 20 mm^3 in human cortex.*)

and $$n = 5000^{\dagger}$$

and $(P \times N) = 1$, that is, there is on average a single neurone in the network with the required convergence.

Then, $$Q = \sqrt[3]{8{,}000} = 20$$

In other words at least 20 neurones must be activated by each component of the speech sound, if their convergent influence is to lead to the firing of just a single neurone which can represent the three components with correct sequence and timing.

If all these values remain the same, but **quadruple convergence** is required (i.e. **g = 4**), corresponding to the representation of a speech sound with four identifiable frequency components, Q becomes substantially larger (fourth root of 16,000,000 = 63.24).

If however, the cortico-cortical connections have NO delay (i.e. **w = 1**), then the amplification in the cortico-cortical link is not reduced by temporal dispersion, and Q has the value $\sqrt[3]{8}$, i.e. only 2. In other words only 2 neurones need to be activated by each of the three components, in order that a single neurone can represent their co-occurrence (regardless of sequence or timing).

Conversely, assume that N = 1,000,000 and n = 5,000 and there is NO delay in cortico-cortical connections, and assume also that Q = 20. In this case we can calculate $(P \times N)$ = 125 neurones (rather than just 1). In other words, if all we require is representation of the

* Braitenberg and Schüz (1991) obtain a value of 90,000 neurones/mm3 in the mouse cortex. Blinkov and Glezer (1968; Table 249) gives values of 30,000 to 106,000 neurones/mm3 in different regions of the human cortex. The higher values are for sensory cortex, especially visual cortex. A value of about 50,000/mm3 is typical of the cortex as a whole.

† The connectivity ratios of cortico-cortical neurones, designated n in the preceding sections has been estimated by a number of workers in various species, as reviewed by Peters (1987). Values lie between 2,000 and 17,000 with no obvious relation to the size of the brain in various species. An initial value of 10,000 is assumed here. However, it is likely that the synapses from a cortico-cortical neurone are more densely distributed close to the parent cell body than over the whole cortical mantle (Braitenberg and Schüz, 1991). For the purposes of the above calculation, it is assumed that the temporal dispersion of signals within the network uses just the distant connections, since the signals are assumed to have been relayed from a region of cortex distant from the assumed 10 mm3 block of cortical tissue. Following Braitenberg and Schüz (1991, Chapter 17) it is assumed that these distant axon terminals comprise 50% of the axonal length and a corresponding 50% of the number of synapses of each neurone, i.e. about 5000.

Table 1 Sample values for the computation of the amplification factor in spatio-temporal neural networks (see text of section 3.4 for interpretation of symbols).

$P*N$	N	n	g	w	Q
1	10^6	5000	3	10	**20**
1	10^6	5000	4	10	**63.24**
1	10^6	5000	3	1	**2**
125	10^6	5000	3	1	20

Figures in bold type are **calculated** values; other figures are values **assumed** as the basis for each calcuation.

co-occurrence of the three components of the speech sound, regardless of sequence or timing, activation of 20 neurones by each component will lead to activation of 125 neurones receiving the required triple convergence. Table 1 summarises the various illustrative calculations made above.

In conclusion, there is, *prima facie*, a significant constraint imposed by the use of the axonal conduction delay principle for representing brief temporal patterns. From the examples chosen, the number of neurones which can represent the whole speech sound will scale inversely with the number of time bins encompassed by the temporal dispersion produced by direct cortico-cortical connections. As the speech sound becomes more complex, with more components to be represented at the single unit level, the number of neurones needing to be activated in order that each component can gain representation increases significantly. Thus there is a fairly stringent limit on the complexity of sounds which can be encoded if one is to use the Hebbian mechanism to write in the associations, in combination with axonal conduction delay to represent the temporal separation of the components. Mechanisms other than these must therefore be involved in sequencing of speech in a longer measure than the single speech sound.

Thinking realistically, the individual components of a speech sound are themselves not single pure tones, but more complex combinations of frequencies. It is likely that we can recognise the different sensory qualities of different simultaneous combinations of pure tones, even if each tone is present for only the fraction of a second involved in uttering each component of a speech sound. This can be guessed from our ability easily to distinguish different timbres or accents of speech, or different individuals speaking. For this task, one needs not so much association between events which occur at slightly different times: One needs more or less simultaneous convergence of signals for all the frequencies which make up a single component of the speech sound. Such simultaneous convergence is likely to be achieved by the use of rapidly conducting cortico-cortical axons.

Thus, there is a role for rapidly conducting axons in simultaneous (or "spatial") convergence of signals, just as important as the role for slowly conducting axons in temporal convergence of signals. Roughly speaking, one would expect cortico-cortical axons with conduction time from soma to synapse of less than 10 msec to play a role in "spatial" convergence and representation of simultaneous ("spatial") patterns; while those with longer conduction times would be expected to play a role in temporal convergence and in representation of temporal patterns. Axons which can relay their messages "instantaneously" (as far as the temporal window of the Hebb interval is concerned) thus have a different function, independent of that of the slow-conducting ones where the conduction delay is longer and its precise value more critical.

3.5 EMPIRICAL EVIDENCE ABOUT TEMPORAL DISPERSION IN NEURAL NETWORKS OF THE CEREBRAL CORTEX

A number of quantitative values were assumed in making the above calculations. Some of these are supported in the footnotes in section 3.4. The most important of them however, was about the range of conduction delays imposed by cortico-cortical connections. This needs to be validated in detail from empirical data. The subject has been briefly reviewed elsewhere (Miller, 1995a).

3.5.1 Methods available

Evidence described in section 3.2 indicated that there can be precise temporal structure in the dynamics of neuronal firing in cortical tissue. The main method used was cross renewal density analysis for simultaneously recorded neuronal spike trains. This method shows temporal structure on the scale of tens or hundreds of milliseconds which is readily compatible with the hypothesis of long conduction delay in single axons. However, this evidence does not directly prove that hypothesis, because of the competing explanation that such long delays could be multisynaptic. An even less-direct method provides information about global (rather than single axon) conduction velocity and conduction time in the cortico-cortical network. It involves *cross-correlation of the EEG* recorded at different loci in the hemisphere (Thatcher *et al.*, 1986; Wright and Sergejew, 1991). By such cross-correlation methodology, it is possibly to show that the EEG can in part be interpreted as travelling waves of neural activity. The wave velocity, so computed, might be regarded as a global measure of axonal conduction velocity. However, there are many assumptions in this interpretation, and there are alternative ways of viewing EEG wave velocity. Another demonstration of the temporal structure in the activity of the cerebral cortex was recently reported by Arieli *et al.* (1995) who made optical recording of neural activity in the visual cortex of anaesthetised cats, using voltage-sensitive dyes, combined with single unit recording. When the optical signal was subjected to spike-triggered averaging (controlled by the action potential trains of single units), temporal structure was revealed over intervals of several hundred milliseconds before and after each action potential. However, the temporal structure so revealed was far less precisely measured in the time axis than in the triple-coincidence method used by Abeles (section 3.2). Taken together, all the above methods can be regarded as no more than corroborative evidence, not as direct proof of long axonal conduction delays. They do not positively identify the contribution of axonal conduction delays to this temporal structure.

The most direct method for assessing axonal conduction delays in central neurones is by measuring *latency of antidromic responses* with single neurone recording. Most studies of antidromic responses in cerebral cortical neurones report values for conduction *velocity* rather than conduction delay. This value is derived from the raw latency data using an estimate for conduction distance (which is only approximate because usually it cannot be measured directly, and often makes the arbitrary assumption that the axonal trajectory is straight). However, the raw data for latency, as actually measured, is the best value available for evaluating the hypothesis about temporal representation. Only a few papers present such unprocessed data. Even this method is beset by inaccuracies and biases which arise in several ways. These biases all lead to an underestimation of the contribution of the more slowly conducting axons. Thus a falsely high estimate of conduction velocities or a low estimate of conduction time is given by all published data. The biases should be carefully considered in considering the plausibility of axonal conduction delay as a means of representation of temporal patterns extended over intervals as long as a few 100 msec:

There are three reasons why empirical estimates of conduction velocity are biased in favour of the more rapidly conducting axons:

(i) Recording techniques may have a sampling bias in favour of larger neurones, which are also likely to have larger-calibre (and faster conducting) axons. This is a major factor when using glass micropipettes. The studies referred to below used tungsten or tungsten-in-glass microelectrodes. These can record from neurones up to 100 μ from tip, and have a lesser bias than glass micropipettes. Nevertheless, even with such electrodes, the smallest extracellular unit potentials, produced by the smaller cell bodies, may produce a signal which is so small that it is indistinguishable from noise in a single trace. (Averaging methods are seldom used in antidromic stimulation studies.) These small cell bodies are likely to project the finer calibre axons with lowest conduction velocities. Thus, even with such electrodes, there may be a bias against detecting the longest-latency responses.

(ii) In addition to biased sampling due to recording techniques, there may be a major sampling bias due to criteria for identification of cortical neurones. Very often in cortical recording, the presence of spontaneous neuronal firing is the identification criterion. However, it is known that in both anaesthetised preparations (Miller, 1975) and in awake ones (Swadlow, 1990; 1994) neurones in lamina V of the cortex have much higher spontaneous firing rates than those in other laminae. Many cortico-cortical neurones, with cell bodies in superficial laminae are essentially silent, and so might not be identified. Since the latter neurones are also those with slowest conducting axonal projections, this would again introduce a sampling bias in favour of the rapidly conducting axons.

(iii) Apart from these biases, antidromic stimulation also may ignore the terminal arborization of an axon, where calibre is likely to be smallest, and conduction is slowest. Alternatively, if stimulation is actually within the arborization, only the most rapidly conducting pathway in that arborization may be detected, any response elicited in the others being cancelled by "collision".

Apart from overestimation of conduction velocity, the conduction delay from soma to synapse may be underestimated because the full conduction distance is not taken into consideration. Antidromic single unit responses have generally not been sought from those cells which, on the basis of anatomical connection studies, would be expected to have the longest trajectories. In addition almost all antidromic stimulation studies have been conducted on smaller-brained animals (rabbit, cat) and seldom on primate species.

More data are available for axons that connect two cortical areas across the corpus callosum than for ipsilateral cortico-cortical connections. What data there are suggest that these two classes of axons are rather similar, and in the section that follows, callosal axons will be regarded as a special class of cortico-cortical axon. Latency histograms for ipsilateral and callosal cortico-cortical antidromic responses are provided in four papers (Swadlow 1974; Miller 1975; Swadlow and Weyand 1981; Swadlow, 1994).

The method of antidromic conduction could be refined, to give a better approximation to an unbiased conduction delay histogram. Signal averaging could be used to enhance the signal-to-noise ratio of the tiniest consistent-latency spikes. There have however been no serious attempts to improve data in this way however.

All these physiological data can be compared with ultrastructural data on axon calibres in cerebral white matter. Since there is a regular relation between axon type or calibre and conduction velocity, one would expect the data on the range of conduction velocities to be consistent with data on the proportion of unmyelinated to myelinated fibres, and their respective spectrum of calibres.

3.5.2 Experimental evidence

3.5.2.1 Maximal latencies

Swadlow (1974) has observed conduction latencies of rabbit transcallosal responses up to 39 msec (see Figure 3.3). A later study in rabbit (Swadlow and Weyand, 1981) found that the longest latencies for transcallosal conduction, or for conduction between visual areas V-I and V-II were around 24 msec. Similar findings in the rabbit motor cortex have also been reported recently (Swadlow, 1994). In the cat Miller (1975) observed maximal latency for conduction between somatosensory areas S1 and SII of 44 msec, and for transcallosal responses of about 34 msec (Figure 3.3). The conduction distance in each of these latter cases was about 10 mm. Bullier *et al.* (1988) found antidromic latencies between area 17 and 18 or 19 in the cat of up to 10 msec, indicating conduction velocities of around 1 m/sec or less. Zarzecki (1989) found callosal antidromic responses in cat motor cortex to have a mean latency of 6.7 msec, the longest latencies being about 15 msec. Expressed as conduction velocities, Swadlow has found that the ipsilateral cortico-cortical axons conduct more slowly than any other class of cortical efferent axon. This is a finding made consistently in visual areas (Swadlow and Weyand, 1981), first somatosensory area (Swadlow, 1989, 1990) second somatosensory area (Swadlow, 1991), and primary motor area (Swadlow, 1994). Callosal axons from these areas have somewhat faster conduction, though still quite slow compared to the descending efferent axons originating in lamina 5 neurones.

3.5.2.2 Shape of latency histograms

Published latency histograms for both ipsilateral and callosal cortico-cortical axons generally show a significant concentration of short latency responses compared with those with long latencies, and otherwise a fairly flat distribution, with gradually falling frequency of occurrence for the longest latencies. For instance, latency histograms in the papers of Swadlow (1974), Miller (1975) and Swadlow and Weyand (1981) indicate that 30–60% of ipsilateral cortico-cortical axons have a latency less than 10 msec (Figure 3.3). The true percentage with conduction times from soma to synapse of less than 10 msec would be lower than this bearing in mind that axon terminal ramifications (where conduction velocity is likely to be slower) may not be included in measured antidromic latency. For instance Swadlow *et al.* (1978), found that antidromic conduction velocity estimates for callosal axons in the monkey was greater when the stimulation site was

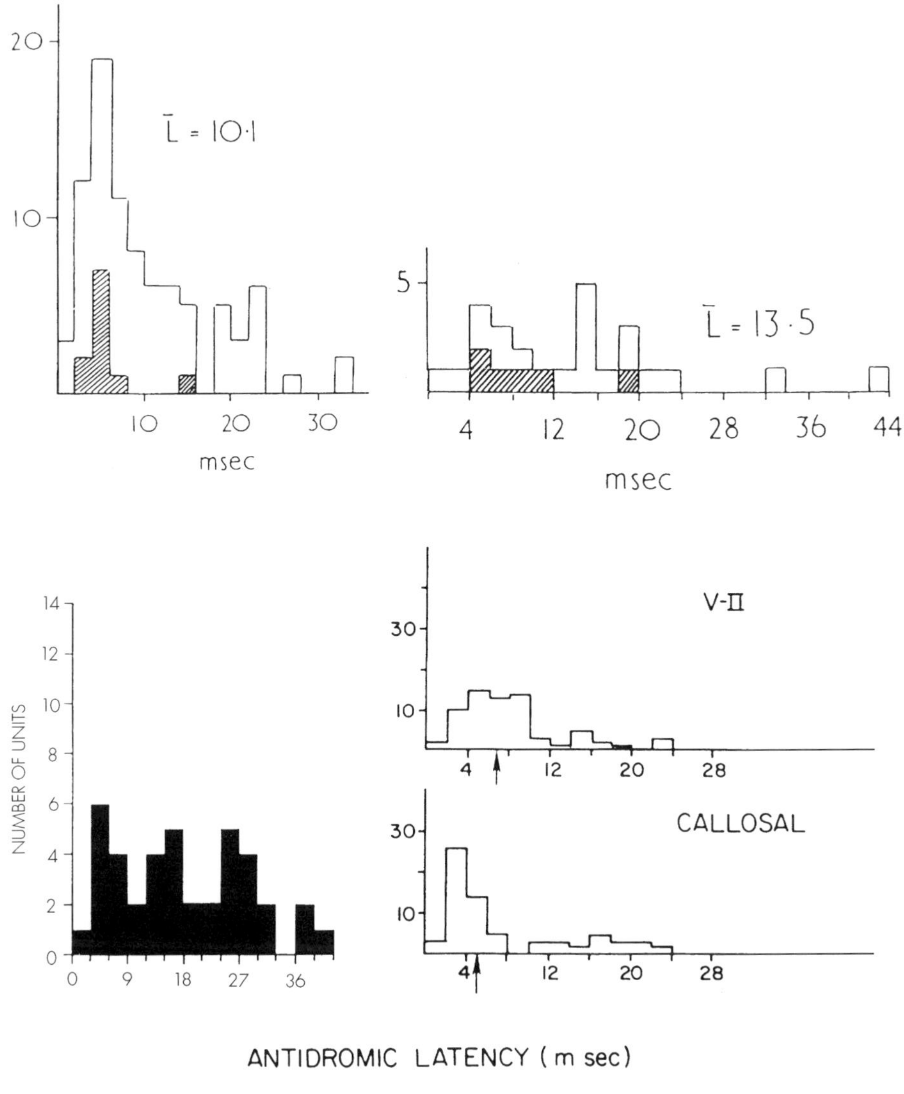

Figure 3.3 Composite Diagram, from Various Sources, Showing Latency Histograms for Antidromic Responses in Ipsilateral Cortico-Cortical and Callosally Projecting Single Neurones. *Upper left:* Callosal neurones in primary somatosensory cortex of cat (from Miller, 1975); *Upper right:* Neurones in primary somatosensory cortex of cat projecting to second somatosensory cortex (from Miller, 1975); *Lower left:* Callosal neurones in rabbit visual cortex (from Swadlow, 1974); *Lower right:* Neurones in rabbit visual area V-I projecting to ipsilateral area V-II (above) and across the callosum (below) (from Swadlow and Weyand, 1981). Reprinted with permission from *Journal of Comparative Neurology* [Miller, R., 1975, *Distribution and properties of commissural and other neurons in the cat sensorimotor cortex*; Copyright © Wistar Institute Press; Swadlow, H.A. and Weyand, T.G., 1981, *Efferent systems of the rabbit visual cortex: laminar distribution of the cells of origin, axonal conduction velocities and identification of axonal branches*; Copyright © A.R. Liss Inc. Reprinted by permission of Wiley-Liss, a subsidiary of John Wiley & Sons Inc.] and from *Experimental neurology* [Swadlow, 1974].) Composite figure reproduced with permission from "Oscillatory event-related brain dynamics," Plenum Press, 1995.

actually in the callosum than when it was at a more distant site in the contralateral cortical grey matter near where the axon terminated. Nevertheless the role suggested in the previous section, for rapidly-conducting cortico-cortical axons (with conduction times from soma to synapse of 10 msec or less) receives support from these empirical data.

3.5.2.3 Comparison with ultrastructural data

Data that are available come from the callosum in various species. The callosum may have a different axonal composition from white matter within one or other hemisphere (although some of the axons would be continuous with those in the corpus callosum). Nevertheless, to a first approximation, hemispheric white matter is probably not greatly different from that in the callosum, so the callosum may be a suitable indicator of the spectrum of axon calibres in cerebral white matter generally. In rat (Seggie and Berry, 1972), rabbit (Waxman and Swadlow, 1976) and cat (Fleischhauer and Wartenberg, 1967; Koppel and Innocenti, 1983; Berbel and Innocenti, 1988) between 50 and 60% of axons are myelinated in the adult. Other studies (e.g. Naito, *et al.* 1971) probably have methodological flaws and overestimated the proportion of myelinated axons. Lamantia and Rakic (1990a,b) produced excellent pictures of white matter from rhesus monkey callosum, and found that in most regions about 90% of axons were myelinated. Aboitiz *et al.* (1992) present data for a single human corpus callosum, in which there were very few unmyelinated axons. However, the quality of electronmicrography possible in human material is poor, which makes it unlikely that the full complement of unmyelinated axons was identified.

The 40–50% of unmyelinated axons in the callosum of smaller species have a mean calibre of about 0.25 μm, ranging from a little below 0.1 μm to over 0.5 μm. Myelinated axons have a mean calibre (for the axon cylinder) of 0.6–0.7 μm. Lee *et al.* (1986) have estimated that central axons of calibre 0.5 μm have a conduction velocity of about 0.5 m/sec. The finer-calibre unmyelinated axons may be expected to have conduction velocities lower than this, and Ferraya Moyano and Molina (1980) have estimated that the conduction velocity of the slowest conducting axons in the olfactory peduncle of the rat is about 0.1 m/sec.

The unmyelinated axons found in the adult callosum are similar to the total population of callosal axons found in new-born animals (rats: Seggie and Berry, 1972; kittens: Fleischhauer and Wartenberg, 1967; Berbel and Innocenti, 1988). Myelination in the 50–60% of callosal axons occurs gradually during development. This differentiation of structure into myelinated and unmyelinated axons may correspond to a specialisation of their functional role within neural networks, such as suggested in the previous section. Empirically it might correspond roughly to the separation of axons with conduction delays above and below 10 msec, as suggested in section 3.4.

3.5.2.4 The possibility of regional or species variation

Swadlow (1991) drew attention to the fact that conduction velocity in visual callosal axons of the rabbit was substantially slower than that in visual callosal axons of the cat (Harvey, 1980) or in callosal projection from the prelunate gyrus of the monkey (Swadlow *et al.*, 1978). All these results were obtained with tungsten-in-glass microelectrodes, so the studies are likely to be comparable, with low to minimal degrees of sampling bias. It is not clear from these results whether the variations are mainly species differences or mainly differences between regions of the callosum. However, on the basis of electronmicroscopy of the cat callosum, Koppel and Innocenti (1983) suspected that there were regional differences within the callosum in both mean axon calibre and ratio of myelinated to unmyelinated axons. Berbel and Innocenti (1988) provide quantitative data on regional variations in the cat callosum for both mean axon density and mean axon diameter. Juraska and Kopcik (1988) find that the posterior fifth of the rat corpus callosum contains about seven times as many unmyelinated as myelinated axons, a higher fraction than in the evidence cited above for the callosum as a whole. Lamantia and Rakic (1990a,b) showed that there were significant differences between different parts of the rhesus monkey callosum in the proportion of unmyelinated axons, which was lowest in the regions of callosum connecting primary sensory areas of cortex, and highest in the parts connecting frontal regions. It is significant that in most regions of the callosum around 90% of axons were myelinated, the number falling to 60–70% for regions connecting frontal regions. The electronmicrographs were of high quality, and it is likely that these estimates were fairly accurate.

Since monkeys are the species phylogenetically closest to humans for which good morphological data are available, this study might be taken as evidence against the idea that very slow conducting unmyelinated axons are of major importance in the human cerebral hemispheres. However, it should be pointed out that monkeys do not have a sophisticated language, nor the dramatic lateralization of function found in humans. Since language is a major function for which representation of brief temporal patterns is essential, it is quite plausible to suggest that one of the evolutionary changes between monkey and human brains is in this cytological feature of cerebral white matter. Moreover there may be significant differences in axon calibre spectrum between the corpus callosum, and the white matter containing predominantly axons which do not cross between hemispheres. It would be very interesting to have data from cerebral white matter, both in the callosum, and in each hemisphere, from higher sub-human primates, for which there is some evidence of cerebral lateralization similar to that in humans, and from whom first class electronmicroscopic pictures could be obtained.

3.5.2.5 Summary and discussion of empirical data

There seems to be no unbiased estimate of the full repertoire of cortico-cortical conduction delays. Antidromic conduction methods so far used are to greater or lesser extent biased in favour of the shorter latency axons. Cross-correlation methods applied to impulse trains or the EEG occasionally show time dependencies with quite long delays, but the interpretation of such indirect methods leaves doubt about the source of the delays.

Swadlow himself (1991), who has produced much of the evidence for long-latency axonal conduction, believes that the interhemispheric conduction time in rabbits for the slower 40% of axons is 20–60 msec. Nevertheless he is equivocal about the possible functional role of these slow-conducting axons. He writes: "It may be that the very slowly conducting components … are *all* modulatory in nature, whereas only the more rapidly conducting (presumably myelinated) elements ... are capable of faithfully driving post-synaptic neurones." He derived indirect support for this notion by comparing the latency histogram for antidromic and orthodromic activation of neurones by callosal stimulation, and finds the mean response latency of the latter to be less than half of the former.

However, there is a flaw in this argument for the relative inefficacy of the slower conducting axons. When electrical stimulation of brain tissue is used, all axons are activated synchronously. The faster conducting axons will then tend to deliver their action potentials to their post-synaptic destinations relatively synchronously, and summation of post-synaptic potentials to the point where a new action potential is generated is likely. On the other hand, the more slowly conducting axons will deliver their post-synaptic effects in a temporally-dispersed fashion, and the chances of summation occurring will be less. Thus the observed difference of the latency histograms for antidromic and orthodromic conduction, to which Swadlow draws attention, is to be expected without making any inference about relative strength of synaptic effects produced by rapidly and slowly conducting axons. Inference of axonal conduction velocities and delays from latencies of orthodromic responses to electrical stimuli is, like other approaches to this question, inherently biased towards the rapidly conducting axons in a population.

Apart from questions about the synaptic efficacy of signals relayed along slow conducting axons, the various biases leading to an underestimation of conduction delay from soma to synapse are rarely taken into consideration by those considering network dynamics in the cerebral cortex. It is often claimed that axonal conduction delay can only play a small role in delays within neural networks in the real brain. Carr and Konishi (1988) consider conduction delay as important for time representation, but only for delays up to a few 100 μsec. Tank and Hopfield (1987) consider it unlikely that delays longer than 10 msec can be explained on the basis of axonal conduction. Those who, using cross correlation methods, have demonstrated long-delay interactions between neurones, likewise dismiss the possibility of axonal conduction playing a major part in the delay. Neither Gassanov nor Abeles believe that their longer-delayed interactions represent single axon conduction times. Instead, Abeles considers that the organised temporal chains of activity which he infers involve very many synaptic links. He appears to be unaware of the magnitude of single axon delays in the slowest conducting fibres. While some synaptic links may be needed to explain his observations of interactions which have delays of as much as 300 msec, the number of such delays is probably very much fewer than he assumes.

Wright and Sergejew (1991) have estimated the velocity of transmission of EEG waves as 0.1–0.29 m/sec. If these figures are to be interpreted mainly in terms of conduction along axons, the velocity is within the range for the unmyelinated cortico-cortical axons, such as seen in the callosum by Swadlow. However, Wright and Sergejew (1991) believe that their "estimate of wave velocity ... is much lower than the velocity of conduction in myelinated or unmyelinated axons and indicates that wave conduction probably depends upon comparatively short range fibre connections, thus introducing many synaptic delays". Nunez (1989) builds a theory for the generation of the human EEG on the assumption that "the bulk of — action potential velocities ... lie in the 6–9 m/sec range". With such an assumption, hypothetical axons passing from frontal to occipital pole of the human hemisphere (axons of such length being unlikely from hodological evidence) would have a conduction delay at the very maximum of 30 msec.

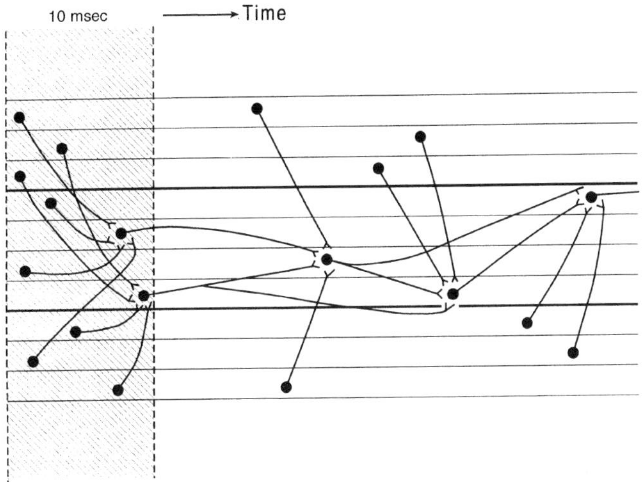

Figure 3.4 Diagram of Division of Function Between Short and Long Latency Cortico-Cortical Connections in a Network Configuration. This diagram compresses the spatial dimensions of a network into the vertical axis, while time is measured on the horizontal axis. Each horizontal row thus represents a single neurone, and a circle is drawn at particular horizontal position(s) to indicate the time(s) when it fires. A temporal chain of activation is shown in the central four neurones, with accessory active neurones in the upper and lower parts of the diagram. Connections show those axonal links which contribute the activation which makes each neurone in the chain fire at a particular time. For this diagram, four-fold convergence is needed to fire each neurone in the chain. Within the neuronal integration time of 10 msec, effective convergence is ample to sustain activity in the network. For longer temporal spans, long latency connections are necessary but are not sufficient to allow faithful transmission. Other (short latency) inputs are necessary to maintain suprathreshold activity.

However, despite the shortcomings in the empirical evidence, the antidromic single unit data presented above, and the simple logical extension of them, make it entirely feasible that single axon conduction times much longer than this are quite common, especially in larger brained animals. Consider a cortico-cortical axon of length 10 cm (quite plausible in the primate), whose conduction velocity is 1 m/s (far from the slowest conduction velocities known). Its conduction time would be 100 msec. With longer axon trajectories and slower conduction (both entirely possible from available data) conduction times from soma to synapse of several 100 msec would be expected in a proportion of axons. In a previous section a suggestion was made that axonal conduction delays in areas of the human cortex concerned with representation of speech sounds could use both rapidly conducting axons for spatial convergences, and slowly conducting axons for temporal convergence. The empirical evidence considered in this section indicates that the assumptions behind such a division of function are quite plausible. In short, neural networks of the human cerebral cortex are likely to have the required temporal dispersion of signals to represent brief temporal patterns such as speech sounds, as well as sufficient concentration of influence at short delays to represent simultaneous patterns with adequate amplification. This is illustrated in Figure 3.4. (There may nevertheless be other cerebral mechanisms involved to amplify lengthy temporal sequences of neuronal activation, which would otherwise fail to transmit as a result of temporal dispersion of the signal.)

3.6 THE CENTRAL HYPOTHESIS: THAT ASSOCIATION CONNECTIONS OF THE LEFT HEMISPHERE INCLUDE A HIGHER PROPORTION OF SLOW AXONS THAN THOSE OF THE RIGHT HEMISPHERE

The final part of this section follows easily from the previous discussion. The patterns of the receptive and productive aspects of speech which need to be represented are temporal ones on the time scale 50–200 msec. In the right-handed majority of people these patterns appear to be represented more effectively in the left hemisphere than the right (a subject discussed extensively in later chapters, especially Chapters 6 and 9). The values for conduction delays in a substantial population of cortico-cortical axons, when projected from animal data to the human cerebrum, can plausibly be assumed to lie within the same range of 50–200 msec, or longer. Such a spectrum of delays could play a major role in encoding the temporal sequences which make up perceived speech sounds, or the motor programs for emitted speech. Therefore, an obvious hypothesis is that *the left hemisphere should contain a richer repertoire of such long delay lines in its cortico-cortical axons than does the right.*

Probably this hypothesis translates to a simpler one, *that there is a higher proportion of unmyelinated fibres amongst the cortico-cortical connections in the left hemisphere than in the right.*

One or other version of the hypothesis may apply *globally*, to comparisons between left and right hemispheres of cortico-cortical connections linking any parts of the hemisphere taken in its entirety. Alternatively, these hypotheses may apply *locally*, to comparisons of corresponding specific regions in each hemisphere, especially the language zones of the left hemisphere and their counterpart in the right. In Chapter 11 correlations between the many different psychological measures of cerebral laterality are briefly considered, which give indications of whether cerebral lateralization is a global attribute of a hemisphere, or one requiring special consideration of each of the several major highways of cortico-cortical projections within a hemisphere. In any case the relevant parts of the left hemisphere should be specialised for temporal convergence and divergence of signals, while the corresponding parts of the right hemisphere should be specialised for "spatial" convergence and divergence.

Despite the simplicity of this hypothesis, its implications for actual cytological experiments are quite subtle and beset by many problems, both technical and ethical. Discussion of these issues is included in the next chapter.

4 Biological Predictions from the Conduction Delay Hypothesis of Cerebral Lateralization

4.1 MORPHOLOGICAL PREDICTIONS

The most direct morphological prediction of the hypothesis just presented, about the difference between the sides in the human hemispheres, involves microscopic study of the white matter in either hemisphere: By such study it is, in principle, possible to derive a histogram of calibres for myelinated and unmyelinated axons in the white matter from either hemisphere, or from specific matched regions in either hemisphere. The simplest prediction from the above hypothesis is that *the cortico-cortical connections in white matter of the left hemisphere would include a higher proportion (numerically) of fine calibre or unmyelinated axons than those in the white matter of the right hemisphere.* A more subtle prediction along the same lines would restrict the differences to cortico-cortical connections arising from selected corresponding regions of each hemisphere.

The technical and ethical difficulties of such morphological experiments are immense. The most important axons in such a comparison are the unmyelinated ones, because they, principally, are the ones with delays longer than 10 msec. They are much more delicate for cytological studies than myelinated ones, and because of their small size, are easily overlooked when intermingled with much larger myelinated ones. Their demonstration therefore requires the very best fixation methods, as well as accurate identification in the electron microscope. Careful unbiased stereological methods would be needed to obtain quantitative data on numbers of axons and their calibre. As part of the stereological methods, tissue samples would have to be obtained in a random but systematic fashion from accurately defined regions of brain tissue, so that there could be no sampling bias. It is unlikely that the ideal experiment in scientific terms could be reconciled with the necessary ethical constraints, since the required experiments would have to be carried out on human brain tissue*.

* Psychological evidence suggesting that cerebral laterality similar to that found in humans also exists in experimental animals is mentioned in Chapter 12. If this evidence is accepted, the difference may have the same structural basis as proposed here for humans. If so, the ideal ultrastructural experiment envisaged here could be conducted on small laboratory animals, such as rats.

Nevertheless, there are ways in which small samples of white matter could be obtained in an adequately fixed state for electron microscopy. For instance, tissue removed at temporal lobectomy operations could be the source of small fragments of white matter, to be fixed by immersion within minutes of excision. The difficulty here would be that there could be important differences in the calibre histograms obtained in different regions of each hemisphere. Therefore, with this approach, there would be major sampling problems in ensuring that data from one small fragment of white matter was representative of the region from which it came, or from its hemisphere as a whole. There is no data at all available using either of these direct approaches.

Another prediction, of a neurochemical nature, follows from the central hypothesis in a less direct manner: If the white matter in the left hemisphere contains a higher numerical proportion of unmyelinated axons than the right, *one might expect there to be a slightly increased concentration of markers for axoplasm (e.g. neurofilament protein) in white matter from the left, and an increased concentration of myelin in that from the right.* It is actually known that the concentration of myelin or myelin markers varies significantly for samples of myelin taken from white matter underlying different parts of the human cerebral cortex (Malone and Szoke, 1985; Johnson *et al.*, 1986). At present there is no data at all with which to evaluate the prediction that there be systematic differences between corresponding parts of the left and right hemispheres. However, this approach, though an indirect one, is feasible both technically and ethically.

There are other indirect structural approaches to the central question which could provide data with which preliminary tests of the hypothesis could be made. Unmyelinated axons take up less volume than myelinated ones, both because the axon has smaller calibre, and because the bulky myelin sheath is absent. If one assumes an equal *number* of cortico-cortical neurones on each side, but finer calibre, and/or a higher fraction of unmyelinated axons on the left, *one would predict that the volume of white matter would be lower on the left than on the right.*

In addition one can assume that cell body size is positively correlated with the volume of axoplasm each neurone has to support. Therefore, if the hypothesis were true, one would expect that a significant population of neuronal cell bodies on the left side would be smaller than their counterparts on the right side. This would mean that, numerically, *the cortex on the left side would have a higher cell density than on the right.* There is actually some evidence for this in the older German literature: Berger (1921) compared cell density in the precentral gyrus of left and right hemispheres in four human brains, studied *post mortem*. In all cases, the number of cells under a unit area of cortical surface was higher (in absolute terms) on the left than the right (by 5–11%, mean 8%). While none of the individual brains gave statistically significant results, collectively the mean cell density on the left was significantly higher than that on the right ($p = 0.012$, paired t-test). This result could be an indication that cortico-cortical projection cells on the left are smaller in volume than on the right, in association with their axons being of finer calibre. There appear to be no modern data re-examining the issue.

If neuronal cell bodies on the left side were smaller than their counterparts on the right, and again assuming equal neurone numbers in each hemisphere, *one might predict that the volume of cortical grey matter in the left hemisphere would be less than that in the right.* Since both grey and white matter are predicted to have smaller volume on the left, *one would predict that the left hemisphere as a whole would have a smaller volume than the right,* but this would depend on the assumption of equal numbers of neurones in the two hemisphere.

In addition, it is very likely, regardless of hemispheric differences, that cortico-cortical neurones or their processes occupy a smaller volume fraction of the grey matter than of the white matter to which these neurones project. Specifically, in the cortical grey matter, cortico-cortical neurones are located mainly in laminae II and III, whereas the neurones projecting to the brainstem and below are located in lamina V. The former have relatively small cell bodies and short dendritic trees, and a large part of their axonal projection is within the subcortical white matter, where they constitute by far the largest proportion of axons. The latter often have large cell bodies and have long dendritic trees, extending to the cortical surface, and much of the length of their principal axon will be below the subcortical white matter. In addition, the grey matter contains a variety of other cell types not found in the subcortical white matter. Therefore, the cortico-cortical axons are likely to form a larger volume fraction of the cerebral white matter than the cell bodies and dendritic and local axonal arborizations of such axons will form for the cortical grey matter. We can then introduce the hypothetical differences between hemispheres: Suppose, as inferred above, that the cell bodies of cortico-cortical axons are smaller on the left, and also their axons are of smaller calibre on the left than on the right. The left-sided volume deficit in the grey matter will have less overall influence on the total volume of the grey matter than does the left-sided volume deficit of the white matter. Therefore one might predict *that the volume ratio of white to grey matter would be lower in the left hemisphere than in the right.*

These predictions are not completely certain, because of uncertainty of assumptions and relative numbers of neurones in each hemisphere, and the length (and therefore the volume) of axoplasm contributed by axons of each calibre. These variables may show different side-to-side variations in different regions of the hemisphere. However, the predictions, such as they are may help to understand empirical data on the relative size of left and right hemispheres, the relative size of corresponding lobes on each side, the volumes of grey and white matter and the ratio of grey to white matter volumes in each hemisphere. What is more, measurements of this sort might give an indication of whether the hypothetical interhemispheric differences apply to the comparison of the hemispheres as a whole, or only to selected portions of them. Data on these questions are reviewed in section 4.2 below.

4.2 EVIDENCE OF MORPHOLOGICAL DIFFERENCES BETWEEN THE HEMISPHERES

4.2.1 Methods

The volume of different components of the forebrain can be measured from post mortem brains, fixed by traditional methods. However, with the development of computer-assisted tomography (CT) and magnetic resonance imaging (MRI) methodologies, similar measurements can now be made *in vivo*. Most of the studies using these methods (tabulated below) were not intended to investigate lateral differences in normal humans, but rather to study various patient groups (especially persons with diagnosis of schizophrenia). However, in such studies data for each hemisphere are also reported for control groups, and such data are used below. In many cases, the published data include the mean, the standard deviation, and the number of cases. Thus, t-test comparisons can be made, even when they were not included in the original published reports. For some of the comparisons, only a minority of the published reports provide statistically significant differences, and yet mean values in all or almost all reports may show differences between left and right side in the same direction. Such an occurrence can be taken as tentative support for there being a real difference between sides, but, since the aim of the study may have been something other than to establish a lateral difference in control brains, the data collected have insufficient statistical power to prove this in a single study.

The ratio of grey to white matter can be derived most directly from *measurements made on slices* of post mortem brains, or from MR images. Such measurements also give absolute values for volumes of grey and white matter, which aid interpretation of ratio values. However, there is also an indirect physiological approach to the question, which relies on the fact that white matter has a lower blood flow than grey: By measuring *clearance of radioactive xenon*, with detectors positioned over the skull, a clearance curve of radioactivity *vs.* time is obtained which can be analysed as the sum of two exponential curves, with different decay rates. The relative proportion of each component in the curve obtained with a particular detector gives a measure of the relative proportion of grey and white matter under the detector. A third method which provides a measure related to the ratio of white to grey matter is the *density of the x-ray signal in a particular region of a CT scan*: White matter attenuates x-rays less than grey matter, and thus shows up as less dense in the image. Taking each hemisphere as a whole therefore, one would expect a higher overall radio-density in the side with the highest fraction of grey matter.

4.2.2 Size of complete hemispheres

Fifteen studies (three quite old) address the prediction that, overall, the right hemisphere should be bigger, in weight or volume than the left (see Table 2). Out of fifteen quantitative comparisons of means, fourteen show an absolutely larger right than left hemisphere, but usually the difference is very small. In only one of the latter 14 comparisons do *volume* measurements show a right hemisphere which is larger to a high level of statistical significance, while in a second there is a strong trend ($p = 0.06$). The one study showing a larger *left* hemisphere also gives a highly significant difference. This result is not unique: De Lacoste *et al.* (1990; not tabulated) measured the area of the medial surface of each hemisphere in 28 male and 36 female brains. In both sexes the area measurement from the left hemisphere was larger than for the right, though in neither case was it a statistically significant difference. In addition to the tabulated quantitative *volume* comparisons, one study reports a higher *proportion* of brains with larger right than larger left hemispheres, to acceptable levels of statistical significance. In two studies, *areal* measurements show highly significant differences, one of these showing significance in male but not female brains. Greater rightward asymmetry in males than females is reported in three other studies. Of the other studies, one shows a trend to larger right hemisphere in brains from hypertensive patients, but not in the control group, and one shows no side-to-side differences. Overall, the vast majority of studies show a larger absolute hemispheric volume or area on the right, but only sometimes does this reach statistical significance. Thus, the prediction that there be a larger hemisphere on the right

Table 2 Lateral differences in the size of normal cerebral hemispheres

Authors	*Method*	*n*	*Right Hemisphere*	*Left Hemisphere*	*Comments*
Broca (1875) see Henderson (1986)	weighing of p.m. brains	19 18	531.3 g 575.4 g	530.8 g 574.4 g	Samples from two hospitals
Crichton-Browne (1880)	weighing of p.m. brains	male female 32 total	618.2 g 558.0 g 588.1 g	616.1 g 556.7 g 586.1 g	No statistical tests
Braune (1891) see von Bonin (1961)	weighing of p.m. brains	100	551.2 g	549.7 g	No statistical tests
Miller *et al.* (1980)	Volumes from p.m. brain slices	18 male brains 21 female brains	R > L in 13 out of 18 R > L in 7 out of 21		$p < 0.05$ $p = 0.07$
Jernigan *et al.* (1982)	CT	32	no significant side to side differences in hemicranial volume		
Schwartz *et al.* (1985)	CT (vol. derived from no. of pixels)	30	477.4 cm^3	463.8 cm^3	No statistical tests possible. CSF excluded
Yeo *et al.*	CT	43	left hemisphere volume was average of 1.3 (±0.9) % larger than right		$p < 0.001$ (1987)
Kelsoe *et al.* (1988)	MRI	14	402 ± 24 (S.E.) cm^3	392 ± 24 (S.E.) cm^3	N.S.
Weis *et al.* (1989)	Stereology in p.m. brain slices	29	538.2 ± 49.2 cm^3	534.4 ± 50.6 cm^3	$p = 0.06$
Kertesz *et al.* (1990)	MRI; linear and areal measurements	98	"Hemispheric area" larger on right by 1%		$p < 0.02$–0.001 for different groups
Dauphinais *et al.* (1990)	MRI; areal measurements from two coronal slices	12 male 9 female	"hemispheric area*" larger on right for anterior but not posterior slices		$p < 0.001$ in males. N.S. in females
Gur *et al.* (1991)	MRI	23 male 20 female	580.7 cm^3 532.8 cm^3	574.6 cm^3 529.3 cm^3	} R > L } $p < 0.0001$
Heckers *et al.* (1991)	Volume of p.m. brains	23	546.6 ± 75.0 cm^3	531.1 ± 79.8 cm^3	N.S.
Salerno *et al.* (1992)	MRI MRI	17 control 18 hypertensives	545 ± 58 cm^3 551 ± 112 cm^3	544 ± 61 cm^3 548 ± 117 cm^3	N.S. N.S.
Murphy *et al.* (1993)	MRI	19 female controls	37.2 ± 0.89†	36.9 ± 0.89†	N.S.

* excluding temporal lobe.
† % total intracranial volume

receives some support, but there is conflicting evidence. It is possible that a real difference is obscured in some studies by lack of statistical power. However, data from individual regions of the hemisphere tells a less equivocal story.

4.2.3 Differences in the ratio of grey to white matter between hemispheres

Table 3 shows data available from nine studies, using any of the methods described above for measuring the relative proportion of grey and white matter in the two hemispheres. None of the three studies on post mortem brain shows a significant difference between sides in grey/white matter ratio. However, two studies using clearance of radioactive

Table 3 Lateral differences between hemispheres in grey/white matter ratio.

Authors	*Method*	*n*	*Right Hemisphere*	*Left Hemisphere*	*Comments*
Aresu (1914) (see von Bonin, 1961)	estimate of volum e from cortical surface area	12	G = 360 cm^3	G = 364 cm^3	No significant difference between sides
Blauenstein *et al.* (1977)	Xenon clearance	34	G = 39.8%	G = 40.4%	$p < 0.05, 0.001$ (two comparisons)
Meyer *et al.* (1978)	Xenon clearance	15	G = 38.8%	G = 44.6%	$p < 0.02$
McHenry *et al.* (1978)	Xenon clearance	15	absolute values for %G higher in L than R		N.S.
Coffmann *et al.* (1984); Golden *et al.* (1981)	CT density	24	Left hemispere denser than right in all 12 blocks		$p < 0.05$ in all 12 blocks; $p < 0.005$ in 7 of them
Coffmann and Bloch (1984)	CT density	22	Left hemisphere denser than right		$p < 0.05$ in many blocks
Reveley *et al.* (1987)	CT density	18	Left hemisphere denser than right		N.S.
Weis (1989)	stereology on slices of p.m. brain	29			No difference between sides
Heckers *et al.* (1991)	stereology on slices of p.m. brain	23	G = 47.4%	G = 47.4%	No difference between sides

xenon show a significantly higher fraction of grey matter in the left hemisphere, and in a third with this method, absolute values showed a non-significant difference in the same direction. Moreover, three studies show that the radio-density of the left hemisphere is greater than that of the right, to a high degree of statistical significance in two of the studies. This was found uniformly, in comparisons between corresponding blocks on the two sides, for all 12 blocks of tissue into which each hemisphere was divided. The explanation of this lateral difference needs discussing. Coffman and Bloch (1984) provided evidence that the left/right differences in radio-density were not due to the partial volume effects or to "beam hardening", the artefactual increase in x-ray attenuation near to bone, which falls off non-linearly with distance from bone. In several studies where much smaller regions of the hemispheres were scanned, defined as entirely white or grey matter, no lateral differences were seen in either tissue type (Largen *et al.*, 1983; Gado *et al.*, 1983; Dewan *et al.*, 1986). However, regardless of side, in the studies of Coffmann (Table 2) radio-density was higher in just those tissue blocks where a higher ratio of grey to white was to be expected on geometrical grounds (the most anterior and most posterior blocks, and the slice closest to the dorsum of the hemispheres). It is likely therefore that the lateral differences also arise from differences in ratio of grey to white, rather than differences in radio-density of either tissue type between sides.

In conclusion, these findings suggest strongly that the interhemispheric difference in density is due to differences in grey/white matter ratio, rather than to difference in density of each tissue type between hemisphere. The ratio appears to be higher on the left. The consistency of these results suggests a genuine difference, supported by the data from clearance of radioactive xenon, despite the failure to find such a difference from measurements of post mortem brains. The latter method may have lacked the statistical power to show the differences demonstrated with CT scanning density.

4.2.4 Lateral differences in the size and grey/white matter ratio for individual lobes

The data in section 4.2.2 gave somewhat equivocal support for a prediction made about difference between sides in overall hemispheric size, though the data in section 4.2.3 on lateral differences in grey/white matter ratio gave more solid support to the corresponding prediction. Significant differences may be hidden by lumping all lobes together in the comparison. Therefore side-to-side comparisons of individual lobes may be a more incisive and detailed test of the predictions.

4.2.4.1 Frontal lobe

The earliest report of the relative sizes of the left and right frontal lobes is from Broca (1875; cited by Henderson, 1986). In groups of 18 and 19 brains from two different hospitals, the left lobe was somewhat larger than the right (by about 4 grams in 240 grams). However, Crichton-Browne (1880) observed larger right frontal lobes, (the difference being greater in male than female brains), in a series of 60 post mortem brains. In absolute terms, the frontal lobe weights in these studies indicate much larger tissue blocks than in modern estimates of frontal lobe volume (see Table 4 below), and it seems that much tissue in addition to the frontal lobe of the cerebral cortex may have been included. Fifteen other recent studies give data on the relative sizes of right and left frontal lobes (Table 4). Amongst this data are five *volume* comparisons, four of which give a larger right frontal lobe. One of these, using post mortem brains, achieves a high degree of statistical significance. The three which fail to reach statistical significance of the difference between sides, are those with the smaller subject numbers. The fifth study, using entirely female subjects, finds no difference between sides, but bases the comparison not on absolute volume, but on percentage of total intracranial volume occupied by the frontal lobe. For *linear* measures, seven studies base the comparison on width measurements from CT or MRI images, two on both length and width measurements. Of eleven comparisons, made in right handers (or non-specified handedness) six give statistically significant larger right-side measurements, and three of the remainder give an absolutely larger right side, or a higher proportion of larger right than left lobes. Asymmetry is not found in left handers (three studies), and is less in females than males (one study). The greater width of the right frontal lobe (found in right handers) is probably an indication of a greater total volume because (in three studies) it was shown to be associated with a right petalia (the concave impression made by the frontal pole on the cranial vault), and in a two more studies both width and length was greater on the right. Thus the larger size of the right frontal lobe is a fairly robust finding, and may represent a substantial absolute volume difference (13% in the one post mortem study).

The ratio of grey to white matter in the frontal lobes is addressed in ten studies (Table 5). Of these, the two studies using clearance of radioactive xenon both find a significantly higher fraction of grey matter on the left. Three studies measure radio-density in the frontal lobe. Two show evidence of significantly increased density on the left, while the third shows a trend in the same direction, for an anterior portion of the frontal lobe only. The higher density on the left is attributed to increased density in "white matter". However, the "white matter" sample in one of these (Daniel *et al.*, 1992) may have been a

Table 4 Lateral differences in the size of normal frontal lobes

Authors	*Method*	*n*	*Right Hemisphere*	*Left Hemisphere*	*Comments*
Le May (1976)	frontal width from CT	174 R handers	R > L in 122	L > R in 15*	p < 0.001
		49 L handers	R > L in 19	L > R in 10*	N.S.
	frontal petalia from CT	206 R handers	62	14	p < 0.001
		58 L handers	11	12	N.S.
Le May and Kido (1978)	frontal width from CT	80 R handers	R > L in 46	L > R in 10*	R > L: p < 0.001
		85 L handers	R > L in 30	L > R in 26*	N.S.
	frontal petalia from CT	80 R handers	23	10*	p < 0.05
		78 L handers	16	16*	N.S.
Chui and Damasio (1980)	frontal width from CT	50 R handers	R > L in 18	L > R in 11*	N.S.
		25 non-R handers	R > L in 12	L > R in 5*	N.S.
	frontal petalia from CT	50 R handers	18	4*	p < 0.01
		25 non-R handers	7	4*	N.S.
Jernigan *et al.* (1982)	frontal width from CT	32 normal controls	R > L in 22	L > R in 3*	p < 0.001
Weinberger *et al.* (1982)	volume from p.m brains	40†	R > L in 32, by average of 13%		p < 0.01
De Lacoste-Utamsing and Woodward (1982)	volume from CT	29	larger right frontal lobe, and right frontal petalia more common than those on left		No statistical tests
Bear *et al.* (1986)	Frontal length from CT	66	mean values: R > L‡		p < 0.05
	Frontal width from CT	66	mean values: R > L‡		p < 0.001
Koff *et al.* (1986)	frontal width (CT)	100	R > L in 39	L > R in 18*	
	frontal length (CT)	100	R > L in 38	L > R in 11*	
Kelsoe *et al.* (1988)	volume from MRI	14	61 ± 6 (S.E.) cm^3	58 ± 6 (S.E.) cm^3	N.S.
Hashimoto *et al.* (1989)	frontal petalia from MRI	18 controls age 2–11	7	1	
Stratta *et al.* (1989)	frontal width from MRI	20	28.17 mm	27.30 mm	p < 0.07
Suddath *et al.* (1989) from MRI	volume	17 ± 14.7cm^3	60.5 ± 13.6 cm^3	57.1	N.S.
Kertesz *et al.* (1990)	frontal width from MRI	98	no difference between sides		
Murphy *et al.* (1993)	MRI controls	19 female	18.4 ± 0.88§	18.5 ± 0.93§	N.S.
Falkai *et al.* (1995)	width from CT images	43 males	12.6 ± 1.5 mm	11.8 ± 1.6 mm	
		113 females	12.8 ± 1.6 mm	11.9 ± 1.6 mm	

* Remainder of n have R = L to within 1 mm.
† including 20 fetal brains
‡ Asymmetry significantly greater in male than female brains
§ % total intracranial volume

mixture of grey and white, since its density was intermediate between that of "frontal grey" and a region definitely identifiable as white, namely the "corpus callosum". The difference between sides could therefore plausibly be explained as a difference in the proportion of grey. This could also explain the similar result of Schlegel and Kretzschmar (1987). Of the remaining studies, one, studying post mortem brain slices, shows significantly thicker cortical grey matter in the precentral region on the left than on the right. Of three studies using MRI, two find a larger fraction of grey in the left frontal lobe, while the third finds an equal ratio between sides. The last study, using CT finds a slight excess

Table 5 Lateral differences between normal frontal lobes in grey/white matter ratio

Authors	*Method*	*n*	*Right Hemisphere*	*Left Hemisphere*	*Comments*
Meyer *et al.* (1978)	Xenon clearance	15	% G higher in left than right frontal lobe		$p < 0.01$
Gur *et al.* (1980)	Xenon clearance	36 right handers	% G was 1–3% higher for left than right prefrontal detectors		$p < 0.02$ for 3 out of 4 detectors
Bondareff *et al.* (1981)	CT density	29	ant. frontal lobe: L > R post. frontal lobe: L = R		N.S.
Thompson *et al.* (1985)	volume from CT images	31, age <16y	G/W = 2.05	G/W = 1.97	$p < 0.05$
Schlegel and Kretzschmar (1987)	CT density	53	"frontal white" more dense on left*		$p < 0.002$
Weis (1989)	stereology on slices of p.m. brain	29	Precentral cortex thicker on left side than right		$p < 0.05$
Suddath *et al.* (1989)	MRI images	15	G = 49.3%	G = 50.7%	No statistical tests
Suddath *et al.* (1990)	MRI images	15	G = 52.5%	G = 52.5%	no difference
Daniel *et al.* (1992)	CT density	20	"frontal white" more dense on left*		$p < 0.05$
Wible *et al.* (1995)	MRI images	15	G = 68.4%	G = 73.1%	

* see text for interpretation

of grey in the right side of the brain, but the subjects were children, aged 1–16, who may have very different radio-densities and proportions for grey and white matter from adults.

Overall there is a measure of support, with only one possible conflicting result, for the prediction made above, that the left frontal lobe has a higher ratio of grey to white matter than the right frontal lobe. Available data leave uncertainty about the contribution of differences in amount of grey or white between sides to the overall differences in the ratio. Since the right frontal lobe as a whole is larger than the left (see above) it could be suggested that the volume of grey is similar between sides, and the overall volume difference is due entirely to there being a larger volume of white on the right. Weinberger's data showed that the right frontal lobe as a whole was as much as 13% bigger than left. While not all studies show such a big overall volume disparity, the volume disparity seems to be much bigger than the disparity in grey/white matter ratios. Therefore, the possibility that the right hemisphere has more grey as well as more white matter cannot be ruled out.

4.2.4.2 Parietal lobe

LeMay and Kido (1978) measured the "central width" in 58 right handers and 41 left handers. This measure was more commonly greater on the right than the left, especially in the right handers. However, the measure of "central width", presumably near the level of the central sulcus, would be a poor guide to parietal volume. One early study (Crichton-Browne, 1880) and two more recent ones compare the volume of parietal lobes on left and right sides. Crichton-Browne (1880) found larger left than right parietal lobes (by about 6.5 g in 125 g), this difference being bigger in males than females. De Lacoste-Utamsing and Woodward, (1982) also report (in abstract only) that the left lobe was larger than the right. In female brains the central sulcus was more anteriorly placed on the left than the right. This made the difference in ratio of parietal to frontal lobe volumes greater on the left in females than males, although the grossly detectable asymmetry (frontal and occipital protuberances) are less obvious in female. Murphy *et al.* (1993) fail to confirm the larger right parietal lobe, but express their result as percent intracranial volume rather than absolute volume (percentages were 9.28 ± 1.08 on the right, and 9.16 ± 0.93 on the left. N.S.)

Three studies comment on grey/white matter ratio in the parietal lobe. Bondareff *et al.* (1981) found no differences between sides in tissue density in CT scans of the parietal lobe. Schlegel and Kretschmar (1987) found that the density of both "white matter" and "grey matter" in the parietal lobe in CT scans was significantly higher on left than on right. Since it is difficult to separate the two tissue types completely, the lateral differences may have reflected higher ratios of grey to white on the left (see previous section). Iwasaki *et al.* (1990) charted the medullary white matter in CT scans, and found that the inferior parietal lobule was more deeply folded on left than right side, in 68% of 276 subjects below the age of 20, and in 62% of 361 adults, with only 6% of each group having deeper folding on the right side. These differences are highly significant.

In conclusion the two studies with volume measurements show a larger parietal lobe on left, although no statistical analyses were presented. The probable larger size of the left parietal lobe is likely to be due more to an excess volume of grey matter, than to that of white matter.

4.2.4.3 Temporal lobe

Crichton-Browne's (1880) early study found slightly larger left than right temporal lobes. Since then thirteen studies, all using MRI, provide information about the size of the temporal lobe on each side (Table 6). These include one study carried out on fixed post mortem brains. Out of fourteen *volume* comparisons, there was a larger *right* than left temporal lobe in them all, the result being statistically significant in 8 comparisons. In a further four "area of slice" comparisons the right side was also significantly larger. The discrepant early result may reflect a difference of definition of the temporal lobe. Taking all the recent studies which report volumes, the right temporal lobe is larger by 2.7 to 16.1%, with a mean (for 197 normal subjects) of 8.17%. In these studies there is usually no way of establishing whether or not the posterior speech region is included in the volume measurements of the temporal lobe. However, the difference in volume between left and right temporal lobes is a robust finding. What evidence there is (four studies) suggests that this asymmetry applies equally in both sexes.

Four studies provide information about lateral differences in grey/white matter ratio in the temporal lobe, one using CT density, and three using MRI (Table 7). In all comparisons, the left hemisphere had a higher proportion of grey than the right, though in none of the studies was this difference shown to be statistically significant. It is possible that a real difference in grey/white matter ratios exists, which has not been proved because of lack of sufficient statistical power in these four studies. Since the right temporal lobe has a larger overall volume than the left, this difference, if it exists, may be due mainly to a smaller volume of white matter in the left. However, further evidence is needed to substantiate this suggestion, and estimate any additional contribution of differences in grey matter volume.

Table 6 Lateral differences in the size of normal temporal lobes

Authors	*Method*	*n*	*Right Hemisphere*	*Left Hemisphere*	*Comments*
Kelsoe *et al.* (1988)	volume from MRI	14	53 ± 3 cm³	49 ± 3 cm³	$p < 0.01$
Jack *et al.* (1988)	volume from MRI	11 females	57 ± 7 cm³	49 ± 5 cm³	$p < 0.05$
	of fixed brains	14 males	60 ± 11 cm³	53 ± 10 cm³	N.S.
Jack *et al.* (1989)	volume from MRI images in vivo	52	"ant. temporal lobe" larger on right by mean of 2.3 cm³ in 60 cm³. (Range: −10.2 to +13.2 cm³)		$p < 0.005$
Suddath *et al.* (1989)	volume from MRI	15	85.6 ± 14.4 cm³	74.1 ± 15.6 cm³	N.S.
Suddath *et al.* (1990)	volumes from MRI*	15	50.92 cm³	47.05 cm³	no statistical tests
Barta *et al.* (1990)	volume from MRI	15	34.1 ± 4.1 cm³	32.5 ± 3.7 cm³	N.S.
Rossi *et al.* (1990)	area of 2 slices from MRI	13	"right temporal lobe area larger than left"		$p < 0.002$
Dauphinais *et al.* (1990)	MRI; areal measurements	12 males	right temporal area larger than left in both sexes,		$p < 0.01$, 0.05 in males
	from two coronal slices	9 females	and in both slices		$p < 0.05$, 0.001 in females
Swayze *et al.* (1990)	volume from MRI	28 males	68.89 cm³	64.00 cm³	$p < 0.05$
		19 females	64.47 cm³	61.64 cm³	NS
		47 (total)	67.11 cm³	63.05 cm³	$p < 0.02$
Blackwood *et al.* (1991)	MRI; areal measurements	30	11.5 ± 1.6	10.7 ± 1.4	$p < 0.05$
Lencz *et al.* (1992)	volume from MRI	14	"right temporal lobe 8% larger volume than left"		$p < 0.005$
De Lisi *et al.* (1992)	volume from MRI	18 men	93.42 ± 8.1 cm³	87.2 ± 10.3 cm³	$0.1 > p > 0.05$
	15 women		78.06 ± 7.8 cm³	71.91 ± 7.8 cm³	$p < 0.05$
Murphy *et al.* (1993)	volume from MRI	19 females	4.91 ± 0.48†	4.78 ± 0.48†	N.S.

* summing values for grey and white matter to give total temporal lobe volume
† % total intracranial volume.

Table 7 Lateral differences between normal temporal lobes in grey/white matter ratio

Authors	*Method Hemisphere*	*n Hemisphere*	*Right*	*Left*	*Comments*
Bondareff *et al.* (1981)	CT density	29	$G = 40.23 \pm 8.1\%$	$G = 41.55 \pm 7.9\%$	N.S.
Suddath *et al.* (1989)	MRI images	15	G = 58.6%	G = 62.5%	no statistical analysis
Suddath *et al.* (1990)	MRI images	15	G = 59.9%	G = 60.4%	N.S.
Harvey *et al.* (1993)	MRI images males females	19 15	G = 73.6% G = 72.6%	G = 75.6% G = 74.0%	no statistical analysis

4.2.4.4 *Occipital lobe*

Crichton-Browne (1880) found that the occipital lobes were larger on the right side. Since then ten studies have investigated the relative sizes of the occipital lobes on each side (Table 8). One of these is a volume study from post mortem brains and finds a *left* lobe which is larger to a high degree of significance. In addition six studies report widths, and two more report both width and length of the occipital lobes. In right handers (or

Table 8 Lateral differences in the size of normal occipital lobes.

Authors	*Method*	*n*	*Right Hemisphere*	*Left Hemisphere*	*Comments*
Le May (1976)	width from CT images	158 R handers	R > L in 26	L > R in 101*	p < 0.001
		50 L handers	R > L in 23	L > R in 11*	p < 0.05
	occipital petalia from CT	269 R handers	24	185*	p < 0.001
		62 L handers	12	35*	p < 0.01
Le May and Kido (1978)	width from CT images	80 right handers	R > L in 7	L > R in 57*	p < 0.001
		85 left handers	R > L in 27	L > R in 29*	N.S.
	occipital petalia	80 right handers	6	60	p < 0.001
		85 left handers	12	36	p < 0.01
Chui and Damasio (1980)	width from CT images	50 right handers	R > L in 10	L > R in 18*	N.S.
		25 non-right handers	R > L in 1	L > R in 16*	p < 0.001
	occipital petalia from CT	50 R handers	10	30*	p < 0.01
		25 non-R handers	5	11*	N.S.
Weinberger *et al.* (1982)	volume from p.m. brains	40†	L > R in 32 cases by mean of 20%		p < 0.01
Jernigan *et al.* (1982)	width from CT images	32	R > L in 7	L > R in 24*	p < 0.01
Bear *et al.* (1986)	Occipital length from CT	66	mean values: L > R‡¶		p < 0.001
	Occipital width from CT	66	mean values: L > R‡¶		p < 0.001
Koff *et al.* (1986)	width from CT images	101	R > L in 19	L > R in 59*	
	length from CT images	100	R > L in 24	L > R in 60*	
Hashimoto *et al.* (1989)	occipital petalia from MRI	18 controls age 2–11	2	10	
Stratta *et al.* (1989)	width from MRI images	20	3.27 ± 0.28 cm	3.32 ± 0.21 cm	p = 0.11
Falkai *et al.* (1995)	width from CT images	43 males	13.6 ± 1.7 mm	14.7 ± 1.9 mm	
		113 females	13.4 ± 1.7 mm	14.7 ± 1.7 mm	

* Remainder of n have R = L to within 1 mm.
† including 20 fetal brains
‡ Asymmetry significantly greater in male than female brains
¶ In left handers, asymmetry is reduced or reversed.

handedness not specified) all six width comparisons give a larger value for the left hemisphere (significantly so in four cases). An indication that the greater width on the left is also an indication of greater volume comes from the finding, in three of these studies, that there is also a significant predominance of left over right occipital petalia, and in the two studies reporting both width and length, the left lobe is larger on both measures. The discrepant finding of Crichton-Browne (1880) may reflect a different definition of the occipital lobe, perhaps including some of the parietal lobe. In left handers, the asymmetry is less, or even reversed. One study reports that asymmetry is less in female than male brains.

Table 9 summarises the data available from four studies on differences in grey/white matter ratio between sides, in the occipital lobe. Two studies (one using clearance of radioactive xenon, the other using radio-density in CT images) provide evidence of a significantly greater fraction of grey matter on the left. The other two studies show either a non-significant trend in this direction (another radioactive xenon study) or no difference (a post mortem study). The fact that the left occipital lobe is substantially larger than the

Table 9 Lateral differences between normal occipital lobes in grey/white matter ratio

Authors	*Method*	*n*	*Right Hemisphere*	*Left Hemisphere*	*Comments*
Weinberger *et al.* (1982)	areal measures on p.m. slices	40	% G did not differ between sides		
Meyer *et al.* (1978)	xenon clearance	15	% G lower on right side		$p < 0.01$
Gur *et al.* (1980)	xenon clearance	36 R handers	% G lower on right side		N.S.
Schlegel and Kretzschmar (1987)	density of CT image	53	Right side less dense than left		$p < 0.001$

right lobe may therefore be due mainly to a larger absolute volume of grey matter on the left. The contribution of differences in absolute volume of white matter to the overall size difference are at present uncertain.

4.2.4.5 Conclusion

The *overall conclusion* from these data is that, *in the right handed majority, the right frontal and temporal lobes exceed left frontal and temporal lobes in volume,* or in other measures likely to correlate with volume. *The left occipital and perhaps the left parietal lobes are larger than right.* The early report of Crichton-Browne (1880) agreed with respect to the larger right frontal and left parietal lobes, but disagreed about the other measures (although regularly finding that the total weight of the three lobes behind the frontal were larger on the left-side). Weinberger *et al.* (1982) found that in 28/40 of the brains they studied both the larger right frontal and the larger left occipital lobes were present together. Quantitatively the volume differences reported are up to 13% for the frontal lobe asymmetries, and up to 20% for the temporal and occipital asymmetries. The differences which exist within individual lobes are more obvious, and the findings statistically more robust than those for the comparisons of the hemispheres as wholes. This may be because the asymmetries in frontal and temporal lobes cancel out those in occipital (and perhaps parietal) lobes. Asymmetries are much less, if they exist at all, in the left handed minority (at least in frontal and occipital lobes, for which studies report data on handedness), and females show a lesser degree of asymmetry than males in most cases where these gender effects have been investigated.

These lateral differences (as well as their interaction with sex) appear to have their origin early in foetal development. de Lacoste *et al.* (1991) found, in brains of gestational age 13–37 weeks, that in male brains the right hemisphere was consistently larger than the left, but in females, the two hemisphere were equal or there was a left side superiority. The prefrontal and occipital regions were those showing such asymmetry most markedly. It is thus likely that further changes must occur later in development in the occipital lobe to give an adult left lobe which is larger than the right one.

These differences in size of corresponding lobes on each side are associated with differences in grey/white matter ratio. The right frontal lobe has a higher ratio of white to grey than the left. Thus, the greater size of the right frontal lobe depends more on a greater volume of white matter on that side than on a greater volume of grey, though probably both contribute. The greater size of the right temporal lobe may also be accounted for mainly by more white matter than the left, but more evidence is needed to prove this. In the parietal lobe there is deeper infolding of cortex on the left, implying a greater volume of grey matter on the left. More evidence is needed. In the occipital lobe the larger overall volume on the left is due mainly to a larger volume of grey matter. The contribution of white matter is uncertain. Thus although there are differences between the lobes in the direction of volume asymmetry, *there is probably a higher ratio of grey to white on the left side, which generalises across all four lobes.*

4.2.5 Differences between sides in gyral pattern, or in size or cytoarchitecture of specific regional areas

The previous subsections have reviewed evidence about differences in volume and in grey/white matter ratios between the two hemispheres, or between corresponding lobes in each hemisphere. This dealt entirely with large scale quantitative comparisons between sides. However, there is another sort of comparison between sides, namely the comparison of size of local regions, defined either by gyral pattern or on cytoarchitectural grounds. Information of this sort could be quite independent of the global measures just reviewed. Investigation of lateral differences in gyral pattern or size of cytoarchitectural areas have naturally concentrated on the anterior and posterior speech areas and their surroundings.

4.2.5.1 Anterior speech area

Comparison of the size of the anterior speech region ("Broca's area) with that of its counterpart on the right side is beset with the problem of definition of the relevant areas. Wada *et al.* (1975) studied 100 adult and 100 infant brains and found no significant differences in size of the frontal operculum between left and right. However proper areal measurements were not made. Galaburda (1980) found, in a study of 102 brains, that there was an extra sulcus in the frontal operculum in 27 cases on the left, and in 13 cases on the right, which was statistically significant ($p < 0.05$). Falzi *et al.* (1982) defined the anterior speech area by gyral pattern in 12 right handed brains, and made areal estimates of the much-folded surface. Mean area on the left side was 27.4 cm^2, on the right side 22.4 cm^2, which was statistically significant ($p < 0.05$). In six of the brains, area on the left was greater than on the right, in three the right side was greater, the remaining three being about equal. Differences could not be detected from surface measurements of the speech area. One study (Galaburda, 1980) measured the area of a cytoarchitecturally defined region. The area concerned had characteristics similar to Brodmann's area 44 in macaques (i.e. large pyramidal cells in lamina IIIc, a narrow lamina IV, and a "crowded" lamina Va). From a sample of 10 human brains this area was larger on the left in six cases (up to 250% larger) about the same in 3, and smaller on left in 1.

Apart from these studies, there are three reports of finer structural differences between the sides in this region. An early study by Mellus (1911) measured the thickness of the cortex in various parts of sections through the inferior frontal gyrus on either side of human brains. He found that in parts supposed to be the anterior speech area, thickness was greater on the left side, both for the complete thickness and for laminae I, II and III individually. (Deeper laminae were not measured specifically. Mellus was probably using a 5-layered description of the cortex in his work, rather than the modern 6-layered description.) This difference may be part of a more widespread thickening of the grey matter in parts of the left frontal lobe, since a similar finding by Weis *et al.* (1989) was made for other parts of the left frontal gyrus (referred to in section 4.2.4.1 above). Scheibel *et al.* (1985) analysed dendritic branch patterns of pyramidal cells in left and right side of various regions of eight human brains. In lamina III pyramidal cells in the frontal opercular region the basal dendrites had richer distal branching on the left than the right sides. Hayes and Lewis (1993) conducted a cytoarchitectural comparison between left and right of Brodmann's area 45 (found in the frontal operculum). In this area there is a population of large pyramidal cells in lamina III. These are significantly larger on the left side than the right (although there is no difference between sides in the smaller pyramidal cells of this region).

Thus, in conclusion there is consistent evidence of various types that the anterior speech area, located in the left frontal operculum is larger in area and probably thicker on the left than on the corresponding region on the right. Since the volume of the frontal lobe as a whole is larger on the right, this implies that expansion of the anterior speech area on the left is atypical for the left frontal lobe as a whole (which is smaller). Likewise, the increased size of lamina III pyramidal cells on the left, reported by Hayes and Lewis (1993) reflects a change opposite to the overall change in cell size postulated in section 4.1 for the left side. Thus the local changes in the grey matter of the speech area appear to be additional to, and quite independent of larger scale differences between frontal lobes, which, quantitatively, are probably accounted for mainly by white matter differences.

4.2.5.2 Posterior speech area

A variety of methods have been used to compare gyral pattern, and size of local areas, on the temporal lobe of either side. One of the easier parameters to measure is the length of the Sylvian fissure which is usually greater on the left side. Asymmetry in this was first noted in early German literature (Eberstaller (1884) cited by von Bonin, 1962) who found a mean difference of 6.4 mm in 55 mm (left longer than right). The result was confirmed by Conolly (1950) and Von Bonin (1962). Yeni-Komshian and Benson (1976) found the Sylvian fissure to be longer on the left than the right in humans and to a lesser extent in chimpanzees, but not in monkeys. More detail on the difference in humans was provided by Rubens *et al.* (1976) who superimposed photographic images of the gyral pattern on left and right in the neighbourhood of the Sylvian fissure. In 25/36 brains the right Sylvian fissure angulated sharply upwards at its posterior end, compared to the left which remained roughly horizontal up to its posterior portion. As a result it necessarily followed that there be smaller areas of grey matter on the right, specifically on the superior surface of the temporal lobe, in the insula, and in the overhanging opercular part of the parietal lobe, with compensatory expansion of the right inferior parietal lobule posterior to the Sylvian fissure. The larger Sylvian fissure on the left has been detected to a high degree of significance in infants aged up to three years (Seidenwurm *et al.*, 1985). Asymmetries in the temporal lobe (which has been most extensively studied) are discussed below, followed by additional comments on the other regions.

Lateral differences in the *superior surface of the temporal lobe* were reported in early literature. The region of particular interest is a triangular area on this surface, called the planum temporale, defined by the surface line of the Sylvian fissure laterally, one of the transverse temporal fissures anteriorly, and the posterior limit of the superior temporal lobe surface. A larger area for the planum was reported by Kakeshita (1925), von Economo and Horn (1930), and Beck (1955). This finding became better known when Geschwind and Levitsky (1968) found that the planum temporale was larger on the left in 65/100 adult brains, and on the right in only 11/100. No data were provided on handedness of the subjects whose brains were studied. The surface edge of planum was 1/3 longer on the left than the right. Since the proportionate enlargement of the planum length on the left is much greater than that of the length of the Sylvian fissure, it is likely that the asymmetry of the planum accounts for most of the asymmetry of the length of the fissure. The greater area of the left planum was confirmed by Teszner *et al.* (1972) who found the surface area of the planum to be larger on the left in 64% of brains, and on the right in only 10%. Wada *et al.* (1975) found, in 100 adult brains that the left planum area was about twice that of the right. Falzi *et al.* (1982) found, in 10 brains from right handed subjects, that the planum area was larger on the left side ($p < 0.005$) by an average of 25%, though the area of planum was only about 1/4 of that of the anterior speech region. The degree of planum asymmetry varied greatly between brains. Pieniadz and Naeser (1984) found that the degree of planum length asymmetry assessed post mortem correlated with the extent to which the left occipital lobe projected further back than the right (as seen in a CT scan, and referred to as "occipital petalia"). Galaburda *et al.* (1987) found that the degree of asymmetry in the planum reflects mainly variation between brains in the size of the right planum, the left planum being more constant in size. Iwasaki *et al.* (1990) found, in the CT images of the brains of 276 children, that the planum was located more posteriorly on the left than the right in 56%, with a reversed asymmetry in 21%. This finding correlates with the description by Rubens *et al.* (1976) of the angulation of the posterior part of the Sylvian fissure upwards on the right but not on the left. Steinmetz *et al.* (1990) confirm, in 10 post mortem brains, that the surface area of the planum is larger on the left than the right side (670 mm^2 *vs* 403.5 mm^2; $p < 0.0078$), but also finds that the buried surface of the temporal lobe posterior to the planum is larger on the right ($p < 0.016$). Leonard *et al.* (1993) also found that, although the leftward asymmetry of the planum applied to the "temporal" part of it, (the horizontal part on the superior surface of the temporal lobe), the parietal part (the upwards angulated part described by Rubens *et al.*) was usually larger on the right.

Definition of the planum is not uniform between studies, since one or more of the transverse temporal gyri lying in the anterior part of the superior temporal surface may or may not be included. Campain and Minckler (1976) found that neither the number of transverse temporal gyri, nor their surface area differed from left to right. Chi *et al.* (1977a), who studied 207 foetal brains, from 10–44 weeks of gestation, found that 54% of brains had more, and larger, transverse temporal gyri on the *right* as well as a longer left temporal planum (see also Chi *et al.*, 1977b). Witelson and Pallie (1976) measured areas of left and right planum defined in a "biased" way, and in an "unbiased" manner (the latter definition not making assumptions about the number of gyri on superior surface of temporal lobe anterior to the planum). The left/right asymmetry was detectable and statistically significant using either method, but was less using the unbiased method.

(The left planum on average had an area 1.6 times that of the right [$p < 0.05$] using the unbiased method and 2.4 times that of the right using the biased method [$p < 0.005$]).

A cytoarchitectural study of four brains from the Yakovlev collection was carried out by Galaburda *et al.* (1978b). They defined area Tpt, which was characterised by lack of a sharp division between lamina IV, and the laminae on either side of it (III or V), because pyramidal cells from the latter two laminae encroached on lamina IV. The volume of this area on the left was 101–254, on the right 35–95 (non-standard planimetric units). The ratio (L/R) of volumes ranged from 7.26–1.14, and correlated with area of the planum temporale.

Apart from questions of the size of corresponding areas on opposite sides, there is one study which draws attention to cytoarchitectural differences in corresponding areas of the temporal lobe. Seldon (1981) studied Nissl stained sections of area TA (= Tpt = area 22), TB and TC (= primary auditory cortex). In these areas, columns of cells were wider on the left than the right. Intervals between columns were wider on the left in TA and TB, implying a larger fraction of neuropil to cell bodies, but this was not found in area TC. Dendritic extent covered the same number of columns on either side, implying that dendritic trees were wider on the left.

Kodama (1934; cited by von Bonin 1962) compared the size of the *insulae* on each side and found the left insula to be 4 mm longer (in 55 mm) and 1.5 mm wider (in 38 mm) than the right insula.

Eidelberg and Galaburda (1984) conducted a cytoarchitectural study of the *inferior parietal lobule*. Three fields were identified in the angular gyrus region. One of these (PG) was larger in the left, but only when the planum temporale was also larger. Another field (PEG) was substantially larger in the right, regardless of size of planum temporale.

As an addendum to these results, Foundas *et al.* (1995) have recently reported, from an MRI study, that the asymmetry of the planum temporale (and the adjacent pars triangularis), typically found in right handers, is absent in left handers.

In *summary*, several studies show, using linear, areal or volume measurements, that, to a highly significant degree, the planum temporale is larger on the left than the right side. Adjacent regions (inferior parietal lobule and insula) also appear to be larger on the left. The expansion of the posterior speech area on the left, like that of the anterior speech region, is against the trend for that lobe as a whole, which is larger on the right (see section 4.2.4.3). As with the frontal lobe it seems that local expansion of the speech areas occurs independently of other more global anatomical differences between hemispheres.

4.2.5.3 Other local differences between sides

It is likely that there are lateral differences in other regions of the hemisphere, about which very little is known at present. Kopp *et al.* (1977) give some clues to lateral differences in parts of the temporal lobe other than the posterior speech area. The "bridge" between the first and second temporal gyri was commoner on the left (especially in brains from male subjects). The second temporal gyrus had a greater maximum width on the right, and the fusiform gyrus (at the base of the temporal lobe) was wider on the left in the majority of cases. The larger planum temporale on the left correlated with the size of the anterior commissure. All these differences were highly significant statistically. Jack *et al.* (1989) found, from MRI studies, that the right hippocampus was significantly larger than the left (mean difference 0.3 cm^3; $p < 0.001$). Weis *et al.* (1989) found from study of post mortem brains that the cortex of the parahippocampal gyrus and in the basal temporal region was thicker on the right than the left ($p < 0.01$) and that the frontal orbital cortex was larger on the right ($p < 0.01$). Kopp *et al.* (1977) comment: "Anatomical asymmetries of the two cerebral hemisphere in man seem to be numerous. The interpretation of these asymmetries is difficult."

4.2.5.4 Conclusion

On the left, both the prefrontal and the temporal speech areas are larger than the corresponding areas on the right, whether the areas are defined by gyral pattern or on cytoarchitectural grounds. According to Falzi *et al.* (1982) there is no relation between the degree of asymmetry seen in the anterior and the posterior speech areas. There also appear to be subtle cytoarchitectural differences between sides in both areas, which fact provokes an attempt at explanation in terms of information processing in language (see Chapter 8). For instance, the larger lamina III pyramidal cells in the anterior speech area, and the wider spacing of neurones in the posterior speech area (compared with the opposite sides) might be a consequence of cortico-cortical cells in the anterior region projecting to the posterior region and having richer terminal arborizations there on the left side than the corresponding region on the right side. In the inferior parietal area, one cytoarchitectural area was consistently larger on the right, another on the left (the latter correlating with the size of the planum temporale). The measurements which have been made of the size of individual cytoarchitectural areas are highly selected data. It is possible that many other lateral differences might be found if they were looked for.

4.2.6 Overall conclusion concerning morphological differences between sides

The well known asymmetries in the anterior and posterior speech regions are not the only ones. There are also more global differences: Frontal and temporal lobes are both larger on the right, while occipital lobe, and possibly parietal lobe are larger on the left. In the frontal lobe (and tentatively in the temporal lobe), the difference appears to be due mainly to a higher volume of white matter on the right, while in occipital (and tentatively in the parietal) the difference arises mainly from a higher volume of grey matter on the left.

The areas of grey matter defined in the left hemisphere as important in speech are quantitatively only a small part of their respective lobes. Differences between the volumes of these areas and the corresponding areas on the right are quite small relative to the size of the whole lobe, or the whole hemisphere. If anything, increased size of the speech areas (on the left) would contribute to an increase in the size of left frontal and temporal lobe, but in fact these lobes are smaller than the right frontal and temporal lobes. These facts suggest strongly that there is a more global difference between hemispheres, which overrides any asymmetry related to speech areas. As a specific example of the global asymmetry, Steinmetz *et al.* (1990) and Leonard *et al.* (1993) point out that parts of the temporal lobe other than the planum temporale have a size asymmetry in a direction opposite to that of the planum. The large-scale asymmetry is related mainly to the grey/white matter ratio. *It is highly probable that there is a global difference between the two hemispheres, with a higher ratio of grey to white matter in all lobes of the left hemisphere than those of the right.*

This result is in accord with the prediction made about the neural basis of some aspects of hemispheric specialisation for language. The central premise was that there be a higher proportion of unmyelinated or fine-calibre axons amongst long cortico-cortical axons in the left side, compared to the right.

4.3 ELECTROGRAPHIC EVIDENCE OF ASYMMETRY BETWEEN HEMISPHERES

The two hemispheres in humans differ in the sort of information they can handle most effectively. Therefore it would not be at all surprising if there were marked asymmetries in event-related potentials or in electrographic activation evoked by stimuli for which one or other hemisphere has a preference. Examples of such asymmetries are included in a review by Davidson (1988). Conceptually this evidence goes little beyond the psychophysical evidence of lateralization (to be discussed in detail in later chapters), although it may give information about which region within a hemisphere is most involved in a particular test. It does not tell us much about the biological basis of such lateralization. However, the central hypothesis that the two hemispheres have different repertoires of axonal conduction delays leads to a number of specific predictions about electrographic activity and evoked potentials.

4.3.1 Evoked potential latency

The latency of the earliest components of an sensory evoked potential are determined largely by the conduction time in subcortical pathways, and there is no reason to expect differences between sides. However, the later components are likely to involve intrahemispheric relay, and so their latency might reveal lateral differences predicted by the central hypothesis. Bentin and Feinsod (1983) presented visual stimuli appropriate for each hemisphere (four letter words, or nonsense patterns), and recorded a late positive component of the evoked potential. With the former stimulus, amplitude was larger over the left hemisphere, and for the latter over the right hemisphere, as might be expected from the lateralized processing of the two stimuli. When this component was recorded from a midline position (the vertex), it had a mean latency of 511.3 msec in the verbal task and 464.4 msec in the non-verbal task, indicating that for this component, there was a faster transmission link in the right than the left hemisphere.

4.3.2 EEG coherence analysis

The EEG is a pattern of electrical activity containing a complex and often irregular mixture of many frequencies. The fact that the activity of many cortical neurones summates to produce measurable gross potentials implies that it is produced in neuronal elements with a high degree of parallel organisation of their dendrites, that is in cortical pyramidal cells. This population of neurones includes the cortico-cortical pyramidal cells. Since these transmit electrical activity from one part of the cortex to another (both locally and distantly), one might expect there to be regular correlations between the EEG recorded at different parts of the cortical surface. The measure of this correlation is known as the *coherence* of the EEG between the two recording sites. EEG coherence between two sites is a complex form of correlation, because, in principle it is specific to the frequency component under consideration (see Bendat, 1986, pp. 172–6). Moreover, once a significant coherence between the signal at two sites has been established, the delay between the signals at a particular frequency at one recording site and those at another can also be investigated. Such a computation allowed Wright and Sergejew (1991) to derive a figure for the wave velocity of the EEG over the cortical surface, as described in section 3.5.2.5. While it is not proven that the axonal conduction delay controls the velocity of spread of the EEG, this is a plausible conjecture. Therefore, our central hypothesis about the neurobiology of cerebral lateralization has implications for data on the coherence and wave velocity of the EEG.

If that hypothesis is correct, it is predicted that, *in the right hemisphere, there be a higher overall coherence than in the left*, because in the left hemisphere, the signal at one site would be dispersed amongst a wide span of time relative to the other, whereas in the right hemisphere, the two signals would be concentrated within a smaller window of time. Moreover, if an analysis of delays is carried out *it is predicted that the correlated signals recorded at two points in the left should exhibit longer average delays than those in the right.* Expressed as wave velocity, *in the left hemisphere, average wave velocity over the cortical surface should be slower over the surface of the left hemisphere than the right.*

The amount of data available to evaluate these predictions is not large, but is very significant. Four studies give data comparing the coherence of the resting EEG between sides. Tucker *et al.* (1986) found that right frontal intrahemispheric coherence was greater than left, in delta band ($p < 0.005$) and in theta band ($p < 0.04$). In temporal sites it was higher on the right for the delta band ($p < 0.04$). Apart from these sites and frequency bands, most comparisons yielded higher absolute values for coherence on the right, though not to a statistically significant degree. Thus, there may be a global difference, although these experiments rarely had sufficient power to show it. Thatcher *et al.* (1986) also reported data on EEG coherence in either hemisphere. For several of the comparisons between left and right intrahemispheric pairs of electrode sites there was a significantly higher coherence on the right side. None of the comparisons between sides gave a significantly higher left side coherence. Merrin *et al.* (1989), in a study of EEG coherence in patients with schizophrenia found incidental evidence in the control subjects of higher right sided than left sided coherence for some electrode pairs. Leuchter *et al.* (1994) gave values for coherence of the EEG on each side, for various regions of the hemispheres. The only interhemispheric difference which was statistically significant at the 5% level (in the frontal lobes) showed a higher right hemisphere coherence. These data all support the prediction about coherence. This prediction is probably the most obvious conclusion to be drawn from the data, although other interpretations are possible, especially if one takes into account the many non-linearities in signal transmission from one point on the cortex to another.

Thatcher *et al.* (1986) also report on phase delays between different recording sites for EEG wave forms. Overall the phase delays were greater in the left than the right hemisphere. The relevance of this to our central hypothesis requires clarification, because the mean absolute phase delay was not constant for different frequencies, but ranged from a few tens of msec for the fastest frequencies, up to 100 msec for the delta frequency band. Phase delay would be expected to stay constant as frequency is varied if the source of that delay was axonal conduction time, unless the different EEG frequencies are generated in different populations of neurones, with projections having different conduction times. It is possible however that factors other than conduction delay are important. Nevertheless the larger delays in the left hemisphere are intriguing, and lateral differences in axonal conduction time could be one part of their explanation. Combined with the higher coherence on the right side, the EEG data provides important evidence in favour of the central hypothesis. Overall, the study of EEG coherence has a great deal to offer in understanding cerebral laterality, and may even have a potential for giving detail of inter-regional interactions and their relays in specific cortico-cortical pathways.

4.3.3 Interhemispheric conduction time in either direction

A small number of papers using quite different methods provide some evidence relevant to the question of differences between the hemispheres in conduction delays in cortico-cortical axons. Neurones which project their axons across the corpus callosum are in many ways similar to those which distribute their axons to distant regions of cortex ipsilaterally. For instance, these neurones have their cell bodies mostly in superficial laminae of the cortex (II and III), and, in studies in animals, the axons of both are amongst the slower conducting long axon projections from the cortex (see section 3.5.2 above). Methods are available for measuring interhemispheric conduction times. Since this can be done in either direction, one has a method of measuring conduction delay in at least one class of cortico-cortical axons whose cell bodies are located in either hemisphere.

A relatively direct method of measuring interhemispheric transmission time uses the latency of evoked potentials in either hemisphere, for lateralized stimuli (especially visual stimuli). Components of evoked potentials recorded in the hemisphere contralateral to the stimulus presumably are a result of direct transmission to that hemisphere, while those recorded ipsilaterally involve the additional transmission via the corpus callosum. The difference between the latencies under the two conditions gives a value for callosal transmission time. Rugg *et al.* (1985) found transmission time using the "P120" wave" from left to right as 17 msec, and only 10 msec from right to left. Similar differences were seen using the later "N160" wave. In a later study, Saron and Davidson (1989) found transmission time (for lateral occipital electrode positions over either hemisphere) of 15.4 msec from left to right, and of 10 msec from right to left.

Another less direct method of estimating interhemispheric transmission time uses reaction time data. Although this is a psychophysical rather than a biological measure, the conclusions it leads to are physiological ones, exactly comparable with those just reviewed, and so will be dealt with here. The experiment again involves comparison of times under conditions requiring interhemispheric transmission with those that do not. If right visual fields and right hand, or left field and left hand are used, in reaction time tests, no interhemispheric transmission is involved. The combination of right visual field and left hand involves transmission via the corpus callosum from left to right hemisphere, while the combination of left visual fields and right hand involves transmission from right to left hemispheres. Reaction time is a measure subject to many influences (as we will see in Chapter 6), and different studies give somewhat different values. Nevertheless, in a meta-analysis of sixteen studies, Marzi *et al.* (1991) found that the average transmission time from left to right hemispheres took significantly longer (7 msec) than in the opposite direction (3 msec). Another recent study confirms this conclusion (Bisiacchi *et al.*, 1994). In a related study Verfaellie and Heilman (1990) found that simple reaction time was equally fast for right field/right hand and left-field/left hand combinations. Performance could be improved by using attentional warnings (indicating the field to be used) or intentional warnings, (indicating the hand to be used). Under such optimal cued conditions, the combination of field and hand involving callosal transmission from right to left hemisphere was improved above baseline more than that for transmission in the opposite direction. While the authors interpret this result in terms of specialisations for perception or motor performance within each hemisphere, the data also fit the present conjecture about differences in conduction time in either direction between hemispheres.

The absolute value of the transmission times reported in these studies is of course much less that the conduction times in individual axons which are emphasised in the central hypothesis of this book. However, the difference, both for reaction times and evoked potentials are consistently in the direction predicted by that central hypothesis. The smallness of the callosal conduction times reported in reaction time studies may be related to the fact that, in a task where speed of response is critical, it is the fastest conducting interhemispheric axons which determine the reported response time. In the case of evoked potential studies, it is likewise the fastest interhemispheric axons which would have a dominating influence, because these are the only ones which, as a population, would be sufficiently synchronous to produce peaks in the evoked potential.

4.4 OTHER BIOLOGICAL EVIDENCE RELEVANT TO THE CENTRAL HYPOTHESIS

A miscellany of studies using other methods provide support for the hypothesis that the right hemisphere is superior to the left with regard to "instantaneous" convergence of signals: Mullan and Penfield (1959) found that subjective illusions produced either by cortical stimulation, or as the aurae preceding an epileptic seizure, were much more common for right hemisphere involvement than for left. This is compatible with the idea that local activation of the cortex is more likely to have a wide enough influence to evoke a subjective illusion if it is in the right hemisphere than if it is in the left. Conceivably the effect of left hemisphere stimulation tends to be dissipated as a result of its temporal dispersion over pathways with different conduction times.

The parameters of brain stimulation in Mullan and Penfield's study are not stated. Later work (Libet, 1982) has shown than the threshold intensity of brain stimulation required to produce conscious sensory experience falls as train length increases, up to a duration of about 0.5 sec. In other words, there can be an integration between duration and brain stimulus intensity, for durations up to about 0.5 sec. In the next chapter, the topic of temporal integration by sensory stimuli is raised, and the prediction is made that the left hemisphere can integrate duration and intensity over a longer period than can the right. If the intensity/duration effect described by Libet for brain stimulation is really determined by the possibility of temporal convergence in pathways of differing delay, one would predict that the time needed for integration of brain stimulus effects to produce conscious sensation would also be longer for the left hemisphere.

Two studies of brain-injured patients suggest that cerebral functions are more highly localised in the left than the right hemisphere. Semmes (1968) showed that brain injury to the left hemisphere is likely to lead to classical syndromes such as sensory loss or motor weakness only when the lesion is in the corresponding cortical area. With right hemisphere lesions however, these impairments occur more commonly, and with lesions beyond the cortical area expected to be primarily involved. Landsell (1968) used tests of left hemisphere function (vocabulary test) and right hemisphere function (face categorisation) in patients with either left or right temporal lobectomy. The vocabulary test scores correlated significantly with the extent of left temporal removal, but the face test did not correlate with extent of right temporal lobe removal. Although these studies are intriguing, their exact significance for our central hypothesis is unclear. There is no reason for thinking that "instantaneous" convergence (the hypothetical speciality of the right hemisphere) and temporal convergence (the hypothetical speciality of the left) should be different in their involvement of long *vs.* short range cortico-cortical connections.

4.5 SUMMARY OF EVIDENCE PERTAINING TO BIOLOGICAL PREDICTIONS

A substantial body of gross morphological evidence suggests that in all lobes in the human, the left hemisphere has a higher fraction of grey matter than the right. In the frontal and probably the temporal lobes, this difference appears to be mainly because the absolute volume of white matter is smaller in left than right, while in the occipital and perhaps the parietal, the difference arises mainly because the absolute volume of grey matter is larger on the left than the right. As a whole these differences are what would be predicted from the central hypothesis, taking the version that the difference in axonal calibre is a global one between hemispheres. However, this morphological evidence could have a variety of explanations and so does not force an uncommitted student towards the central hypothesis. Comparisons of white matter composition between hemispheres would help to test that hypothesis, but again this method is not sufficiently direct to force a conclusion.

A much smaller body of relevant electrophysiological evidence is available and again addresses the issue of lateral differences in conduction delay in an indirect fashion. What evidence there is supports the central hypothesis, but does not compel one to adopt it. The most suggestive evidence is that concerning the left-right asymmetry of callosal transmission time, supported also by reaction time studies. If measurement of the wave velocity for the EEG (as carried out by Wright and Sergejew, 1991) were to be done separately for each hemisphere, a further very incisive biological test of the hypothesis would be possible.

Study of gross morphology and cytoarchitectonics has clearly defined two regions in the left hemisphere which are, in most individuals, larger than the corresponding region in the right. These correspond in location to the classic anterior and posterior speech regions, though they may not be the only individual regions which are asymmetrical between sides. Moreover, these local asymmetries are likely to be independent of the global asymmetries in size and grey/white matter ratio.

Apart from biological tests of the central hypothesis, the vast majority of evidence about cerebral lateralization uses psychological test. Again it is an indirect approach to the central hypothesis, but it is a wonderfully rich and diverse body of information. We move now to this area.

5 Perceptual Aspects of Lateralization: Theory and Predictions

5.1 INTRODUCTION

From the morphological evidence reviewed above, the hypothesis that there be a richer repertoire of delay lines between different cortical areas in the left hemisphere compared to the right is at least plausible, though critical tests of the morphological aspects of the hypothesis have not yet been conducted. The evidence, such as it was, implied a global difference between the hemispheres, not confined selectively to the areas known to be primarily involved in language. This in turn has another major implication: It encourages us to think that the functional difference between the sides can be demonstrated not only with respect to language functions but with respect to a wide variety of other functions (both sensory and motor) which depend on the same lateralized differences in morphological substrate. Therefore, two further important questions are posed: What is the full range of functional differences between left and right hemispheres? Can they all be explained as direct or indirect consequences of differences between the hemispheres in the degree of temporal dispersion of signals within each hemisphere?

In the present chapter, predictions will be made about the differences between the hemispheres in processing *sensory* signals, and in the following chapter corresponding evidence is reviewed. Admittedly, the hypothesis of the structural basis for interhemispheric differences also lends itself to predictions about differences in programming of motor output, and has implications for the higher cognitive functions of the forebrain. However, it is much easier to design experiments to evaluate temporal aspects of sensory processing than those of motor programming or cognitive processes. Discussion of the lateral differences in motor programming is therefore deferred to Chapters 8 and 9, when they can be placed in an appropriate theoretical framework. Laterality of higher cognitive functions of the forebrain is discussed in Chapter 10.

The basic theme of the discussion here, closely related to the hypothesis about lateral differences in cortico-cortical connections, is that the left hemisphere has a superior ability to represent temporal patterns (roughly in the range 50–200 msec) compared to the right. There are however a variety of other more exact predictions, derived from the hypothesis of cortico-cortical axonal conduction delays, and involving several variables in the sensory input. The effect of any one variable may be small, and hidden, unless one can control for several other relevant variables. Therefore, the predicted effect of all variables, singly and in interaction is derived in the next sections. There follows, in Chapter 6 a number of reviews of the empirical data about each variable in turn, also subdivided according to the sensory modality.

5.2 PREDICTIONS OF THE DIFFERENCE BETWEEN LEFT AND RIGHT HEMISPHERES IN PROCESSING SENSORY INPUT

5.2.1 Time/intensity integration

In sensory psychophysics there is a well recognised relation between time of exposure to a stimulus, objective stimulus intensity, and perceived intensity: For exposure times up to about 150 msec, lengthening stimulus exposure time increases the perceived intensity, just as does increase in objective intensity. Beyond about 150 msec, further prolongation of the stimulus does not increase subjective intensity, and is reflected instead as increase in perceived duration. In fact, for both the auditory and visual senses, perceived intensity is determined by an approximately linear integration of time and objective intensity, for stimuli less than the critical duration of about 150 msec. In visual psychophysics, this relationship is known as Bloch's law (see also Blumenthal, 1977). This law applies in a different way to the light-adapted and the dark-adapted eye, since in the light adapted eye, linear integration does not hold for durations beyond about 30 msec, whereas for the dark adapted eye the "saturation" duration is 100 msec or more (Aiba and Stevens, 1964). In part Bloch's law arises from the properties of photochemical reactions occurring in the retina. However, subjective intensity of brief light stimuli can in part be affected by temporal summation between corresponding parts of the left and right retinae (e.g. Battersley and Defabaugh, 1969), which indicates summation at the cortical level. Moreover, single unit electrophysiological studies in the cat visual cortex (Duysens *et al.*, 1991) reveal cortical cells which can perform linear time/intensity integration for durations well exceeding 100 msec, properties which are not found at either the retinal or the geniculate level. As far as the auditory sense goes, Zwislocki (1960) has shown linear integration of intensity over stimulus durations between 10 and 200 msec. (Zwislocki, 1960; 1965; Figure 5.1). In a review (Zwislocki, 1969) he writes: "There is compelling evidence that the linear temporal summation takes place in the central nervous system ... The apparent time constant is on the order of 200 msec near the threshold of audibility, and decreases to about 100 msec at moderate and high suprathreshold levels. ... From the neurophysiological point of view an experimental search for a temporal integrator with a time course on the order of 200 msec is called for." Thus, we have reason to believe that time-intensity integration involves, in part at least, cortical mechanisms.

It has been suggested before (Miller, 1981 p. 89) that the temporal integration referred to by Zwislocki (corresponding also to the "concentration in time" of Tank and Hopfield (1987)) is the converse implication of the temporal dispersion which can occur along axonal delay lines in the human hemispheres. In principle, it should be possible to derive the shape of a time/intensity integration curve, such as that shown in Figure 5.1 from assumptions or empirical data about the spectrum of axonal conduction delays.

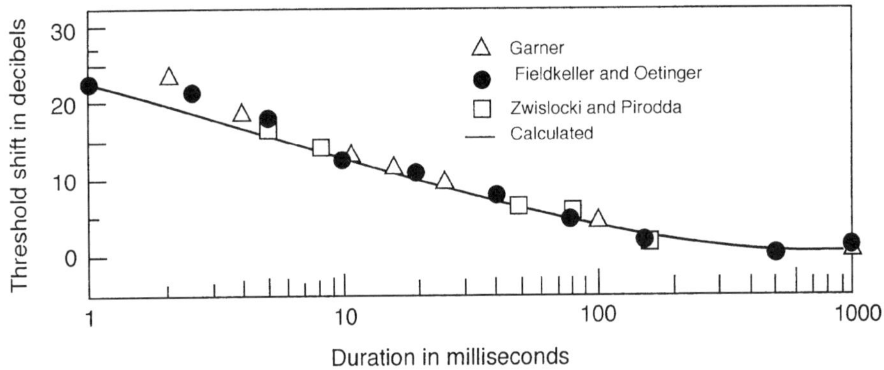

Figure 5.1 Time/intensity Integration at Threshold According to Zwislocki (1965; Figure 30). Time axis is logarithmic, to match the logarithmic (decibel) scale of sound intensity. With this transformation, there is a linear integration of intensity at threshold, over stimulus durations ranging between 10 and 200 msec Reproduced with permission.

For instance if one assumes a flat distribution of conduction delays (say, between 0 and 150 msec), linear integration might be expected, provided the relevant summations within individual neurones are also linear (see Figure 5.2). If, with such a flat distribution, cortico-cortical axonal delays are distributed over a longer time period they will integrate perception more slowly than if the same number of axons have their delays distributed over a shorter period. This difference is illustrated in Figure 5.2 (left *vs.* right).

Although few data are available from which to estimate the true histogram of axonal conduction delays, one can make the above model correspond better to those data by a making it slightly more complex. It is known that a sizeable fraction of cortico-cortical axons is myelinated, and therefore these axons have relatively short conduction delays (possibly 10 msec or less, as suggested in sections 3.4 and 3.5.2.2). From this, one might expect that there be *two* components to time/intensity integration. One component, mediated by the unmyelinated (or other slow conducting) axons achieves the linear integration, as already shown in Figure 5.2. The other, mediated by myelinated axons, achieves very rapid integration, which occurs more-or-less instantaneously (since the conduction delay is no longer than the period in each neurone when EPSPs can summate — see section 3.1 and 3.3). Such integration will occur continuously, throughout a maintained stimulus, giving a constant value to perceived intensity, added to the time/intensity product achieved by the first component. The composite integration is shown in Figure 5.3, and corresponds closely to the space-time neural network shown in Figure 3.4. Figure 5.3 implies that time/intensity integration for stimuli of short duration (of the order of 10 msec) should occur more rapidly than for longer ones. Thus, the intensity at threshold with such very short stimuli would be smaller than would be predicted from the time intensity curve derived using longer duration stimuli. To put it another way, as illustrated in Figure 5.3C, the value of the product T × I, would be constant for stimuli of different duration if there was truly linear integration; but, if there is an added component of *immediate* integration, this product would fall below the constant value as the stimulus duration was shortened towards 10 msec.

Such are the predictions from theory. Can such an analysis be used in practice to relate morphological and psychophysical differences between the hemispheres? Not in a

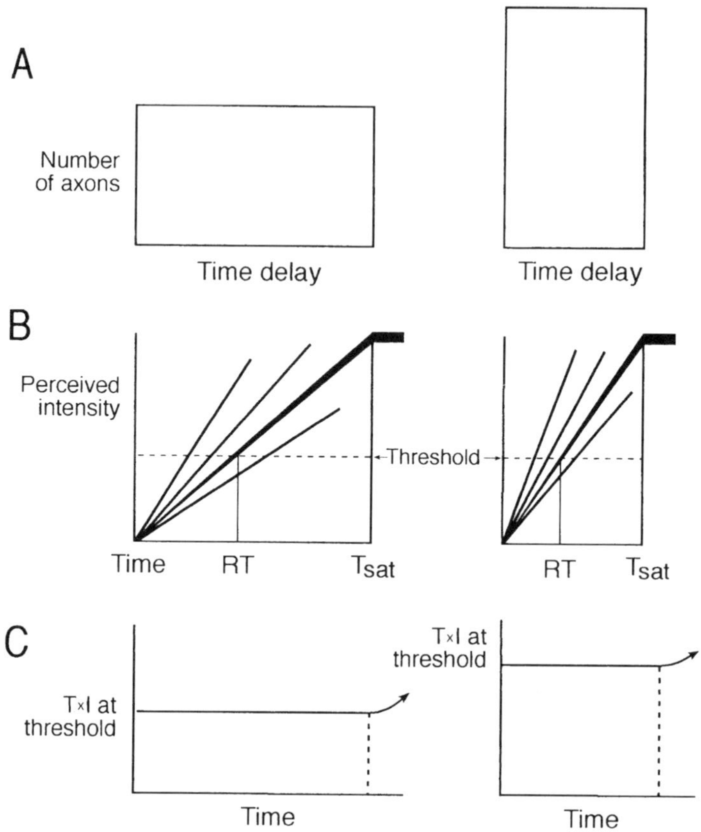

Figure 5.2 Linear Time/Intensity Integration and the Repertoire of Cortico-Cortical Delay Lines. *A:* Histograms of axonal delays. Vertical axis: number of axons; horizontal axis: delay. Rectangles on left and right have equal areas (indicating flat distribution of equal numbers of axons), but the axons are distributed over a longer span of delays on the left than on the right. *B:* Linear integration of time (horizontal axis) and objective intensity (corresponding to flat distribution of axonal delays) to give perceived intensity (vertical axis). Figures on left and right each give a family of linear slopes corresponding to different objective intensities of stimulus. For equal stimulus intensities (e.g. *bold* lines on left and right), the axon delay distribution on left gives slower integration than that on the right. Horizontal dashed lines: Constant value of perceived intensity corresponding to perception threshold. The time at which the integration line crosses the threshold may be critical in determining reaction time (RT), and is longer when the integration rate is slower (left) than when it is faster (right). T_{sat}: Stimulus duration, beyond which perceived intensity becomes independent of duration. *C:* In an experiment where the time of the stimulus is the independent variable, and the intensity at threshold becomes the dependent variable, the product of time and intensity is constant (given linear integration). This constant product is higher for rapid integration (right) than for slower integration (left).

quantitative sense. The data available about the spectrum of axonal conduction delays in the human is minimal. Even on the basis of the fragmentary morphological evidence that is available, there are discrepancies between what would be predicted and what is observed in psychophysical experiments. For instance, the departure from linear integration for short-duration stimuli, illustrated in Figure 5.3 has not been observed (Stevens, 1966; Stevens and Hall, 1966). Indeed, it is commonly found that the stimulus intensity of short duration stimuli needed to match the subjective intensity of longer stimuli is *greater* than would be predicted by linear integration (see Figure 5.1), rather than less. Presumably

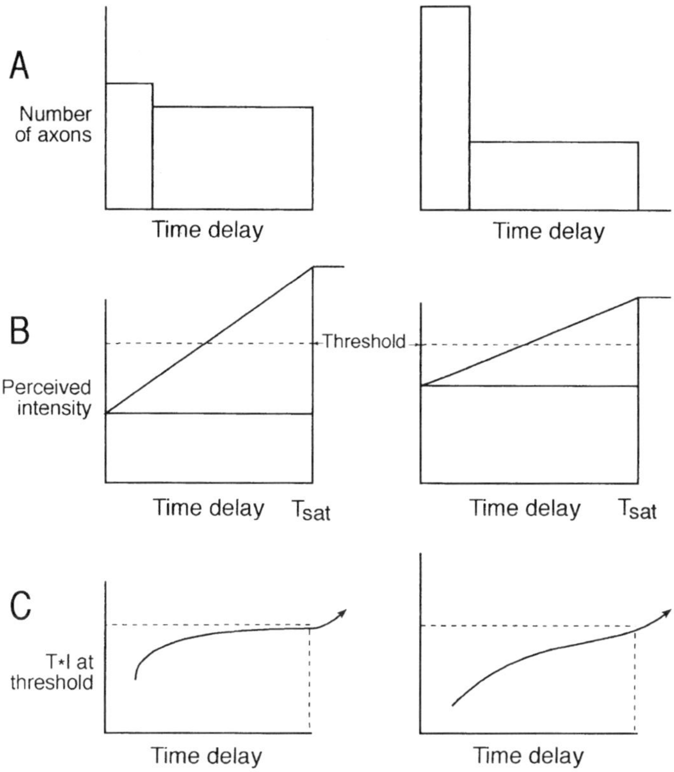

Figure 5.3 Linear Time Intensity Integration, with an Added Constant Value for Instantaneous Integration. *A:* Histograms of axon conduction delays, as in Figure 5.2A, except that there are two categories of axon with flat distribution, one with short delays (<10 msec) and relatively abundant, the other with longer delays and less abundant. The short-delay category constitutes a larger fraction of axons in the right than in the left diagrams. *B:* Time/intensity integration, to give perceived intensity, as in 5.2B. In this case there is a linear component to integration added to a constant value of instantaneous integration. T_{sat}: Stimulus duration, beyond which perceived intensity becomes independent of duration. *C:* The time/intensity product at perception threshold for the axonal delay distributions depicted in *A*.

other factors than integration via networks of cortico-cortical axons come into play for short duration stimuli. Moreover, it must be remembered that the psychophysical data is imprecise. In particular, published information usually presents averages between subjects, even though there appear to be significant differences between subjects, especially in the "saturation" duration (T_{sat} in Figures 5.2B and 5.3B), beyond which perceived intensity becomes independent of stimulus duration (Small *et al.*, 1962). In view of the imprecision of the data it is not surprising that there are discrepancies between what would be predicted from the hypothesis that axonal conduction delays are the basis of temporal integration of perception, and the psychophysical evidence available. The derivation of time/intensity integration from morphological premises will therefore not be pursued further in a strictly quantitative sense. Nevertheless, the psychophysical data do lend themselves, in many ways, to interpretation in terms of the axonal conduction delay hypothesis in a qualitative sense. In the following section schematic graphs are presented from which certain important qualitative predictions can be made.

5.2.2 Lateral differences in time/intensity integration

Of particular interest is the idea that the "time/intensity integration" can be tested separately for either hemisphere, i.e. using stimuli lateralized to either hand, ear, or visual field. The diagrams on the left side of Figures 5.2 and 5.3 may then be possible versions of the temporal integration in the left hemisphere, while those on the right side correspond to possibilities in the right hemisphere.

The detailed anatomical requirements for such an identification are worth discussing: If the hypothesis of a richer repertoire of long axonal time delays in the left hemisphere than the right of humans is true, one might suspect that the duration beyond which no further temporal integration is detectable, would be less for left visual field and left ear presentations than for those to the right sensory surfaces. This is depicted in Figure 5.2, where the value T_{sat} is greater on the left side than the right.

However, there is no reason to expect that the unmyelinated cortico-cortical axons in the right hemisphere should be any different from those in the left, except in that their relative number should be less. Given this, the critical duration beyond which no further temporal integration occurs should, strictly, be the same in each hemisphere. Therefore in Figure 5.3 the value for T_{sat} is the same for both hemispheres. However, suppose that *detectable* integration ceases when it is nearly complete, say 90% complete. Since the slope of integration is steeper in the left hemisphere (according to the premises in Figure 5.3), **detectable** integration may appear to cease at shorter stimulus durations in the right than the left hemispheres. Moreover, the assumption of a **flat** distribution of conduction delays in the unmyelinated population, though convenient, is unlikely to be strictly true. It is more likely that the fraction of axons falling in each "delay bin" decreases gradually to zero as one approaches the longest delays. This is depicted in Figure 5.4. The consequence of such a distribution of conduction delays for temporal integration is that, for long duration stimuli, temporal integration slows down gradually until it ceases, rather than showing an abrupt transition from "linear integration" to "no further integration". If one also assumes again that temporal integration escapes detection once it is almost complete (90% complete is illustrated in Figure 5.4) this limit of detectability will be reached sooner in the right than the left hemisphere.

Although an accurate derivation of the lateral differences in temporal integration cannot be made from anatomical data, it is worthwhile illustrating the above relationships in a more detailed manner, using psychophysical rather than anatomical variables. The crucial aspect of the following section is to define (qualitatively) the conditions under which a *reversal* of hemispheric advantage occurs as stimulus duration or intensity is changed.

Let T = duration of stimulus

I = objective intensity of stimulus

P = perceived intensity of stimulus

T_{sat} = the "saturation duration", i.e. the stimulus duration beyond which no further temporal integration is detectable.

C = the starting level for the time/intensity integration. This is determined by the immediate integration (hypothetically achieved by short latency/myelinated cortico-cortical fibres).

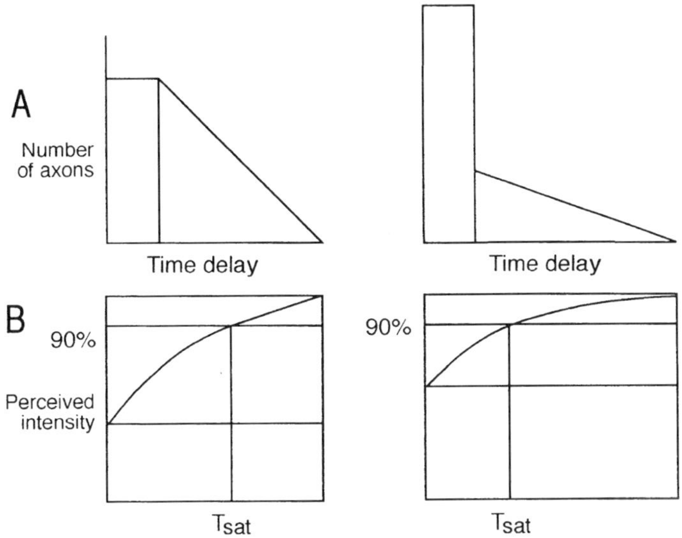

Figure 5.4 Time/Intensity Integration with Fraction of Axonal Conduction Delays Declining Progressively as Delay is Increased. *A:* Quasi-realistic histograms for axon conduction delays, hypothetically in left hemisphere (left) and right hemisphere (right), including both the component of rapid conducting axons (introduced in Figure 5.3.), and the sloping distribution of long-delay lines. *B:* Corresponding temporal integration curves for each hemisphere. If it is assumed that integration is no longer detectable for the last 10% of the curve, for a given objective stimulus intensity, the effective "saturation duration" (T_{sat}) becomes smaller for the right than the left hemisphere.

If there is linear time/intensity integration, then for stimulus durations less that T_{sat},

$$P = C + (k \times T \times I) \qquad \text{Equation 1}$$

where k is the gradient of the linear part of the time intensity integration curve.

Three parameters, which may differ between hemispheres, will determine hemispheric advantages in perceptual tasks: These are:–

(i) The relative magnitude of **C** in each hemisphere (designated as $\mathbf{C_r}$ and $\mathbf{C_l}$).
(ii) The relative values of the gradient **k** in each hemisphere (designated as $\mathbf{k_r}$ and $\mathbf{k_l}$)
(iii) The relative value of the saturation duration $\mathbf{T_{sat}}$ in each hemisphere. (These values are designated as T_{satr} and T_{satl}, in the section below.) Various combinations of relative values are shown in the graphs of hypothetical time intensity integration curves shown in Figure 5.5.

In Figure 5.5(a) $C_r > C_l$, and $k_r = k_l$. In this case there will be a right hemisphere advantage in perceived intensity whatever the stimulus duration. If however, as in Figure 5.5(b) it is also true that $T_{satl} > T_{satr}$, there may be a reversal in hemispheric advantage from right to left as stimulus duration increases. In Figures 5.5(c) and 5.5(d) it is assumed that $C_r = C_l$, but $k_l > k_r$. In these cases there will be a left hemispheric advantage for all durations, provided that $T_{satl} \geq T_{satr}$. In Figures 5.5(e) and 5.5(f), it is assumed that $C_r > C_l$, and that $k_l > k_r$. In these cases there is likely to be a reversal of hemispheric advantage from

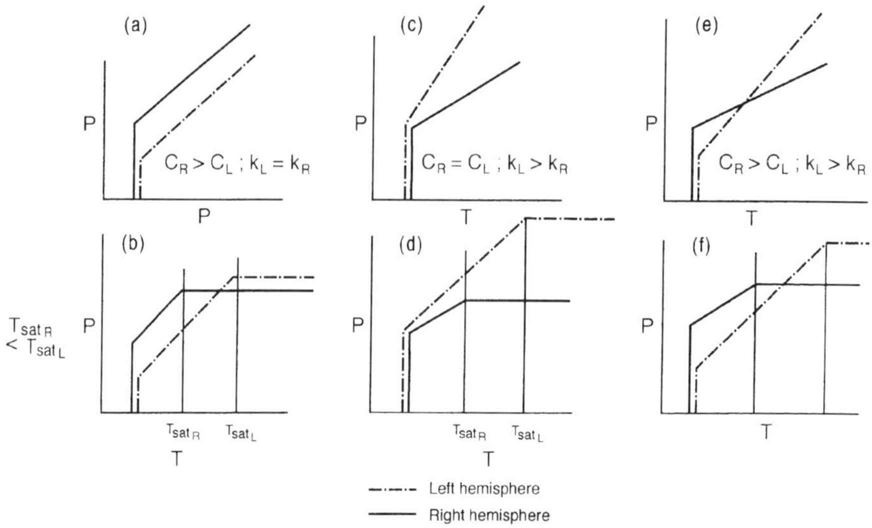

Figure 5.5 Possible Scenarios for Left/Right Differences in Temporal Integration Curves. In the upper three diagrams, the saturation duration is assumed to be equal for the two hemispheres; in the lower three, it is assumed to be longer in the left hemisphere. 5.5(a) and (b): Instantaneous integration is greater in right than left hemispheres, but gradient of time/intensity integration is equal. 5.5(c) and (d): Instantaneous integration is equal in the two hemispheres, but the gradient is steeper in the left hemisphere. 5.5(e) and (f): Instantaneous integration is greater in the right hemisphere, and gradient is steeper in the left hemisphere. Reversals of hemispheric advantage from right to left as duration of stimulus is increased occur under the conditions of (b), (e) and (f).

right to left, as duration is lengthened, and this can occur even when there is no lateral difference in T_{sat}.

In other words, if there is to be a *reversal* of hemispheric advantage (in measures of perceived intensity) from right to left as stimulus duration is lengthened, two conditions must be met:

(i) $C_r > C_l$

and

(ii) Either (a): $T_{satl} > T_{satr}$
or (b): $k_l > k_r$
(or both)

In other words, by this model, if a reversal of hemispheric advantage from right to left occurs as stimulus duration increases, it implies both: (i) a higher starting point for temporal integration by the right hemisphere and (ii) a faster rate of integration by the left hemisphere or a greater saturation duration for the time intensity integration in the left hemisphere (or both).

Although one cannot translate these conclusions into implications for the morphology of cortico-cortical connection in a quantitative sense, qualitative implications can be made. The first of the conditions implies an absolutely larger number of short (immediate) delay lines in the right than the left hemisphere. Either of the two versions of the second condition could be accounted for by an absolutely larger number of long delay lines in the left compared to the right hemisphere.

The reversals of advantage may be sought either in measurements of threshold intensity, or in lateralized reaction time measures. In either case, one assumes that a similar perceptual intensity is needed in either hemisphere to detect the stimulus, or to initiate the speeded reaction. For a continuous stimulus, this will be attained sooner for the hemisphere which has the higher value at a particular duration in the graphs of Figure 5.5. Therefore there will be a similar pattern of hemispheric advantages for reaction time measures as for perceived threshold intensity.

The above analysis made the implicit assumption that numbers of axons remained constant across hemispheres, although their conduction delay (determined by their calibre and/or whether or not they were myelinated) may vary. Although this may be approximately true, it is worth considering the consequences of an inequality in axon numbers. If there are different numbers of cortico-cortical neurones (and their axons) in the right and left hemispheres, the number of instantaneous connections (delay less than 10 msec) can vary independently of the number of delay lines (delay greater than 10 msec). In this case, a superiority in immediate integration need not be bought at the expense of inferiority in integration over time. The two specialisations then become independent of each other, and in principle one hemisphere could have a superiority with respect to both functions. However, this would inevitably mean that the superior hemisphere was the larger of the two. It would truly be the "dominant" hemisphere, both in structure and in all functions.

It should also be pointed out that all synapses, whatever the delay imposed by their axons, are assumed to have equal influence on post-synaptic neurones. As a result the total time-intensity integration possible for stimulus durations longer than T_{sat} in either hemisphere should be the same, provided that the numbers of cortico-cortical axons in the two hemisphere were equal. This is depicted in Figure 5.2A, where although the two rectangles have different shapes, they have equal areas. This represents the fact that in each scenario there are equal numbers of axons, despite different distribution of delays. In Figure 5.2B we see the consequence of this, that total perceived intensity, when integration is complete, is the same for both distributions of axonal delays. The empirical implication to be made from this is that *if, for stimuli long in relation to* T_{sat}*, one hemisphere is found to have greater sensitivity than the other, it must have the greater number of cortico-cortical axons.*

The preceding discussion dealt with reversals of hemispheric dominance as stimulus duration is changed. However, one can also ask another question: Under what conditions could reversal of hemispheric advantage occur from right to left as *intensity* is changed? Change of objective intensity can be assumed to cause proportional changes in both the starting level and the gradients for temporal integration. In the graphs of Figures 5.5 (a)–(d) this should merely expand or compress the graphs in the vertical dimension. There should be no change in the time at which the lines in 5.5(b) cross over, and therefore no effect of intensity on lateral advantage. In the case of the graphs in 5.5(e) and (f) however, the situation is different. Versions of these graphs with two intensity levels are depicted in Figure 5.6. Consider the perceptual intensity when the lines cross each other, and is therefore equal in left and right hemispheres. This is designated as P_{eq}, and the time at which this occurs as T_{eq}.

$$P_{eq} = (T_{eq} \times I \times k_l) + C_l = (T_{eq} \times I \times k_r) + C_r$$

From this is derived

$$T_{eq} = (C_r - C_l)/[I \times (k_l - k_r)]$$

Since values of **C** and **k** scale equally with intensity, the ratio $(C_r - C_l)/(k_l - k_r)$ does not change with intensity. Therefore the time at which perceptual intensity is equal in

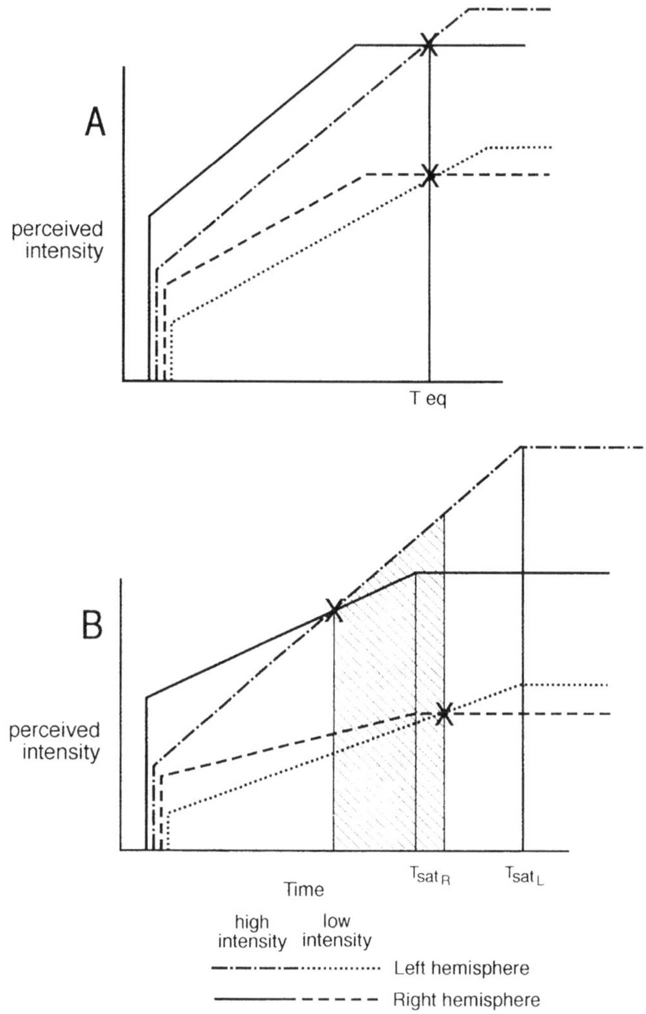

Figure 5.6 Influence of Change of Objective Intensity on Hemispheric Advantage. *A:* Time/intensity integration as in Figure 5.5(b), for two different intensities of stimulation (see key). There is exact scaling of the lines as intensity is changed, and the stimulus duration at which perceived intensity is equal between hemispheres ("X"s at intersection of lines) does not change. *B:* Time/intensity integration as in Figure 5.5(f) for two different intensities of stimulation (see key). Stimulus duration at which perceived intensity is equal in the two hemispheres ("X"s at line intersections) increases, as intensity decreases. Thus, for the intensities of the two pairs of lines, within the duration range indicated by oblique shading, increase of intensity (duration held constant) will produce a shift from right to left hemisphere advantage.

each hemisphere is inversely proportional to the objective intensity, **I**. Therefore, for stimuli of equal duration, one may expect a shift of lateral advantage from right hemisphere to left as objective intensity rises. Conversely, if such a reversal *is* observed, it implies that the graphs of Figure 5.5(e) and (f) are a better description of the differences between hemispheres than is that of Figure 5.5(b). In other words, of the conditions for advantage reversal spelt out above one would have to choose (ii)(b) rather than (ii)(a).

In summary, if one adopts the model depicted in Figure 5.2, that is, simple linear integrations at rates which differ between hemispheres, the hemisphere with highest integration rate will always have the advantage, whatever the duration or intensity, and there will be no reversal of advantage. Likewise if one assumes lateral difference in starting levels for integration and no lateral difference in the added rate of linear time/intensity integration, the hemisphere with the highest starting point will have the advantage at all durations and intensities. *The existence of advantage reversals as stimulus duration is changed, is a critical, albeit qualitative, test of the central hypothesis of this book.* If advantage reverses from right to left hemispheres as stimulus duration increases it has *two* implications: (i) The left hemisphere has a higher number of slow conducting axons, giving it a superiority in temporal integration; and (ii) that the right hemisphere has higher number of fast conducting axons, giving it a superiority in spatial or "quasi-spatial" integration. Such findings do not however, indicate that the left hemisphere has larger numbers of axons with particular ranges of delays. It is possible that the distribution of delays is the same between hemispheres, but, there is an added population in the left hemisphere with delays beyond the range of any of those in the right. If however, there are advantage reversals from right to left hemisphere as objective *intensity* is increased it provides more detail about the difference: It implies that the left hemisphere has a faster gradient of time/intensity integration than the right instead of (or as well as) a longer duration before saturation is reached. This would mean that even for delays lower than the saturation time in either hemisphere, the left hemisphere has an absolutely larger number of axons with such delays. The final implication of this section depends on identifying roughly the saturation time for each hemisphere: For stimuli longer than this time, a greater absolute sensitivity in one hemisphere would imply that there is a greater total number of axonal connections participating in the integration in that hemisphere than in the other.

5.2.3 Time taken to identify brief stimuli

According to the central hypothesis, the left hemisphere is supposed to allow greater temporal dispersion of a signal than the right. Hence when a constant sensory stimulus is relayed to the cortex, it should require to be maintained longer to stabilise its representation in the left hemisphere than in the right. In the right hemisphere, where a higher proportion of cortico-cortical connections are hypothetically of brief conduction times, messages will more readily converge on postsynaptic neurones within a short time after the signal arrives at the cortex. "Temporal integration" to the point where the threshold for conscious perception is reached will take place more quickly in right than left hemisphere. Using the term of Tank and Hopfield (1987), it should take a longer time for the signal to be "concentrated" in the left hemisphere than the right, sufficient to exceed the threshold which permits its detection or identification. This prediction should apply only for short stimulus durations, and should be reversed at longer ones. Given this, it should apply not only to detection of simple stimuli, like points of light or brief sounds, but also to more specific sensory features or qualities (e.g. for visual stimuli, line orientation or colour). For these, absolute thresholds for detection in briefly presented stimuli, should be lower when the stimulus is presented to the right hemisphere than to the left (for which temporal dispersion will reduce amplification).

This prediction can be put to experimental test in two different ways. (i) The subject is presented with a very brief test stimulus on one side (e.g. one ear, or one visual field), and is asked to "locate" it in time with respect to a standard sequence of stimuli. The latter can be presented either contralaterally, or, if another sensory modality is used, bilaterally. If the prediction is correct, brief right sided test stimuli (i.e. those dealt with by the left hemisphere) should be located later with respect to the standard stimulus sequence than the left sided ones (dealt with by the right hemisphere). (ii) Reaction time measurements can be compared for brief stimuli presented on either side. If the hypothesis is correct, reaction times using the right sensory surfaces (left hemisphere) should be slower than those using those on the left side (right hemisphere).

5.2.4 Definition, contrast and retinal eccentricity

Other predictions follow as corollaries of the basic lateral difference in time/intensity integration: Sensory systems generally are specialised to enhance differences between spatially adjacent parts of the stimulus. The result of this "contrast enhancement" or "spatial differentiation" is that spatially uniform stimuli will produce lower intensity sensory signals than stimuli of equal intensity close to an "edge". Therefore, the variables of contrast and definition in the stimulus should be equivalent to that of absolute intensity in a spatially uniform stimulus. Consequently, for stimuli shorter than the saturation of the time/intensity curve it should require a stimulus with greater definition and/or contrast for recognition by the right sensory surfaces (left hemisphere) than by the left sensory surfaces (right hemisphere). In addition, it is known, for visual stimuli, that contrast sensitivity declines sharply as retinal eccentricity increases (Pointer and Hess, 1989). Thus, at high eccentricities one might expect to find a left field ascendancy, because stimuli are relayed to the cortex with lower degrees of contrast than those close to centre of the fovea, and are then better perceived by the right hemisphere.

5.2.5 Speed of encoding information

In section 5.2.2 time/intensity integration of sensory stimuli to give perceived intensity was described, and gave some predictions about lateral differences in intensity thresholds or reaction times. However, when a stimulus is just perceivable (i.e. stimulus intensity is at threshold), the situation also defines a unit of information acquisition. Presumably if time or intensity is increased beyond those obtaining at the threshold combination, a correspondingly larger number of such units of information can be acquired. The experimental conditions in which this could be measured would be when the dependent variable was "number of items reported" rather than speed of response. Given this, the number of items encoded should be related to the product ($T \times I$), in the same way as is perceived intensity. Therefore, *one would predict there to be a trend to right hemisphere advantage in amount of information reported at low intensities and short durations, which reverses to a left hemisphere advantage at higher intensities and longer durations.*

5.2.6 Duration of persistence of sensations arising from brief stimuli

Once the perceptual and neural trace of an event is established, it will have a more enduring influence if it is spread within the hemisphere by connections having long and temporally dispersed conduction delays than if it is spread by those having shorter and more synchronous ones. According to the hypothesis of slower conduction velocities in cortico-cortical axons in the left hemisphere, and given adequate stable representation of a stimulus, *the perceptual image should last longer after stimulus termination for stimuli presented to the right side (right ear or right visual field) than for those presented to the left.* The terms used to designate the brief persistence of perceptual traces are *iconic* or *echoic* storage (in the visual and auditory modalities respectively). These images of a stimulus should persist longer in the left hemisphere than in the right. One method for assessing the persistence of perceptual traces is by measuring the reaction time for detection of the cessation of a stimulus. There are however other ways of identifying the time of awareness of stimulus offset. In such experiments it is necessary that stimuli should have a supraoptimal duration, otherwise the time/intensity integration in each hemisphere, rather than persistence of a fully integrated percept may determine the measure of persistence.

5.2.7 Stimulus size or complexity

The cerebral cortex is commonly conceived as having a modular construction, with each module being activated selectively in response to a particular sensory feature or abstraction, or when a particular aspect of motor performance is being initiated (e.g. Mountcastle, 1978). Sometimes the different modules of the cortex are laid out in spatial array corresponding to the sensory surface they respond to. In other instances, further away from the sensory periphery, there may be no such spatial mapping. According to the hypothesis of this paper, the right hemisphere uses its connections more for "spatial" divergence and convergence while the left is used for temporal convergence. Thinking of the modular construction of the cortex, one must then envisage that the left hemisphere can more easily co-ordinate the activation of several modules selected as a sequence, the right hemisphere can more easily co-ordinate the activation of several modules when they occur simultaneously (or nearly so). From this, two predictions can be drawn: If we are dealing with arrays of cortical modules which correspond spatially to a sensory surface (such as the retina) *the right hemisphere would be expected to be superior at identifying simultaneous patterns that occupy a large portion of that surface (e.g. of the visual field).* Likewise, in the somatosensory modality, the *right hemisphere might show a superiority in complex tactile discriminations, such as those used in reading Braille.* If we are dealing with collections of cortical modules which represent the quasi-spatial variable of auditory sense, namely frequency, we would expect that *the right hemisphere would gain an advantage over the left as the complexity of the spectral characteristics of a sound increased* (e.g. as the number of combined tones, sounding together in harmony increased, or as the number of harmonics in a sound increased). Such right hemisphere superiorities might be discernible in choice reaction time tasks. However, they would also be evident in measures of accuracy of report, with time for response unlimited.

5.2.8 Accuracy of temporal order judgements

Quite apart from speed of acquiring information, there are predictions to be made about accuracy of acquiring information about temporal order. According to the central hypothesis of this book, the "repertoire" of axonal delay lines is richer in the left hemisphere (for delays up to about 100–150 msec) than in the right hemisphere. Temporal patterns in sensory input spanning such lengths of time can therefore be more accurately represented in the left than the right hemisphere, because in the left there is a greater likelihood of there being axons with the delay appropriate to encode the temporal separation of different components of the pattern. This is illustrated in Figure 3.2(B). In the right hemisphere however, a larger fraction of the axons impose a delay which falls within a single integration time of a neurone (about 10 msec). Such neurones cannot represent *temporal* relations between different elements whose influence coincides upon each neurone, but can represent the "spatial" association of features (Figure 3.2(A)).

One can extend this argument to polysynaptic relays. Abeles (1982b) put forward the concept of "synfire chains" (as mentioned in section 3.2). These are organised chains of neurones which tend to fire repeatedly in the same sequence and with the same timing. Such chains of neurones might exist in either hemisphere. In the left hemisphere however, the chains will, according to the central hypothesis of this book, contain fewer sequential synaptic links, on average, but longer conduction delays between successively activated neurones, than in the right. It follows that, even for polysynaptic chains, the left hemisphere can represent temporal sequences more faithfully than in the right. This is illustrated in the three parts of Figure 5.7, which show very similar chains of sequential neuronal firing. In 5.7(A) the neuronal integration time is relatively small compared to axonal delay times. There is then little tolerance for variation in axonal conduction delay (relative to the scale of average delays) and there can therefore be highly specific representation of temporal patterns. In 5.7(B) conduction delays are compressed into one eighth of the time taken by those in 5.7(A), neuronal integration time remaining the same however. Thus, the neuronal integration time is very long compared to axonal delay times and there would be a great deal of latitude for variation in delays (on the scale of average delay times in that network). The network shown in 5.7(B) is thus not a plausible one: The first network could realistically have been selected by the Hebbian mechanism from a much larger number of other connections (not shown) with different delays. The second network could not have been selected in this way: A large number of other connections (not shown), whose synaptic activation fell within the same integration time as that of the firing neurones actually shown, would equally well serve to link the various nodes in this network. A more plausible network using the time scale of 5.7(B), is shown in 5.7(C). This collapses together all synaptic activations which would summate within a single neuronal integration time. In this version of the network, neuronal activation would rely mainly on the "spatial" aspects of connectivity of synapses converging on each neurone, regardless of conduction delay (because most delays are less than the neuronal integration).

Admittedly, in a network scaled as in 5.7(B) neurones *could* be activated in sequence many tens or even hundred of milliseconds apart. However, this would inevitably involve transmission along many sequential steps, with synaptic relays between each. Synaptic relays are the stage in neural signal transmission where continuous variables operate (rather than the quantised transmission via all-or-none action potentials). A wide variety of parameters could influence the quantitative aspects of synaptic relay, which might make subtle differences, determining, for instance, exactly how many converging active afferents were needed to fire a neurone. Since there would be fewer synaptic relays (for a given temporal span of the chain) in networks such as A than those such as B or C, the former has an inherent capacity to represent temporal structure more robustly than the latter. Networks such as 5.7(C) would however be better for representing "snapshots" of the information available at each moment in time, without "chaining" a succession of them together along a temporal metric. They would also have the compensatory advantage that amplification of signals at each step is higher than in those such as 5.7(A).

It follows from this analysis that, if the central hypothesis is correct, the left hemisphere would be likely to have a "finer grain" for resolution of temporal patterns than does the right, and should be able to abstract more information about timing within a stimulus than can the right hemisphere. This should apply not only to temporal patterns compressed within the time spanned by the repertoire of single axonal connections, but also to patterns considerably longer in time (as in the depiction of synfire chains in Figure 5.7(A)). This is because, once a mechanism is present for representing temporal patterns, of 100–150 msec, that mechanism can be concatenated to create longer temporal patterns. Admittedly, the right hemisphere is expected to show an advantage for brief stimuli in rate of encoding of perceptual information (see section 5.2.5). However, the information so gained is spatial or quasi-spatial, and represents little about sequence, let alone about timing. Therefore, provided one limits oneself strictly to considerations of temporal order judgements, there should be no right-hemisphere dominance, even for brief stimuli, and therefore no reversal of dominance as duration is reduced. *The left hemisphere dominance for representation of temporal information should apply at all intervals long enough for sequence to be detectable.* Even if intensity and related parameters limit the accuracy of representation, it should not lead to an absolute right hemisphere superiority in temporal order judgement, as it does in some of the other paradigms discussed above.

Specifically, *where recognition of a stimulus depends on registering the sequence of the components from which it is composed, there should be a right field, right ear, or right hand advantage.* This could be tested with just two components (involving discrimination between sequences AB and BA), or it could involve longer and more complex temporal sequences. Reaction time measures with brief stimuli do not give information about temporal discrimination within a stimulus, and are not appropriate to test this prediction. They are likely to give right rather than a left hemisphere advantages, under some conditions, as described above.

The fine limit of temporal resolution of the left hemisphere, as of the right, should in theory be related to the neuronal integration time, either for EPSPs or for co-operative interaction between synapses in the Hebbian mechanism for synaptic strengthening. As already discussed, this is about 10 msec. Corresponding to this, Hirsh and Sherrick (1961) found that temporal order judgements, for all sensory modalities, could be made with similar temporal constraints: A criterion of 75% correct was met with separation of different elements by about 20 msec. Thus although there may be lateral differences in

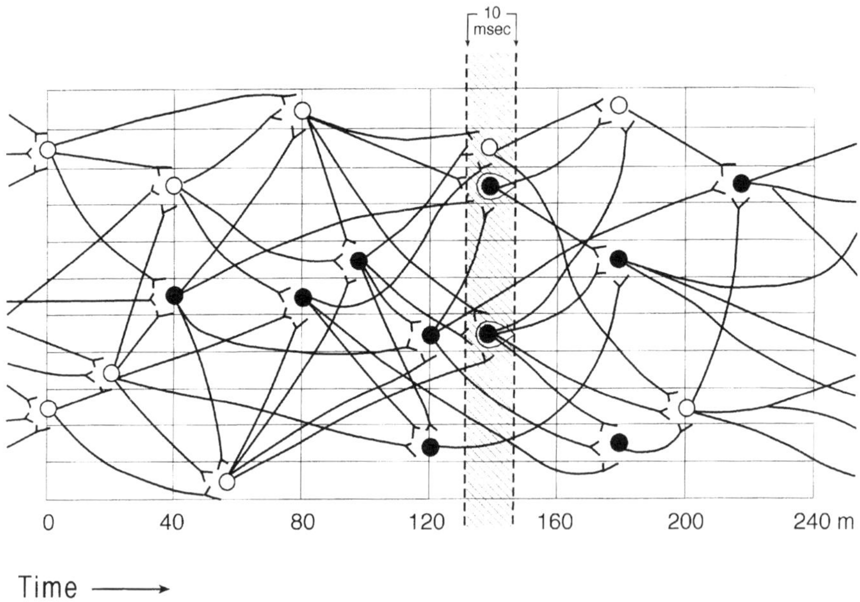
10 msec
0
40
80
120
160
200
240 m
Time

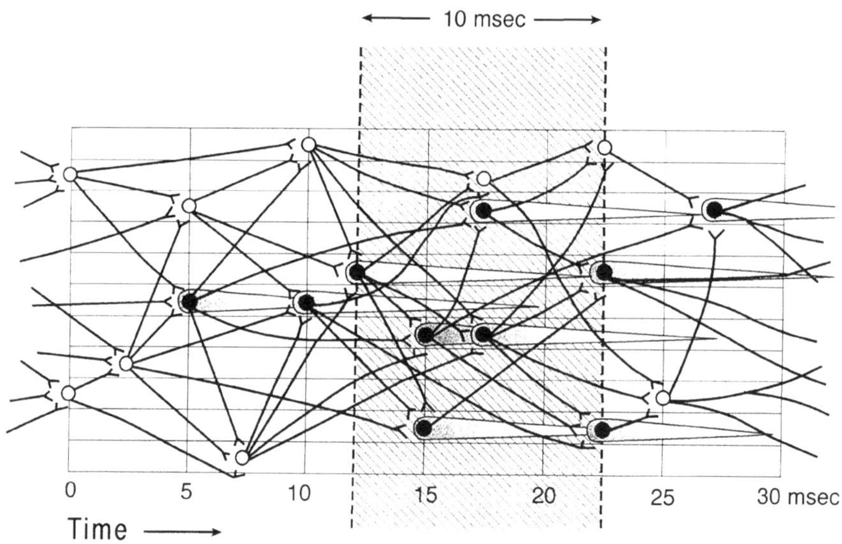
10 msec
0
5
10
15
20
25
30 msec
Time

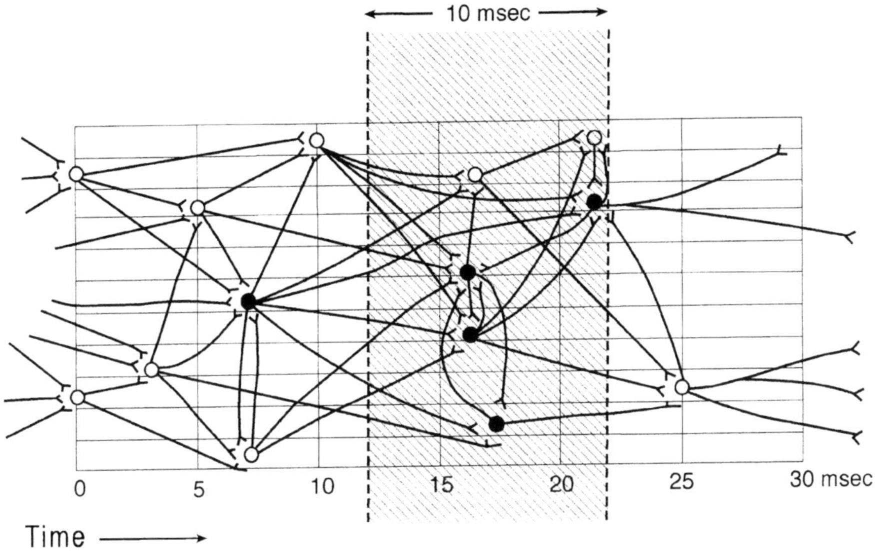

Figure 5.7A, B and C Three Versions of Abeles' Synfire Chains. Each diagram compresses the spatial dimensions of a network into the vertical axis, while time is measured on the horizontal axis. Each horizontal row thus represents a single neurone, and a circle is drawn at particular position(s) on the horizontal axis to indicate the time(s) when it fires. Connections show those axonal links which contribute the activation which makes each neurone fire at a particular time. The latency of each connection is indicated on scale below (suggested as msec, but actually arbitrary). In 5.7(A) triple convergence of active afferents is necessary for a neurone to fire. Connectivity in *A* and *B* is similar in form, but *delays in B are compressed to one eighth of those in A.* The bridge across the network (oblique lines and dashed vertical lines) indicates the time during which integration of EPSPs and the Hebbian process takes place. This is the same in both A and B, namely about 10 msec. The scale of axonal delays compared to neuronal integration time thus differs eightfold between A and B. Filled circles in rows 3, 5, 6, 7 and 10: In 5.7(A) these represent instances where the same neurone fires on successive occasions, the timing of each firing being specified quite accurately by the conduction delays in the three connections which activate each neurone on each occasion. In 5.7(B) however, scaling down the conduction delays brings the successive firings of these neurones within the influence of a single neuronal integration time. Therefore each of these neurones will, on these instances, receive convergent activation by more than three coactive afferent synapses. The temporal structure of firing of this chain would thus be partially lost. *The scaling of 5.7(A) to produce 5.7(B) therefore gives an implausible synfire chain.* The more plausible version of the scaled chain of activity is shown in 5.7(C). In this figure, each time a neurone fires successively within 10 msec, the position on the horizontal axis is represented as the average of the two positions in 5.7(B), and the converging active afferents at the two times in 5.7(B) are drawn at the single time in 5.7(C). This involves drawing some connections as if they have a "negative delay", but this really is a consequence of identifying neuronal activation as a precise point in the temporal axis, when it is really drawn out over a time longer than the conduction delays concerned. As a result of the transformation in 5.7(C) some neurones received 4, 5 or 6-fold synaptic convergences. More complex simultaneous convergences can be represented in the chain shown in 5.7(C) than in that of 5.7(A). Moreover transmission is more secure in 5.7(C) than 5.7(A) (average amplification factor is higher): If the level of activation in the chain falls, transmission will be less severely affected than in the chain shown in 5.7(A).

the temporal information abstractable from stimuli longer than this, neither hemisphere should have a finer temporal resolution than this.

Some paradigms are difficult to allot to the most appropriate predictions. Tests of flicker or click fusion frequency, or of detection of a brief temporal gap in a stimulus could be thought of as tests in which performance (i.e. apparent temporal resolution) is inversely related to perceptual persistence. According to section 5.2.6 (above) performance should then be better with right than left hemisphere. On the other hand, if viewed as temporal resolution in the sense discussed in this subsection (accuracy of abstraction of temporal information) a left hemisphere advantage would be predicted. The latter formulation is probably the correct one: Performance in tests of flicker or click fusion, and gap detection are not determined by the *complete* decay of *all* perceptual trace between stimuli. Rather performance is a function of detection of the least unevenness in the perceptual trace over time, that is, the least perceivable decay of the perceptual trace. According to the central hypothesis of this book this would be achieved by convergence in single neurones from other cortical neurones which represent four slightly staggered events:

(i) the initial continuous stimulus;
(ii) the stimulus offset;
(iii) the succeeding stimulus onset
(iv) the later continuous stimulus.

The hypothesis of temporal representation by a repertoire of delay lines can deal with such a pattern of temporal convergence, just as it does the sequencing of sounds which make up a stop consonant (see Figure 3.2(B)). Therefore this task need never involve the duration of perceptual persistence. Thus, it is argued, such tests access the same sort of information as discussed in this subsection. Measures obtained of how "fine grained" the temporal analysis can be in each hemisphere should have similar significance. One would therefore predict that left hemisphere critical frequencies for flicker fusion would be higher than those for the right hemisphere, and that smaller temporal gaps should be detectable with the right visual fields. The same might be demonstrable in the auditory modality under some circumstances, but since the auditory sense has other means of detecting unevenness in perception (e.g. by transformation into perceived pitch, or perceived position of a sound in space) it is unlikely that auditory evidence will be very useful.

This chapter ends with Figure 5.8, a summary diagram, which captures as a somewhat caricatured form of space/time diagram, the essential difference between right and left hemispheres. At present this diagram is to be applied to lateral differences in the process of perception. However, the same diagram can equally well be applied to other lateralized functions, such as motor preplanning or short term memory.

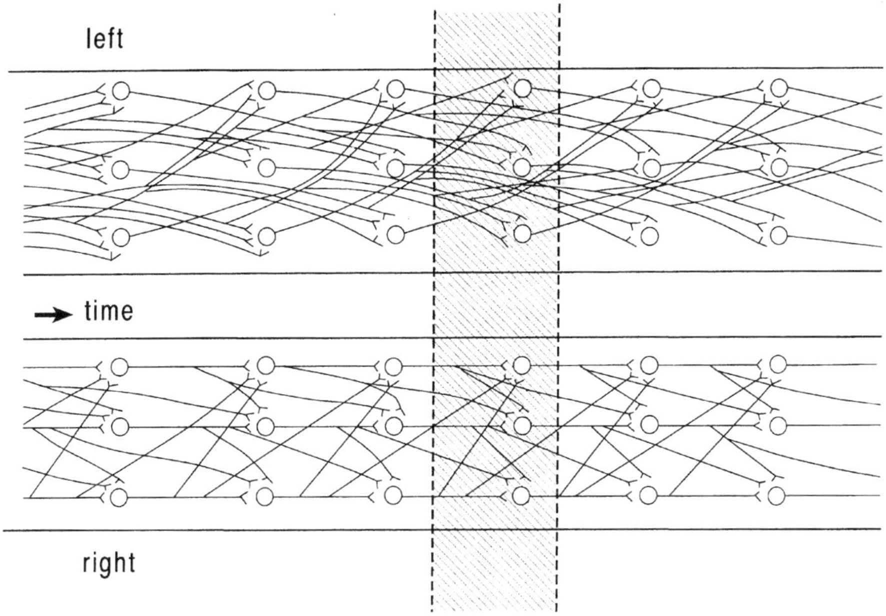

Figure 5.8 Schematised Space-time Diagram of the Difference Between Left and Right Hemispheres. For both the upper and lower staves of this diagram, neurones and their connections are layed out as a geometrically repeating pattern along the time axis. The density of neurones and their connections is the same in each, and the convergence ratio of connections upon neurones in each stave is the same (i.e. threefold). The difference between the two is that connections in the upper stave ("left hemisphere") have longer delays (one each of 2, 3, and 4 time units for each projecting neurone) than those in the lower stave ("right hemisphere") (two of 1 unit of time and a third of 2 units, for each projecting neurone.)

6 Empirical Evidence on the Difference Between Left and Right Hemispheres in Perceptual Processes

As far as the sensory psychophysical differences between the hemispheres are concerned, there is an important constraint to be borne in mind: In establishing the functional difference between the hemispheres, experiments using non-linguistic stimuli, whether auditory, visual or tactile, are more relevant than those using linguistic ones. This is because we already know that the left hemisphere, for whatever reasons, is specialised for language functions. Differences between the hemispheres found using linguistic stimuli may thus originate from the inherent linguistic superiority of the left, even if that superiority has a quite different basis from that hypothesised here. However, if non-linguistic stimuli are used it will be possible to evaluate much more specifically the hypothesis that the left hemisphere has a superior capacity for temporal resolution than the right. If the hypothesis is supported, it would imply that linguistic superiority is only one of the differences that derives from this fundamental cytological and cybernetic difference.

In practice one cannot rigorously exclude linguistic stimuli, because even pure tones or abstract visual shapes have *something* in common with the speech sounds, or written words and letters. Therefore one must recognise a progression amongst stimuli of decreasing affinity to linguistic symbols. In the visual mode words are more linguistic than letters, letters more than abstract shapes. In the auditory mode, words are more linguistic than elementary speech sounds, and these are more linguistic than those made from pure tones. In considering the evidence, support for the hypothesis would be stronger if it generalised amongst different types of stimuli in these lists, and experiments using the later items in each list will permit more definite conclusions, with regard to the central hypothesis, than those earlier in the lists.

One of the most important predictions of Chapter 5 was that there should be a *reversal* of hemispheric advantage from left to right as duration or intensity of the stimulus increases. This is an important aspect of the evidence discussed in this chapter. There are no accurate estimates of the duration at which the reversal takes place, nor can this value be derived on theoretical grounds. Nevertheless, the evidence that is available on reversal of dominance suggests that it takes place between stimulus durations of roughly 50 and 100 msec. Thus, for stimuli less than (roughly) 50 msec in duration, right hemisphere dominance is usually found, while for those of greater than about 100 msec, left hemisphere dominance is usually found. This conclusion is reached in the summary to this chapter (section 6.7), and in Table 10, when all the evidence presented earlier in the chapter is assimilated. This generalisation provides a test of the predictions made in section 5.2.2 which is additional to and stronger than those provided by each section of the evidence taken by itself, where often data on cerebral dominance at only one stimulus duration have been obtained.

6.1 TIME TAKEN TO IDENTIFY BRIEF STIMULI

Evidence is available for auditory and visual modalities, as well as a smaller body of information about somatosensory stimuli. Auditory experiments use mainly the first design mentioned in section 5.2.3, (temporal "location" of a brief stimulus in relation to a standard one), while visual experiments use mainly the reaction time paradigm.

6.1.1 Evidence from the auditory system

Most of the available evidence deals with estimation of the time of delivery of a click in one ear, with respect to a temporal sequence presented to the other ear. Fodor and Bever (1965) carried out such an experiment in which clicks (of duration about 25 msec) were to be located with respect to a sentence (i.e. they had to say during which syllable the click occurred). It was found that clicks to the left ear are perceived as occurring earlier than those to the right ear, a result confirmed later by Bever *et al.* (1969a,b). Bertelson and Tissegre (1972) carried out a more detailed study of this effect, and replicated the original findings. Comparing the condition where the clicks went to the left ear, and sentence to the right, with that of a symmetrical presentation of both stimuli binaurally, the click to the left was advanced by an estimated 70 msec, compared to the symmetrical condition. Gregory (1978) carried out a related experiment, except that the clicks in one ear had to be located with respect to a *musical* sequence in the other. The same asymmetry was detected as in earlier studies. The mean apparent delay for right ear (left hemisphere) clicks was 26 msec greater than for left ear (right hemisphere) clicks.

These results uniformly suggest that the time to integrate a brief sound is less in the right than the left hemisphere. However, in the above papers there are a few other reported findings which make the interpretation more difficult. Bertelsson and Tissegre (1972) found that, averaging between the cases where the click was delivered to the right and left ears, the click was judged to occur earlier in the sentence than it in fact occurred. Gregory (1978) found, in contrast, averaging between conditions, that clicks were *delayed* with respect to the musical sequence. Moreover, Bertelson and Tissegre (1972) found that binaural stimuli (whose perceived position in space would be intermediate between the two asymmetrical conditions) had a delay intermediate between these two conditions. If temporal location of a brief stimulus depended on the hemisphere in which critical suprathreshold convergence from the click occurred first, one would expect that binaural stimuli would be detected as rapidly as by the hemisphere with fastest response (right hemisphere/left ear). This was not the result obtained. The authors suggested that the crucial difference between sides might not be the ear of entry of the stimulus, but the perception of the relative *position of the stimulus in auditory space*. Subsequent research has never reconsidered this point. The interpretation of these additional data is not easy, and will not be dealt with further here.

Six studies compare reaction time to sounds presented in each ear. Four of these involve a choice reaction time. Kallman and Corballis (1975) used pairs of sounds (distinctive notes produced by different musical instruments) presented dichotically, with subjects asked to respond to one of the sounds if presented in either ear. A significant left ear (right hemisphere) advantage (of about 70 msec) was found in the first block of 32 trials, but this was not sustained into subsequent blocks. Rastatter and Gallaher (1982) also obtained a left ear reaction time advantage for judgements of the timbre of different musical instruments. Gates and Bradshaw (1977) measured the reaction time for detection of change in a rhythmic sequence. The left ear had shorter reaction time than the right. Sidtis (1980) required subjects to discriminate the pitch of sine waves or complex tones. For both classes of stimuli reaction time with the left ear was faster than with the right ear. Two other reports compare hemispheres using the simple reaction time method, for detection of tone onsets (Emmerich *et al.*, 1976; 1981). Neither finds any significant difference between ears, and therefore provide no support for the prediction of this section. It should be noted however that, according to theory, perception of onset of a pure tone may be very rapid if intensity is well above threshold, and this may obscure a lateral difference. This qualification may not apply to the tasks in the first four studies, in which the reaction was discriminated to stimuli with specific characteristics. Here simple integration of time with intensity may not apply, but rather recognition of particular, quite transient features. In particular, in the studies of Kallman and Corballis, and Gates and Bradshaw responses were discriminated to a sound with a particular "timbre" or rhythmic feature. Musical "timbre" depends very much on the distinctive sequences of frequencies at the start of each note (the "attack" lasting a few tens of msec). A right hemisphere advantage is predicted for short stimuli (see section 5.2.2), and the stimuli in this task should probably be equated with short stimuli. In both these cases recognition of a temporal pattern is involved, which is very different from integrating intensity over time (see section 6.2 below).

6.1.2 Evidence from the visual system

Several papers compare hemispheres with respect to simple reaction time for identification of familiar visual stimuli in normal subjects. Jeeves and Dixon (1970) found that reaction time to onset of bright point sources presented for 2 msec, was 5–8.5 msec faster for stimuli going to the right hemisphere than for those going to the left, regardless of the handedness of subjects. Jeeves (1972) showed that this difference also occurs in 10-year old children, though it was less in magnitude and found only in right handers.

Supporting evidence comes from Kappauf and Yeatman (1970), using a different method. Subjects were required to judge the simultaneity of onset of two dots (turned on and off, alternately for 2 sec periods) presented almost synchronously to left and right visual fields. In right handed subjects, the onset of the stimulus in the left visual field was judged to be sooner after its objective onset by 7 msec, compared with that in the right visual field. The opposite result was found for left handers. While this result suffers from the disadvantage that it was a test of nasal and temporal hemiretinae in the right eye only, the finding is similar in direction (though smaller in magnitude) to some found using auditory stimuli (section 6.1.1).

Several other papers study visual reaction times with other types of stimulus. Pring (1981) presented four letter words and non-words for 50, 100 or 150 msec, with subjects required to report which category each stimulus belonged to, as fast as possible. There was a left field advantage in reaction time of 20–30 msec, for both words and non-words, provided the exposure duration was 50 msec. At longer durations there was a right field advantage. Rizzolatti and Buchtel (1977) measured choice reaction time to faces previously learned in a practice session. In female subjects no asymmetry was detected. In males however, for exposure times of both 20 and 100 msec there was a significant left field advantage. As exposure time was increased from 20 to 100 msec, the improvement in left field reaction time was quite small (450 to 438 msec) but was very substantial for right field reaction time (600 to 475 msec). Clearly the left hemisphere can benefit more from longer exposure than can the right hemisphere. Sergent (1982a) presented pictures of faces for durations of 40, 120 or 200 msec, which were to be categorised as male or female. At 40 msec there was a left field superiority of 20–30 msec, at 120 msec performance was about equal with either visual field, and at 200 msec there was a right field superiority. (Overall effect of exposure time was significant ($p < 0.001$).) In another experiment luminance was varied reciprocally with exposure duration, so that stimulus energy was kept constant. The same shift of laterality occurred as exposure time was increased. Thus, an exposure duration of 40 msec is suboptimal for the right field (left hemisphere), while the left field (right hemisphere) is closer to optimal performance. A study by Diehl and McKeever (1987) required speeded identification of famous faces, or familiar objects. In this case a right field advantage was found for object naming and a left field advantage for face recognition, and there was no interaction with exposure time (40 msec. *vs.* 120 msec). However, performance in this task may have depended on higher-level processing than in the other tasks discussed in this subsection, as indicated by the large absolute values for reaction times (650–800 msec).

Overall, these reaction-time experiments conform to the prediction of section 5.2.2. Results can be explained by the greater capacity of the right hemisphere for "simultaneous convergence", so that this hemisphere can come to a decision more quickly, when presentation time is limited. The results discussed in the previous paragraph also conform to the prediction in section 5.2.2, that with longer duration exposure, the left hemisphere gains the advantage. Various stimulus types were used in the experiments, and when words were used as stimuli, a left field advantage was still detected at short exposure times. This suggests that the right hemisphere advantage depends on a basic perceptual processing difference between the two hemispheres rather than differences in linguistic processing, which would be expected to produce a left hemisphere advantage.

In addition to these studies, several studies have measured reaction time for subjects deciding whether two letters were the same or different. Use of letters may involve the special linguistic processing by the left hemisphere. However, this factor can be assessed by comparing same/different judgements of letters identical in name (e.g. A,a) or physically identical (e.g. A,A), the latter comparison presumably being more closely related to the basic perceptual properties of each hemisphere. According to Cohen (1972), for physical identity judgements there was a 51 msec advantage for left field presentations. Name identity judgements, which should rely on linguistic processing, were slower, and showed a right field advantage of 43 msec. This result has been confirmed by Geffen *et al.*

(1972), and by Davis and Schmit (1973). However, this demonstration of laterality seems to be a rather capricious effect, sometimes shown only in special circumstances. Hellige (1976) showed a right field advantage for name identity judgements only after some practice at the task; Ledlow *et al.* (1978) found this only if the field in which the stimulus was located, and the type of match (name or physical identity) was unpredictable. Moreover, three studies show different results. Simion *et al.* (1980) found a right field reaction time advantage for both name and physical identity judgements; and Hellige and Webster (1981) found a left field advantage for both types of judgement. Sergent (1983) conducted a complex experiment in which reaction times were measured for physical and name identity judgements, and in addition three variables were manipulated: exposure duration, retinal eccentricity and stimulus size. Although laterality effects were seen with regard to the latter three variables, no difference between hemispheres could be attributed to the difference between physical and name identity matching. No clear explanation of the discrepancies between the various results mentioned above can be found in terms of the values of the several variables known to influence laterality. The finding of opposite hemispheric advantages for the two tasks is thus not very robust.

6.1.3 Evidence from the somatosensory system

Several studies have compared simple reaction time to tactile stimuli delivered to either hand (Peters, 1983; Chentanez *et al.*, 1988; Davidson *et al.*, 1990a), but none has detected significant laterality effects. Most of the visual or auditory reaction time studies showing a left field advantage involved choice reaction time, and it may be that this design, rather than a simple reaction time such as has been used in tactile studies is necessary to show the predicted laterality effect. Reaction times for tactile *recognition* of objects or shapes have been measured but the response is much slower than for visual or auditory sense, and so is not relevant to the basic hypothesis at present under evaluation.

6.1.4 Summary

Experiments using the auditory mode have consistently found right hemisphere advantages for relative timing of brief stimuli, and those using reaction time also find a right hemisphere advantage, provided that a choice reaction time is used and stimuli with temporal structure are to be recognised. Experiments in the visual modality generally report a faster simple reaction time for brief stimuli using the right hemisphere. Choice reaction time, in two studies, shows well the predicted reversal from right hemisphere advantage for short duration stimuli, to left hemisphere advantage for longer ones. Overall the predictions of section 5.2.2 are supported.

6.2 INTENSITY, TIME-INTENSITY AND CONTRAST/DEFINITION DETERMINANTS OF PERCEPTION

6.2.1 Comparison between hemispheres of absolute auditory intensity discrimination

Two studies compare intensity discrimination between hemispheres. Wexler and Halwes (1981) used pure tones of 250 msec duration, and found that right ear tones were perceived to be louder than left ear ones (in right handed subjects). Emmerich *et al.* (1988) tested the absolute auditory intensity threshold for tones of 200 msec duration and also found that the right ear was more sensitive. If it is assumed that temporal integration is essentially complete in both hemispheres after 200 msec, these studies agree in suggesting that the left hemisphere is capable of a greater degree of total time-intensity integration, given unlimited stimulus duration. The implication of this is discussed in section 6.7.

All other evidence on lateral differences in time/intensity trade-off has used visual stimuli. Therefore, as far as auditory (or tactile) stimuli go, the detailed predictions of section 5.2.2 remain for future investigations.

6.2.2 Comparison between hemispheres of stimulus intensity equivalencies for brief visual stimuli

Two older papers, suggest that stimuli of equal objective luminance, presented for about 100 msec, are perceived as brighter or clearer when presented to the left fields than to the right ones (Curtis and Foster, 1917; Dallenbach, 1923). Davidoff (1975) presented subjects with rectangles of varying shades of grey (stimulus duration 100 msec) to right, left or centre field. Right handed subjects saw the stimulus to the left fields as subjectively lighter. Likewise, Davidoff (1976) demonstrated a right hemisphere superiority for discrimination of colour, and Umilta *et al.* (1974) showed the same for recognising line orientation, in each case for stimuli of about 100 msec. Since these results study only one duration and intensity, they do not critically evaluate the prediction of section 5.2.2 that there be laterality reversals as the duration and intensity increase.

6.2.3 Comparison between hemispheres of time-intensity trade-off for visual stimuli

There appear to be no systematic studies explicitly comparing the time-intensity trade-off between the two hemispheres, using visual stimuli. However, more fragmentary evidence exists in three papers. Davidoff (1977) presented small dots for varying periods of time, and found that the shortest time needed for their detection was significantly shorter in the left fields (29 msec) than in the right (31 msec). Sergent (1982b) presented right handed subjects with faces for a duration of 120 msec, which they had to categorise as male or female. With left field presentations high or low luminance had no effect on reaction time. Implicitly, the 120 msec stimulus duration is greater than the saturation time for temporal integration with the right hemisphere. With right field presentations, increasing luminance accelerated reaction time, such that there was a left-field advantage at low luminance and a right-field one at high luminance. The difference was quite substantial: 572 msec at low luminance and 511 msec at high luminance (compared with about 535 msec for left field presentations, at either luminance). Thus the left hemisphere seems to be limited in temporal integration for a 120 msec stimulus if intensity is low. The evidence supports the idea that the time/intensity integration saturates at shorter times for the right hemisphere than for the left hemisphere. However hemispheric comparisons with shorter stimuli would be needed to amplify this. Sergent (1983) performed a complex series of experiments using letter stimuli, with manipulation of three variables: exposure duration, retinal eccentricity and stimulus size. Two exposure durations (20 and 150 msec) were used with luminance held constant. At 20 msec, left field exposures gave a faster reaction time (by 22–30 msec in different experiments) than did right field ones. This difference disappeared at 150 msec exposure. Thus the right hemisphere appeared to synthesise the visual pattern better at short time exposures, the left one being then limited by the brevity of exposure. Glass *et al.* (1985) presented faces for 100 or 190 msec, for a recognition test, and found a left field advantage only for the shorter exposure durations. Christman (1988) employed a temporal integration task, in which a digit had to be synthesised from two asynchronous components, each lasting 8 msec, the interval between them being 55 msec. The dependent variable was an accuracy-of-report measure. Under low illumination the left fields did considerably better than the right ones*. This suggests that the right hemisphere can form a stable representation of a stimulus even if it is presented for only 8 msec, and this image persists for the 55 msec of the interval. On the other hand the left hemisphere, because of its greater temporal dispersion produces a less complete representation for an 8 msec duration, and therefore the persisting image 55 msec later is also less complete. Two studies involved speeded judgements of which of two numbers was greater. With reaction time as a measure, Katz (1980) presented the numbers for only 50 msec and found a left field advantage, while Besner *et al.* (1979) using a 150 msec presentation found a right field advantage. Taken together, the above results provide considerable support for the idea that temporal integration saturates at shorter time intervals in the right hemisphere than the left.

6.2.4 Comparison between hemispheres for identification of brief stimuli differing in contrast or definition

Five reports compare performance by the two hemispheres for stimuli varying in the degree of stimulus degradation or blurring. Hellige (1976), using letter stimuli presented for 100 msec, found that stimulus degradation favoured the left fields for reaction time in name-identity judgements. Jonsson and Hellige (1986) using letter stimuli, and Michimata and Hellige (1987) using non-linguistic patterns, presented in each case for 150 msec, found that blurring impaired performance when stimuli were presented to the right but not the left visual fields. Sergent (1985) presented images of faces for 100 msec, the images having been passed through either a low-pass spatial filter (0–2 cycles per degree), or a broad-pass filter. The low-pass filtered images were better recognised by the left fields, while broad-pass filtered pictures, or the original high resolution pictures, were either non-lateralized or better recognised in right fields.

Four papers examine the effect of varying the degrees of contrast on visual laterality differences with tachistoscopic presentations, with rather discordant results. Davidoff (1977) used a task where small rectangles, presented for 10 msec, had to be detected. There was an overall left field advantage in accuracy of report, with a non-significant trend towards a larger advantage as contrast was lowered. Two other papers (Previc, 1982; Kitterle and Kaye, 1985) using tachistoscopic presentations (100 msec) of grating patterns failed to find a left field advantage at low contrast. This may have been because of

* Luminance increase worsened performance in both left and right fields, but more so in the left, so that, at highest luminance, performance in two fields was about the same. The author comments that "the finding of better performance with decreasing luminance reflects the inverse relation between luminance and duration of visible persistence." (See Bowen et al. (1974), and Haber and Standing (1969)). It is not clear why persistence should decline as stimuli became brighter, nor why this should be more evident for left field stimuli. However, it seems likely that these effects depend on a quite different mechanism from the time-intensity integration occurring while the stimulus is actually presented.

the long duration of the stimulus. Sergent (1984) found a right field advantage in a task involving speeded naming of words flashed for 50 msec, which was increased with reduction of contrast. This result is in opposition to the prediction: A 50 msec stimulus should be well suboptimal for the left hemisphere, but closer to optimal for the right hemisphere, and so one might expect a left field advantage. However, since the experiment used words as stimuli, the perceptual difference may have been overridden by the right field advantage due to use of verbal stimuli.

A number of experiments have studied the effects of varying the eccentricity of the stimulus on the retina. McKeever and Gill (1972b), using letters presented for 50 msec, found that left field reaction time deteriorated much less as eccentricity was increased than did that of the right field. Polich (1978) obtained a similar result, with both reaction time, and accuracy measures. Marzi *et al.* (1979) used sets of nine letter or shape stimuli, arranged in a 3 × 3 array, presented in left or right visual fields for 100 msec. Subjects were required to report the stimuli in a specified column, as indicated by an auditory cue given at various times (0–4000 msec) after stimulus termination. Completeness of report declined as the time of the cue was delayed. In addition however, averaging across delays, for the shape stimuli, there was a significant right field advantage for the inner columns and a significant left field advantage for the outer columns. In this experimental design, where several images had to be assimilated in a limited time, there was likely to be a limit on performance set by brevity of stimulus presentation, as well as by cue delay. Therefore it is plausible to attribute the left field advantage at higher retinal eccentricity to a superior ability of the right hemisphere to represent low-definition visual information. Sergent (1983) found that when stimuli were presented at high retinal eccentricity a left field advantage (of 20–30 msec in different experiments) appeared, not present for small retinal eccentricities. Hirata and Bryden (1976) presented strings of letters to right or left of the fixation point, with the number of letters reported accurately as the dependent measure. A right visual field advantage was seen, which was less for the longer strings (i.e. those which extended to regions of higher retinal eccentricity). Not all studies show the interaction between retinal eccentricity and laterality however. No such effect was seen by Lefton and Haber (1974), or Ericksen and Schultz (1977), and a clear result was also not obtained by Worrall and Coles (1976). Moreover experiments using verbal stimuli (words rather than letters) do not show the effect, probably because of the strong right field advantage likely with this sort of stimulus, overriding the basic perceptual capacities of each hemisphere (see review by Christman, 1989a).

In summary, the results using varying degrees of blurring consistently support the predictions made in section 5.2.2. Lowering of definition by blurring can be thought of as equivalent to lowering intensity, and so should impair the left hemisphere performance more than the right. The left-field advantage for stimuli with high retinal eccentricity, in most studies, also fits the predictions, since these parts of the retinae convey their representation of stimuli to the cortex with lower definition. However, the results involving variations in contrast are inconsistent and difficult to interpret.

Christman (1989a) notes that high spatial frequencies in visual stimuli are less well represented as retinal eccentricity increases. The evidence is then used in support of the hypothesis that the right hemisphere is specialised for low spatial frequencies in visual input. Sergent also interprets her results as evidence that the left hemisphere is better for high spatial frequencies and the right for low spatial frequencies. However, enhancement of contrast or definition has a similar effect to mechanisms filtering for high spatial frequencies. Hence the two hypotheses have closely similar predictions. Nevertheless, with respect to the central hypothesis of this book, it is more useful, in relating results to neurobiological substrates, to emphasise the similarity of variations in intensity to those of definition, rather than to spatial frequency.

6.2.5 Summary

Overall the evidence on the interaction of laterality with variables such as time, time-intensity product, definition, and retinal eccentricity support the predictions made in section 5.2.2 and 5.2.4. There are several clear examples of the important prediction that there be reversal from left to right hemisphere dominance as intensity or duration is increased. The data reviewed support the idea that the "saturation time" for temporal integration in the left hemisphere is longer than that in the right. In detail however, there is disagreement about the implied saturation time for the left hemisphere. In some of the studies referred to, 150 msec is supraoptimal for right field presentations. In others, there is still an effect on reaction time produced by variations of intensity or definition for stimuli of this duration. Sergent (1983), who has conducted the most thorough investigation of these relationships notes that the effects of stimulus duration and retinal eccentricity were not strong ones: Any single variable was usually insufficient to produce a significant laterality effect, though combinations of two could do so.

6.3 THE RATE OF ENCODING OF INFORMATION FROM BRIEF STIMULI

Most of the above studies used reaction time (or other measures of time of perception) as the dependent measure. For assessing rate of encoding, unlimited time for response is required, though the time of actual presentation of the stimulus may yet be sharply constrained. Two of the relevant studies use backward masking and recognition of visually presented letters. In this paradigm, a bright mask is presented shortly after the stimulus, and can reduce the amount of information reported about the stimulus. It is assumed that this is because the perceptual trace of a stimulus does not cease with the stimulus, but lasts for a fraction of a second after it, during which time further assimilation of information is possible, unless the mask is presented. Laterality differences in this perceptual trace are discussed in section 6.4. Here we concentrate on other aspects of results obtained with this paradigm.

McKeever and Suberi (1974) presented letters very briefly (20 msec) and masked them shortly (0–120 msec) later by a bright ring presented also for 20 msec. For right field presentations the deficit in percentage correct recognitions produced by the ring was less severe than for left field presentations, especially for delays of 30–90 msec (see Figure 6.1). The method of percentage correct responses, with no time constraint for responding measures the amount of detail that can be assimilated over the exposure plus delay time rather than the time at which a particular stage of synthesis is achieved (as for reaction time measures). The fact that right field presentations were impaired less by the mask than left field ones suggests that the left hemisphere assimilated information faster during the time 50–110 msec after stimulus onset, and reaches a ceiling before the mask appears. The right hemisphere on the other hand assimilates information more slowly and its operations are curtailed by the mask. Interestingly, at the shortest mask delays (0 and 20 msec) the data presented shows a *left* field superiority. This result, not analysed statistically by the authors, suggests that there may actually have been a reversal of cerebral dominance from right to left hemisphere as duration of stimulus plus delay was increased. This would be in accord with the prediction advanced in section 5.2.5.

Cohen (1976) presented 6 capital letters (all at once) for 100 msec, with backward masking of the stimulus following 20 msec later. Data for the masked condition allowed calculation of the number of items encoded per second by each hemisphere. This value was greater for the right field presentations (22 items/sec) than for the left field ones (18 items per second) ($p < 0.001$), again in accord with the prediction.

A study by Wilkins and Stewart (1974) involved the presentation of two letters in succession, for physical identity matching. The first was presented in a central position, the second was to one or other side, and was separated from the first by an interval of 50 or 990 msec. For the shorter interval, a left-field position for the second letter gave a smaller error score, while for the longer interval right-field positions gave the advantage. Assuming that performance in this test is determined by the accuracy of perception of the first stimulus, the right hemisphere can abstract more information than the left from the initial 50 msec presentation, but the left surpasses the right with long presentations.

A study which appears to give a result opposed to those just mentioned is that of McKeever *et al.* (1973). Two letters were presented sequentially each for 30 msec, with their onsets either 30 or 130 msec apart. In a group of right handers without left-handed family history, the initial letters were perceived better by the right fields, when the second letter appeared immediately afterwards, but not when it was delayed for 100 msec.

Three studies of this type use visual stimuli other than letters or words. Gardner and Branski (1976) presented squares, in one or other visual field, and these either did or did not have a gap in the lateral side. These stimuli were presented for about 50 msec (individually determined to give a 30–40% correct identification of the gap). Accuracy in detection of the gap was greater for left than right field presentations. Nicholls and Cooper (1991) presented subjects with an inverted U configuration, whose descending limbs could be either the same or different lengths. Subjects had to report whether the limbs were equal or unequal, this being an "accuracy of report" measure with time for responding unlimited. By varying the exposure duration it was possible to determine the "inspection time", at which a 90% correct assessment could be made. Inspection times for left field presentations were about 20 msec longer than right field ones. Plotting the percentage of correct responses against duration of exposure showed that this right field advantage applied between 40 and 80 msec. The graph is quite similar to that obtained by McKeever and Suberi, except that no left field advantage emerged at the shortest exposure duration (20 msec). This is however easily explained, because accuracy scores at this exposure duration were quite close to the "floor" of 50% correct by chance alone. In a later study (Nicholls and Atkinson, 1993) with the same task, clearer results were obtained: No field asymmetry was seen for an inspection time of 140 msec, but at times of 65 msec or less a right field advantage progressively emerged. This latter result is in serious conflict with the prediction made in section 5.2.5. The authors suggest that "the tendency for the left hemisphere to out-perform the right hemisphere most likely reflects the left hemisphere's superior capacity for detection of fine temporal events." However, the task involved was a spatial and *not* a temporal order judgement. The predictions of Chapter 5 are that speed of acquisition of spatial information should reverse from left to right hemisphere advantage as stimulus duration is decreased. The meta-analysis of data (section 6.7 and Table 9) produces strong evidence in favour of this prediction. The result of Nicholls and Atkinson (1993) remains as a defiant exception to these predictions.

The evidence in this section is not overwhelmingly in favour of the prediction made in section 5.2.5. Two out of seven results are clearly in disagreement with the prediction. Of the remaining five, three do not use testing conditions in which the predicted reversal with change of stimulus duration would be expected. Two studies do however show that information can be encoded more rapidly in the right hemisphere in the first 50 msec (approximately), but with longer stimuli, the left hemisphere encodes faster. The three other studies making observations on only one side of the reversal point, collectively, support this pattern.

6.4 EMPIRICAL EVIDENCE OF PERSISTENCE OF SENSATIONS ARISING FROM BRIEF STIMULI

Evidence of persistence of perceptual traces in each hemisphere is available for auditory and visual modes, but not for tactile stimuli. The prediction of section 5.2.6 is that the perceptual trace should last longer in the left than the right hemispheres.

6.4.1 Persistence of echoic trace

Two reports using auditory stimuli support this prediction. Emmerich *et al.* (1976) estimated the reaction time to offset of tones presented to either ear. Subjects responded more slowly to tone offset than to tone onset (see section 6.1.1). Moreover, in contrast to the symmetry between ears in reaction latency to onset of the tone, offset of the tone was detected about 30 msec later for right ear presentations than for left ($p < 0.01$). The result was confirmed by Emmerich *et al.* (1981). In this paper the shorter persistence of the echoic trace in the right hemisphere was also shown in another experiment where subjects were required to adjust the time of a light flash until it was judged simultaneous with tone offset. With the left ear, tone offset was detected about 23 msec earlier than with the right. Since this experiment involved no rapid response, the difference must have been on the sensory rather than the motor side of the response.

6.4.2 Persistence of iconic trace

Two reports using visual stimuli support the prediction that the perceptual trace of a stimulus should last longer in the left hemisphere than in the right. Brown *et al.* (1984) assessed simple reaction time to detect the termination of visual stimuli, which were presented for 1–2 sec. Termination of a visual stimulus in the right visual fields elicited significantly later responses than those in the left fields (by 10 msec when the left hand is used, by 6 msec when the right hand is used). Polzella *et al.* (1977) presented patterns of 1–5 dots for 16, 25 or 40 msec in random sequence. Subjects had to estimate the duration of each pattern on a 5-point ranking scale. There was a significant effect of laterality, with right field presentations being judged longer than left field ones ($p < 0.001$). Although the results cannot easily be translated into real time units, the data provided suggests that, on average, right field presentations were perceived to be 55% longer than left field ones. These two reports should be compared with that of Rose (1983) who found no lateral difference in contrast sensitivity, using gratings on an oscilloscope screen with subjects given unlimited time to make their decision.

Another experiment by Christman (1989b) uses the temporal integration task described in section 6.2.3, in which a digit has to be synthesised from two asynchronous components (each lasting 8 msec). The partial images were separated by either 30, 55, 80 or 105 msec. As found in Christman's 1988 paper (see section 6.2.3 above) for the shorter two intervals accuracy of report was better with left visual field presentations. This result is probably explained by the superior integration in the right hemisphere during the presentation of the first partial image (rather than by laterality of visual persistence). However, for the 80 msec separation interval, a significant *right* field advantage was found ($p < 0.025$), which may indicate a lengthening of persistence of the icon in the left hemisphere which is sufficient to override the superior initial representation in the right hemisphere. At the longest separation interval this effect was not found, but an adequate explanation of this was that accuracy of report at this interval had fallen so low as to be mainly based on guessing from what was seen of the second partial image, rather than on any lateralized persistence of the first.

Several other papers provide evidence seemingly refuting this laterality effect. Kappauf and Yeatman (1970) found that the latency for detection of termination of light spots was marginally longer for left than for right field presentations in right handers (by 2.9 msec) and was longer for right field presentations in left handers (by 7.3 msec). Although the design of their experiment was mainly similar to that of Brown *et al.* (1984), an important difference was that the comparison between hemispheres was made on the basis of stimulus presentations to the nasal and temporal fields of the right eye only. Since there are perceptual differences between nasal and temporal fields of the retina (Rains, 1963; Rutschmann, 1966), their result cannot be attributed solely to hemispheric differences.

McKeever and Suberi (1974) carried out an experiment, mentioned above, in which subjects were asked to identify single letters presented for 20 msec, with a mask presented following intervals of 0 to 120 msec. The dependent measure was accuracy of report. Since the mask would erase the persisting trace from which the report was made, time dependent deficits associated with the presence of the mask were thought to reveal the duration of the perceptual trace. The results are presented in Figure 6.1. As the mask delay was reduced from 120 msec, the left visual field stimuli suffered a decrement sooner (at about 90 msec delay) than did right field stimuli (at about 60 msec), with overall smaller performance deficits as a result of the mask for right field presentations than for left. According to the authors, this result implied that the perceptual trace endures for a *shorter* period in the left hemisphere than the right, apparently a clear contradiction of the prediction of section 5.2.6. However, there are hints that this anomalous result arose because laterality effects of perceptual persistence were not separated from those of initial temporal integration. An additional source of misinterpretation is that "accuracy of report" measures, with unlimited time for response, give different information from reaction time measures, where speed is critical. Measures of "percent correct" with no time constraint for responding is likely to measure the amount of detail that can be assimilated over the exposure plus delay time (50–110 msec), rather than the time at which a stage in synthesis is achieved (as for reaction time measures). Assume, according to prediction, that for durations in this range, the left hemisphere can assimilate information more rapidly than the right (see section 6.3 for evidence). One would then expect that, given unlimited time for report, more information could be abstracted from the trace with right field presentations than with left. This is in fact the result obtained, with the prominent exception of the longest delay intervals, where there is no lateral difference. The data for delays between 30 and 90 msec can thus be explained on the basis of faster encoding in the left hemisphere. On the other hand, with stimulus presentations as short as 20 msec, such as used by McKeever and Suberi, a consistent *left* field advantage has been shown in reaction time experiments due to its faster temporal integration (discussed in section 6.1.2, especially the results of Pring (1981) and Sergent (1982a)). For accuracy of report measures, such as used by McKeever and Suberi, the same should be true provided no use can be made of the perceptual trace, that is when the mask occurs immediately after stimulus offset. The results of McKeever and Suberi actually show this left field advantage at the shortest mask-delay intervals of 0 and 20 msec (see Figure 6.1). The fact that, at the longest intervals there is not the right field advantage predicted by the speed-of-encoding argument might arise because of a "ceiling effect", similar between hemispheres, limiting the total amount of information which can be abstracted from such a brief stimulus. Thus one can explain McKeever and Suberi's results on the basis of principles already advocated, without resort to an anomalous right hemisphere superiority in visual persistence, which conflicts with much other evidence.

Cohen (1976) also used the masking paradigm. She presented displays of six letters in a 3×2 configuration either for 100 msec without a subsequent mask, or with a mask appearing 20 msec after the display terminated. Subjects were required to report as many letters as possible from each display. In the second condition, in which the sudden mask obliterated any persisting perception, this measure gave an estimate of the number of items

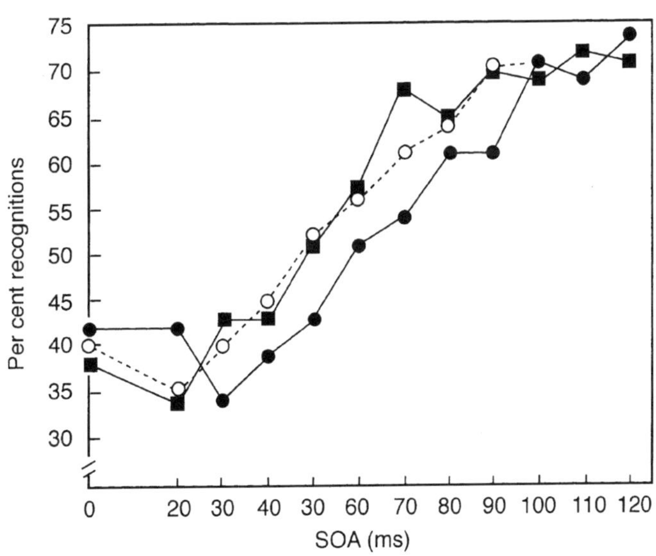

Figure 6.1 Data from McKeever and Suberi (1974), Reproduced with Permission of Academic Press. The experiment is described in the text. SOA: stimulus onset asynchrony. Filled squares indicate recognition scores for RVF (obtained), filled circles indicate those for LVF (obtained). The symbols o-o (LVF adjusted) are not relevant for the interpretation offered in the text (This curve refers to the authors' interpretation that the curves for the two sets of observed data can be superimposed by a simple translation in the horizontal axis.)

encoded per second. For this there was a right-field advantage. (The reason for this is discussed in section 6.3) The improvement in performance which accrued when there was no mask could then be expressed as the duration of visible persistence. It was found that visible persistence, so calculated, was 57 msec with left field presentations, and only 34 msec for right field ones. This result again seems to contradict the prediction of section 5.2.6, although the 100 msec presentation is long enough for the confound of the previously discussed paper to be avoided. However, it is possible that the apparently longer persistence in the left fields arose because of a "ceiling effect" for right field presentations. Since a right field advantage in speed-of-encoding of letters was observed, right field performance may have reached a limit within the first 120 msec, and so could improve little further if the mask was omitted. Encoding from left field presentations however would not have reached this limit at 120 msec, and so could improve given a longer time to examine the perceptual trace. This limit is not actually the maximum number of letters presented, because only two or three out of the six letters could usually be reported even with the right fields. Nevertheless, given the large size of the arrays there may have been another smaller limit in this task, related for instance to use of the more eccentric parts of the retina. Thus, the greater impediment produced by the mask on the left side compared with the right may have had nothing to do with the true duration of persistence in each hemisphere.

Di Lollo (1981) used a task in which a pattern of dots was displayed sequentially over a period of 50–150 msec, subjects being asked to report which of the dots was missing. No laterality effect was observed, but this is not surprising because the very brief duration of presentation of each dot (1 μsec) would by no means saturate the time/intensity integration mechanism.

6.4.3 Summary

The evidence using auditory stimuli is easier to comprehend than that using visual stimuli, and supports the prediction made in section 5.2.6. The issue of lateral difference in iconic storage has been reviewed by Wurst and Long (1983), who conclude that a laterality effects is as yet not proven, with many factors influencing the results, in different ways in different tests. While this may be true overall, those studies in which experimental conditions are most appropriate to test the present hypothesis do in fact give it support, and there is no decisive result against its prediction.

6.5 STIMULUS SIZE OR COMPLEXITY AND LATERALITY EFFECTS

Evidence is available with respect to the variables of complexity and stimulus size for auditory, visual and somatosensory stimuli. The evidence is reviewed in relation to the prediction of section 5.2.7.

6.5.1 Complex patterns of auditory stimuli

A number of papers give information comparing perception of musical chords presented to either ear. All of them use temporally extended chords (in range 0.5–2 sec in most cases), rather than brief presentations comparable to tachistoscopic visual stimuli. Such experiments have been conducted in various ways. Gordon (1970, 1978a), Doehring and Ling (1971) and Peretz and Morais (1979) employed matching of test stimuli against a sample chord. These studies find a left ear advantage in accuracy of matching. In the case of Doehring and Ling (1971) this was found only in musicians, for matches of an arpeggio sample with a chord test. Peretz and Morais (1979) showed the left ear advantage in non-musicians. Preisler *et al.* (1989) found a left ear advantage in accuracy of tuning of a two-note chord with a joystick. Gaede *et al.* (1978) and Burton *et al.* (1989) required subjects to judge the number of notes in a chord of 1–6 notes. There was a left-ear advantage in non-musicians, which was reduced or reversed when subjects with musical experience or aptitude were used. Tramo and Bharucha (1991) studied the improvement of accuracy of detecting mistuning in a chord which arises when primed by an earlier-presented chord harmonically related to the first. This effect was obtained only for left ear presentations.

Another way in which auditory stimuli may be varied in complexity is by altering the number of harmonics added to a basic sine-wave tone. Sidtis (1980) compared the accuracy of recognition of the pitch of tones which were either pure sine waves, sine waves with the first, or the first and second harmonics added, or square-wave tones (with accentuation of the odd harmonics). Right ear performance was unaffected by this progression of increasing complexity in stimulus characteristics. Left-ear performance gradually improved through the sequence, to give a clear left ear advantage with the square-wave tone. In a later paper (Sidtis and Volpe, 1988) it was shown in brain-damaged subjects that right hemisphere lesions (but not left hemisphere ones) impaired the recognition of the complex square wave tones. Sometimes, the right hemisphere advantage in representing spectral complexity depends on musical experience. Kellar and Bever (1980) presented their subjects with three tones in sequence two of which were related to the other either as exactly the intervals of a "fourth" or a "fifth" or approximately so. Following this sample a test series of twelve tones (500 msec each) was presented and subjects had to assign each to the correct one out of the three sample tones. In non-musical subjects there was a significant right ear advantage, but this was replaced by a left-ear advantage in musicians. The authors comment that the musicians could solve this task by remembering Gestalts ("fourth", "fifth") whereas the non-musicians remembered the relation of the individual tones to one another. Samson and Zattore (1988) also studied brain-damaged patients, in a test of detecting the altered note in a three-note chord. There was no significant difference between patients with right and left temporal lobe excisions, although the right-lesioned patients made a greater absolute number of errors. Zatorre (1988) conducted a more subtle test of the ability to integrate related frequencies sounded together. They made use of the fact that pitches of related frequencies can be combined in such a way that they are common harmonics of a particular fundamental frequency. For instance, the frequencies 600 Hz, 800 Hz, 1000 Hz and 1200 Hz, sounded together, imply a fundamental frequency of 200 Hz, whereas those of 600 Hz, 900 Hz and 1200 Hz imply a 300 Hz fundamental. The two combinations do not differ greatly in the overall distribution of power in different frequency ranges. Nevertheless, in normal subjects, a transition from the first combination to the second is perceived as an increase in pitch. Zatorre (1988) showed that discrimination between combinations of this type was impaired in patients with right (but not left) temporal lobe lesions, especially if the lesion involved the primary auditory cortex. In similar tests when the fundamental frequency of each combination was present rather than implicit, no impairment was found. One paper (Gates and Bradshaw, 1977) failed to find lateral differences in perception of changes in two-note chords, but this test appeared to assess analysis of musical sequence as well as chord tuning. Overall these results demonstrate that the right hemisphere is superior in recognising auditory patterns whose components occur simultaneously.

6.5.2 Retinal size or complexity of visual stimuli

Several papers address the prediction that the right hemisphere should increase its superiority over the left as size of the stimulus is increased. Two of them use stimuli which are large letters themselves consisting of arrangements of small letters. The subject is asked to focus attention on either the large or the small letters. Martin (1979) presented such stimuli for 100 msec either in right, left or centre field, with a mask occurring immediately on termination of the stimulus. Reaction time or error rate was the dependent measure. When attention was drawn to distinguishing the local letters there was a right field advantage (especially when local and global letters conflicted). When attention was drawn to the global letters, there were no consistent field differences. Sergent (1982c) presented stimuli for 150 msec or 62 msec, using reaction time as a measure. With the longer-duration stimulus there was a left field superiority for large letters and right field superiority for small letters. When stimuli were presented for the shorter time, only the large letters could be identified, and this faster with left than right fields. Taken together, these results support the idea that when the right hemisphere performs rapid temporal integrations, it can take advantage of widely extended stimuli better than can the left. The left hemisphere advantage for small letters seen at 100 or 150 msec may reflect a superiority which accrues to that hemisphere because of the long duration of the stimuli.

Additional support to the prediction comes from the results of Sergent (1983). Subjects were required to match pairs of letters, with letter size as one of the variables (0.75 or 3 degrees). Reaction time was the measure used. For large letters, and high retinal eccentricities, there was a strong left field advantage (about 50 msec; $p < 0.01$), but this disappeared with the use of small letters. This effect was larger at the shorter exposure duration (20 msec) than the longer (120 msec). For small retinal eccentricities a right field advantage was seen for small letters (the usual condition for reading). This combination of findings is compatible with a superior capacity for wider spatial convergence in the right hemisphere than the left. At exposures longer than the saturation point on the time/intensity curve, it is reasonable to envisage that there be no additional advantage to the right hemisphere from wide spatial convergence. Likewise at low eccentricities, where contrast sensitivity is high, large images should give no further advantage to the right hemisphere.

Taylor and Hellige (1987) presented two vertically orientated bars, one above the other, either aligned or offset by 0.5 degrees. Presentation time was 150 msec, probably supraoptimal for both hemispheres, and subjects had to respond "same" or "different". The retinal width of each bar was 0.5, 1.0 or 2.0 degrees, and reaction time was used as a measure. At all widths, left field presentations gave slower reaction times than did right field ones, for reasons that are not clear. Decreasing the width of the bars slowed reaction time for "different" responses, and this effect was more pronounced for right than left field presentations.

Christman (1987) used a temporal integration task in which a digit had to be synthesised from two components, separated by 60 or 100 msec. The response measure was the proportion of correct identifications. For 60 msec presentations and small letters the two fields gave similar performance. As the size of the digit was increased, right field performance decreased at 60 msec to give a left field advantage. No change in performance with left field presentations occurred as a function of size at either interval. In other words, decrease in digit size at 60 msec improved performance by the left but not the right hemisphere. The 60 msec time of separation may be regarded as suboptimal for the left hemisphere, because full integration (of the first component) might not have occurred at that time. Larger sizes impaired left hemisphere performance, presumably because the pattern required integration over a wide spatial area. The same times of separation of the components were however supraoptimal for the right hemisphere, and so there is no additional effect of size for this hemisphere.

Pring (1981) used four letter words and non-words, with exposure durations set at 50 msec. Three sizes were used: 38′ × 2.24 degrees; 48′ × 3 degrees; 1.08 × 3.39 degrees. The effect of manipulating stimulus size was complex: For words, there was a right field advantage for medium sizes which disappeared or reversed for either smaller or larger stimuli. For non-words there was a left field advantage for medium sized stimuli and a negligible difference between sides for smaller and larger stimuli. The author suggests that his results can be generalised as a right field advantage for combinations of sizes and stimulus types that are easy to decipher and a left field advantage for sizes that are difficult to decipher. This may be so, but the results are not easily incorporated into the hypothesis put forward here about interhemispheric differences in performance.

Apart from evidence about retinal size of stimuli, there is a little evidence about laterality in perception of stimuli varying in complexity. Fontenot (1973) presented shapes varying in complexity (6 pointed, *vs.* 12-pointed) for 10–25 msec, measuring accuracy of identification. There were no differences between fields for the simpler shapes, but increase in stimulus complexity worsened performance by the right fields, so there was a left field advantage for the more complex shapes. In a similar task Edwards (1979; cited by Besner *et al.*, 1979) also, found a left field advantage for more complex figures, but Hannay *et al.* (1976) failed to find significant interaction between complexity and lateral differences.

In summary, the evidence does not point unanimously in favour of the prediction made in section 5.2.7. Nevertheless there is a balance of evidence (five out of eight comparisons) in favour of right hemisphere superiority for large stimuli, or an improvement in right hemisphere performance with respect to left as stimulus size is increased. Some evidence also suggests that the left fields may have an advantage for more complex stimuli. The evidence is sometimes complex and problematical, and discrepancies with the prediction cannot be easily explained.

6.5.3 Laterality and complex patterns in the somatosensory modality

Hermelin and O'Connor (1971) made a chance observation that a blind child who could read Braille was totally unable to do this when he had injured what he called his "reading" (left) hand. They followed this with an experiment in which it was shown that blind children (8–10 years) read Braille more quickly with the left hand than the right. Blind adults showed no difference between hands in speed of reading, but their accuracy was better with the left than the right middle finger. Two other studies of normal subjects confirm these results. Rudel *et al.* (1974) taught Braille characters to normal sighted children using the "paired associates" method (i.e. a tactile-verbal association). Although this was in part a verbal task, it was performed better with the left than the right hand.

Three studies used a "delayed matching to sample" task with tactile perception. Witelson (1976) used as subjects boys and girls aged 6–13 years who had to palpate two different nonsense shapes, one in each hand, for 10 sec. They were then asked to choose a match to each from a visual display of 6 shapes. Boys showed a left hand advantage, while girls up to the age of 13 performed equally with either hand. Milner and Taylor (1972) tested commissurotomy patients, head injury patients, and normal controls in a task where wire figures had to be recognised by touch. The task was very easy for normals and those with head injury but intact commissures (the figures were quite large compared to Braille characters), and in these groups showed no asymmetry of performance. In the commissurotomy patients, performance with either hand was inferior to that shown by the other groups, but was significantly better in the left than the right hand. The authors suggested that both hemispheres normally contribute to performance of this task, communication between the hemispheres therefore being important. Comparing single hemispheres in isolation however, the right played a dominant role. Bottini *et al.* (1995) have confirmed, in brain-injured subjects, the right hemisphere advantage in matching nonsense objects by shape. They add the contrasting fact that tactile matching of objects of the same category, but different shapes (e.g. different-shaped rings) gives a left hemisphere advantage, presumably mediated by an implicit verbal key to the category.

Smith *et al.* (1977) confirmed the left hand advantage in Braille reading in sighted subjects, using error rate as the response measure. However they found that when music was played to the left ear (but not the right) this asymmetry was reversed. In this condition, the right hand was not only better than the left, but was as good as the left hand in the "no-music" condition. This result suggests that the left hemisphere has a potential to read Braille as efficiently as the right, but normally this potential is not realised, unless the right hemisphere is engaged in other tasks. Hemispheric interactions are thus implicated in this laterality effect. Smith's conclusion, which is somewhat discrepant with the other evidence on Braille reading, will be discussed in a later section dealing with lateral differences in relation to attention. In view of Milner and Taylor's result with commissurotomy patients, it is likely that the right hemisphere does indeed have some innate superiority over the left in Braille reading, which does not arise simply as a result of interhemispheric interactions.

A recent study of a different sort, using brain-injured patients, supports the idea that the right hemisphere is superior to the left in representing complex somaesthetic Gestalts. Leonard and Milner (1991) required subjects to move a joystick to a specific position on a dial, and then to repeat the same position, from different starting points, immediately or after a delay. Patients with large right frontal lesions were impaired on this kinesthetic memory task. Small frontal lesions or temporal lobe lesions produced no impairment.

These experimental findings also receive corroboration from clinical facts, well known to neurologists, that damage to the right hemisphere (usually the right parietal lobe) often results in an impairment of a person's image of their own body. In extreme cases the patient may feel that part of their body is alien to them, not part of themselves. This may be interpreted as the patient having awareness of part of the body as perceived exteroceptively with the intact left hemisphere and the corresponding right hand, but the proprioceptive body image normally perceived with the damaged hemisphere is lost. The complex collection of proprioceptive sensations required to form an image of one's own body is typical of the complex Gestalts dealt with better by the right hemisphere. It is thus not surprising that this strange syndrome is generally found with right rather than left hemisphere damage. Carpenter *et al.* (1995) recently reported a graphic illustration of the same principle in patients with intact brains. Neurosurgical patients were being evaluated with the sodium amytal test, in which each hemisphere is subjected to anaesthesia by injecting the short-lived barbiturate into the corresponding carotid artery. When the right hemisphere was anaesthetised, patients had no awareness of the weakness of the contralateral arm, nor any memory of that weakness afterwards. With anaesthesia of the left hemisphere, both these function were intact. Presumably the normal right hemisphere, but not the left could represent arm proprioceptive sensation, as a complex Gestalt.

6.5.4 Summary

The major predictions of this subsection receive considerable support: The right hemisphere is superior to the left in recognition tasks where there is complexity of simultaneous components of the stimulus, or a large spatial extent to the stimulus. This is shown clearly in the auditory or the somatosensory modalities. Visual recognition with widely extended stimuli, directed at the right hemisphere is better than that for such stimuli directed at the left in the majority of studies, though there are some conflicting results. Since the stimuli in most of the studies reviewed are temporally extended ones, it is clear that the right hemisphere advantage gained for complex stimuli overrides the left hemisphere advantage usually seen for stimuli longer than about 100 msec. This generalisation is specifically confirmed by the result with Sergent (1982c) where a left field advantage with large stimuli is seen at both 62 and 150 msec exposures.

6.6 TEMPORAL ORDER JUDGEMENTS

This section reviews an aspect of the predictions made in Chapter 5 on which there is the richest collection of evidence, concerning the abstraction of temporal information from stimuli. The largest part of this evidence is about auditory perception, as might be expected, since human language, the most significant lateralized function of humans, is in origin an acoustic form of communication. However, there is also a significant body of information about temporal order judgements in the visual modality. The predictions in this section derive from section 5.2.8. In dealing with evidence about abstraction of temporal structure from a stimulus sequence, sequences may vary in duration from a small fraction of a second up to many seconds. Generally speaking integration within sensory input achieved over many seconds is considered in this book as an operation of "short term memory" (or "working memory"), while that achieved over a fraction of a second is considered as a perceptual process. A somewhat arbitrary division is made between the two processes: Integration within intervals less than one second is considered under the heading of perception and is dealt with in the section below. That occurring over intervals longer than one second is considered in Chapter 10, under the heading of short term memory.

6.6.1 Experiments using auditory stimuli

In the auditory mode the most obvious stimuli for which temporal order judgements within the relevant duration range are made are speech or speech-like stimuli. In particular, consonants, and especially stop consonants consist of rapid changes of frequency compressed into an interval of 50–100 msec (rapid "acoustic transients"). This is exactly the sort of sequence which could be represented by the axonal conduction delay mechanism postulated in this paper (see section 2.1, and Figure 3.2(B)). Liquid consonants (/l/ or /r/) and especially vowels have less complex temporal structure, and so their recognition should be less likely to show a left hemisphere advantage. Other sorts of auditory stimuli can be used for testing the hypothesis however. In the section below, experiments are grouped according to whether speech-like or non-speech stimuli are used, and also according to whether normal or brain-injured subjects are used. Temporal order judgements may require the subject to make an explicit statement of the perceived order of stimulus components. Alternatively the measure of the subject's temporal order perception may be based on accuracy of recognition of a stimulus defined by temporal order, without an overt statement of the perceived order.

6.6.1.1 Speech-like stimuli: Normal brains

Studdert-Kennedy and Shankweiler (1970) presented pairs of synthetic syllables, one to each ear, and each consisting of consonant-vowel-consonant (CVC) sequences, with total duration about 400 msec. In each pair, only one component differed, and subjects were asked to report what they heard in a specified ear. A right ear advantage was found for initial consonants, a weaker one for final consonants and non-significant effect for vowels. Haggard and Parkinson (1971) confirmed the right ear advantage for stop consonants, but found that other factors (e.g. emotion implicit in a sentence) could reverse this effect. Day and Vigorito (1973) used pairs of syllables, which could start with a stop consonant, a liquid consonant or a vowel, the onset differing between ears by 50 msec. Subjects made a judgement of which started first. For accuracy scores there was a right ear advantage for stop consonants, a reduced right ear advantage for liquids and a left ear advantage for vowels. Cutting (1974) also used tests in which different CV or V sounds were given to each ear, differing in onset by 50 msec, and subjects had to match the first-delivered sound with a diotically presented sample. Accuracy of report for each ear and each sound type was the response measure. There was a right ear advantage using speech-like sounds (which had a full frequency range, whether or not there were rapid transients). This was strongest for stop consonants, and progressively less for liquids and vowels. By comparison, stimuli made up of pure sine waves showed a non-significant left ear advantage. In addition, sounds with rapid transients (sounding like consonant-vowel rather than vowel) gave a right ear advantage whether or not they were speech-like in terms of spectral characteristics. Those without rapid transients gave no lateralization. A right ear advantage was also shown for non-phonetic (unsayable) sounds provided they had rapid transients. Schwartz and Tallal (1980) used synthetic stop consonants with acoustic transients spread over 40 msec or over 80 msec, presented dichotically with subjects required to report from both ears. There was no overall difference in accuracy between the 40 and 80 msec transients, but there was a right ear advantage, which was much better for 40 msec than 80 msec transients.

6.6.1.2 Speech-like temporal patterns: Unilateral head injury

Three papers report laterality effects with speech-like stimuli in head injury patients. Tallal and Newcombe (1978) required subjects to recognise syllables consisting of stop consonants and vowels, with the formant transition lasting 40 msec or 80 msec. At

40 msec only 3/10 patients with left hemisphere damage reached criterion, though 9/10 of those with right hemisphere damage did so. With 80 msec transition, 7/10 left hemisphere patients reached criterion. Clearly there was a left hemisphere advantage for consonant-like sounds with rapid transients. Lackner (1982) used a task in which subjects had to identify the order of repeated sequences of speech-like syllables. These could be repetitions of either CV, V, CV, V; or V,V,V,V, or CV,CV,CV,CV. Each syllable lasted 200 msec. Left hemisphere-injured patients were impaired on this task, equally for left or right ear presentations of the sequences, and as much for the vowel sequences as for those involving consonants. Right hemisphere-lesioned patients were similar to controls on most tests, who showed no lateral differences. Thus the left hemisphere advantage can be demonstrated even with vowels if the conditions are right. This result supports the idea that the left hemisphere is involved in resolving temporal sequences as such, and not just consonant-like sounds (for which the advantage could arise from some linguistic feature, other than temporal complexity). Sidtis and Volpe (1988) also found that left hemisphere patients were impaired in discrimination of speech stimuli.

6.6.1.3 Non-speech-like temporal patterns: Normal brains

Stimuli where temporal patterning is reduced to its barest essentials are those consisting of clicks or Morse-like acoustic patterns. These will be dealt with first in this section. The temporal patterning in musical melodies introduces further complexities, and is discussed later (Chapter 10).

Mills and Rollman (1980) used click stimuli to each ear, with the temporal separation between them varied in order to estimate the threshold for temporal order judgements. The left-ear first click had to be further in front of the contralateral one than did the right-first click, in order to be perceived as separate. The difference was between 3 msec in 35 msec and 6 msec in 60 msec, depending on the experimental method used.

Studdert-Kennedy *et al.* (1970) performed a related experiment. They presented different consonant-vowel syllables to each ear (i.e. dichotically) and slightly staggered in onset time, with subjects required to report both syllables. Overall the advantage in accuracy of report accrued to the lagging syllable. For steady state vowels presented in the same way, or monotic presentation of staggered CVs, the leading syllable had the advantage. The authors suggest that the effect with CV syllables occurs because the trace of the first syllable cannot be formed if it is interfered with by the second syllable. A second, more interesting result was found that the leading CV syllable was disadvantaged more if it was delivered to the left ear rather than the right, for equal time separations.

There are two potential approaches to interpreting the laterality effects in these experiments. It could be suggested that the left hemisphere is so much better at making temporal judgements, or in representing speech sounds, that it is used in either case. Then, with left-ear presentations (projected to the right hemisphere) there is an additional time needed for conduction to the left hemisphere via the corpus callosum. Therefore, the left ear stimulus in Mills and Rollman's experiment would have to be advanced further than in the reverse condition. This is what was found. In the case of Studdert-Kennedy *et al.*'s experiment, one would expect more interference of the leading stimulus if it is delivered to the left ear, than to the right. This is because the initial left ear stimulus would be delayed by callosal conduction time and so would be interrupted at an earlier stage in its integration by the lagging right ear stimulus, than in the reversed condition. This also is what was found.

Alternatively it might be suggested that either hemisphere could accomplish temporal judgements, but with different efficiencies. The one which is activated first then determines overall performance. On the basis of results discussed above we may assume that the right hemisphere would detect a click more rapidly than the left. Therefore, a left-ear-first click would need to be advanced in time less than a right-ear-first click, to be judged separate. This is the opposite of the result obtained.

For Studdert-Kennedy *et al.*'s result one could suppose, according to evidence presented earlier, that the perceptual trace from the first stimulus is active longer if it is directed to the left hemisphere than to the right, and so is susceptible to interference by the second stimulus for a longer time after the first stimulus is presented. If the right hemisphere is activated by the first stimulus, its trace will disappear more quickly, and will not interfere with a following stimulus so strongly. The prediction from this argument is that the leading ear disadvantage will be bigger if the right ear stimulus leads than if left ear stimulus leads. This prediction again does not fit the observed results.

Overall the explanation of both these two experiments in terms of callosal conduction time is more plausible. The degree of asymmetry in timing reported by Mills and Rollman (1980) is quite small, and similar to the asymmetries reported by Efron (1963a) in similar experiments using visual or somatosensory stimuli, also interpreted in terms of interhemispheric conduction time. If the explanation in terms of callosal conduction time is accepted, its corollary assumption, that the left hemisphere is superior to the right in judgements of temporal sequences is also supported.

Three papers use Morse-like temporal patterns: Robinson and Solomon (1974) tested 24 right handed subjects, using stimuli consisting of 4–7 short (50 msec) or long (150 msec) 900 Hz tones, with 50 msec between each tone. Two such rhythms were presented simultaneously, one to each ear. After each pair, four rhythms were presented successively, two of which were new, and two corresponding to the pair just heard. Subjects had to say which of the four sample rhythms were the ones initially heard. Recognition was significantly better for patterns presented to right ear ($p < 0.01$). Natale (1977) also found a right ear advantage in this type of task, correlating in degree with the subject's degree of dexterity. Papcun *et al.* (1974) presented different Morse-code sequences to each ear. Experienced Morse code users recognised more with the right ear than with the left for all sequence lengths ($p < 0.001$). Naive subjects did better with the right ear for sequences less than 7 units long, but for longer sequences there was a left ear advantage. In a loose sense, it could be suggested that the naive subjects dealt with the longer patterns "holistically" rather than "analytically". However, such terms beg the question of what the operational difference between them is. Regardless of this issue, the finding of a left-ear advantage for longer Morse-like signals in naive subjects has similarities to those found for perception of melodies in musically naive subjects (see section 10.2).

Two papers deal with temporal patterns on a slightly longer time scale. Halperin *et al.* (1973) used dichotic presentation of different sound sequences, each up to 3 segments long, with segments of either high or low frequencies, or long or short in duration. For high/low patterns each segment was 600 msec long, while for long/short patterns they were 400 and 200 msec duration. Accuracy of report was the response measure. It was found that, as the number of frequency or duration transitions increased from 0–2, ear superiority shifted from left to right. When there were zero transitions (i.e. three tones each of similar length or similar pitch), a right hemisphere advantage emerged for identifications based on pitch, but not for those based on tone duration, in which case right and left ear performance was similar.

Divenyi and Efron (1979) used a dichotic task in which a sequence of two tones was delivered to each ear, each of 100 msec duration, with 450 msec separating them. To the left the sequence was "AB", while to the right the sequence was BA. Subjects were asked to report whether they heard "AB" or "BA". The stimuli could be pure tones or vowels, or they could have rapid transients. Two effects were seen. For pure tone or vowel stimuli, there could be either a right or a left ear advantage (which was similar for both classes of stimuli for an individual subject). For stimuli with rapid transients (similar to "CV") a right ear advantage was superimposed which could override the ear advantage characteristic of each individual. Thus for stimuli with both rapid transients and pure tones, there could be a reversal of the advantage shown with pure tones but no rapid transients, if a subject was left-ear dominant for the latter.

6.6.1.4 Non-speech-like temporal patterns: Unilateral head injury

Asymmetries between the hemispheres in temporal order judgements have been found in humans with unilateral head injury, and are very much greater than those found in subjects with intact brains: Efron (1963b), for instance, presented 10 msec tones of low and high pitch binaurally, with a temporal order judgement required. Non-aphasic head-injury patients (presumed to have a right hemisphere injury) achieved 100% correct when the temporal separation was little more than 100 msec. Aphasic patients (with presumed left hemisphere injury) required a much longer separation than this, under some conditions two to three times as long. Curiously, the expressive aphasic patients were more impaired on this test than the receptive aphasic ones, though both subgroups were clearly impaired compared to the non-aphasic patients. Swisher and Hirsh (1972) used a task in which two stimuli were presented to the same ear and subjects had to report their order. Brain-damaged patients were impaired in telling their order, compared to age matched controls. The time difference needed to tell the order was greater in fluent dysphasics (with presumed left temporal lesions) than in nonfluent dysphasics, or right lesioned patients. Normal control subjects could sequence the two tones if they were 33 msec apart, nonfluent dysphasics needed about 76 msec, fluent dysphasics needed 300 msec or more. The auditory deficit was the same whichever hemisphere the stimulus was presented to. The descriptions given by patients add information about the role of the right hemisphere in detecting simultaneous auditory patterns. Right lesioned patients reported that "they had difficulty telling the tones apart when they were close together in time". Aphasics on the other hand (with presumed left hemisphere lesions) could "discriminate the sounds but could not tell which came first." These reports suggest that the right hemisphere mediates discrimination of a spatial (i.e. simultaneous) pattern, while the left is better for temporal patterns. Tallal and Newcombe (1978) tested 10 patients with right hemisphere lesions, 10 with left hemisphere damage, and 10 controls. Two pure tones of different frequencies were presented in sequence, of duration 75 msec, with the inter-tone intervals either 428 msec or less (8–305 msec). Subjects were required to recognise which of the four possible sequences had occurred. With short inter-tone intervals the left hemisphere patients but not the right hemisphere ones were impaired. Chedru *et al.* (1978) used a task involving auditory "micropatterns". These were sequences of two pure tones, 20 msec each, with rise and fall time of

7 msec, and separation of 10 msec. They were presented so close together that they were not heard as a sequence but as differences in pitch, differing according to which sequence was given. Two such patterns were given in sequence and subjects had to make a same/different judgement. Fluent aphasics were significantly impaired compared to controls ($p < 0.002$) and subjects with massive right hemisphere lesions had a trend to impairment ($P < 0.1$). Non-fluent aphasics were not impaired compared to controls.

6.6.2 Experiments using visual stimuli

6.6.2.1 Studies of subjects with normal brains: users of American sign language

American sign language (ASL) is a visual language, used by deaf persons, where hand signs are used in rapid succession for communication. A growing number of papers analyse the neurology and psychophysics of this form of language. In studies of lateralization of this sign language, there are a number of possible sources of confusion. For instance, subjects who became deaf after acquisition of a spoken language should not be used, since the sign language may then have been "grafted" on to a language already strongly localised to the left hemisphere. Apart from this, there are difficulties in using static pictures of ASL signs, rather than the natural moving signs, even when the former are presented tachistoscopically. Tests of static ASL signs have been reported to produced either left or right hemisphere advantages. Which of these occurs may be expected to depend partly on the conditions of the experiment in ways predictable from the psychophysical evidence already reviewed, regardless of the lateralization of this form of symbol when used as a language.

For instance, Virostek and Cutting (1979) tested signers (both those who were congenitally deaf, and those with intact hearing) with a variety of patterns including American sign language, illegitimate hand shapes, and visual patterns which were not hand shapes. All stimuli were presented statically for 100 msec, and the dependent measure was percent correct identifications. All users of the sign language (whether or not they were deaf) showed a right field advantage ($p < 0.002$) for legitimate ASL signs, but not for any of the other stimulus types used. McKeever *et al.* (1976) reported a similar study, also using accuracy of report as the dependent measure, but with somewhat shorter exposure duration (75 msec). Here there was no difference between visual fields. We have already seen (section 6.3) that where time for response is unlimited, the left hemisphere can encode information presented for durations about 100 msec faster than can the right hemisphere, but this asymmetry tends to reverse as stimuli are shortened. Together, these two reports do not conflict with what would be predicted from psychophysical data described earlier in this chapter. Basic perceptual processes are probably the reason for the right field advantage found by Virostek and Cutting (1979), and may have nothing to do with the lateralization of ASL *per se*.*

In contrast to this paper, Poizner and Lane (1979) found, for congenitally deaf subjects whose first language was ASL, a substantial left field advantage in recognising ASL signs. The signs were presented statically for 200 msec, and reaction time was the dependent measure. Under these conditions, the psychophysical evidence discussed in section 6.2.3 would predict a right field advantage. That this was not found suggests a tendency for ASL signs to be dealt with better by the right hemisphere.

* The reason why the right field advantage does not also apply to non-familiar signs presented for the same duration is not clear. It may however have a similar explanation to that offered by Carmon (1978; see section 6.6.2.1 below) that increasing complexity confers a left field advantage.

Poizner *et al.* (1979) compared the lateralization of ASL perception between static pictures (presented for 167 msec) and simulated moving ones (three critical frames of moving picture sequence, shown within 167 msec) For the static signs there was a slight right hemisphere advantage, while for the moving picture version there was an equally balanced hemispheric involvement. The authors suggested that for ASL there is less strong tendency to left hemisphere lateralization than for spoken language, probably because of the greater spatial content of ASL signs.

Neville and Lawson (1987) compared normally-hearing children with the children of the deaf (some of whom were deaf, others hearing, and all of whom would be familiar with ASL). The task involved detection of direction of movement. For the normal children there was a left field advantage, but for the children of the deaf, this became a right field advantage. This result suggests that there is something about the left hemisphere which predisposes it to use of language even in the absence of spoken language, and this encourages plastic changes in that hemisphere which can affect basic perceptual processes. However, the distinguishing feature of the left hemisphere may not be specific to language *per se*, but could rather be the ability (if necessary) to detect rapid sequences of perceptual input (whether auditory of visual).

Emmorey and Corina (1993) showed other influences on the lateralization of ASL. Static signs presented for 220 msec showed a right field advantage when the signs used designated abstract notions ("idea", "law") but this switched to a left field advantage for signs of imageable concrete objects ("flower", "egg").

Studies in brain-injured deaf signers have been reported as indicating left lateralization of ASL as for spoken language (Bellugi *et al.*, 1989). However, the numbers of subjects with damage in an appropriate region in this study are very small (one patient with damage to the left temporo-parietal region, three with damage to the right temporo-parietal region) and so do not overwhelm the evidence from subjects with intact brains.

Overall the evidence suggests that ASL, especially when presented under the natural conditions of movement, is more symmetrically dealt with than is spoken language. It should be emphasised however, that the conclusion that ASL is bilaterally represented applies to perception of the signs. Production of them may have different lateralization, and is considered in Chapter 9.

Investigations by Emmorey and Corina (1990) show that the spatio-temporal aspects of ASL perception are different from those of spoken language. ASL signs are recognised more rapidly (within 240 msec) than spoken words (within 330 msec), and this is achieved when only 34% of the sign is detected as opposed to 83% for spoken words. These authors conclude that ASL is mainly "non-concatenative", that is, each perceived component is spatially complex but simultaneous, as opposed to spoken language where there is more complex temporal structure in what is perceived. This analysis corresponds very well to the distinction between two sorts of "synfire chain" depicted in Figure 5.7(A) and 5.7(B) or (C), (or in the two part of Figure 5.8), one linked together by sequences of connections where conduction delay is of critical importance, the other where spatial rather than temporal aspects of simultaneously activated connections is more important. Since this distinction is also the distinction hypothesised between the left and right hemispheres, the analysis of Emmorey and Corina (1990) provides an explanation for the fact that ASL is less strongly lateralized to the left than is spoken language. One other interesting finding supports this interpretation. Wilbur and Allen (1991) asked subjects to tap out the perceived rhythm of spoken sentences and of ASL sentences. This was easily accomplished in the former case but not in the latter. The authors conclude that ASL lacks the temporal structure, with periodic "peaks of perception", found in spoken language. According to Figure 5.7(A) those "peaks of perception" may be the circumscribed times at which conduction delays of several pathways of the synfire chain coincide within single neurones. For the sort of synfire chain envisaged for ASL (Figure 5.7(B) or (C)) these nodes may be so numerous and uniform over time that the percept has no subjectively discernible temporal structure.

6.6.2.2 Flicker fusion frequency

As discussed in section 5.2.8, the critical frequency at which a repeated visual stimulus appears to stop flickering is another paradigm for assessing the amount of temporal information which can be abstracted by either hemisphere. Goldman *et al.* (1968) estimated critical flicker fusion frequency (CFF) in normal subjects and in those with left or right temporal lobectomies. Restricting ourselves to subjects who had no visual field defects (which would confound the question at issue), patients with left lobectomies had lowered CFF ($p < 0.05$) (at least when tested in nasal fields) than those with right lobectomies, or normal subjects. This result suggests that CFF is higher in the normal left hemisphere than the right, and thus supports the prediction of finer grained temporal resolution in the left hemisphere. In this study, no visual field differences could be found in normal subjects. However, recent studies (Nicholls, 1994a,b) have found that the right visual fields can perceive two flashes as separate for briefer separations than can the left fields. The right field advantage was shown in percentage of correct judgements for separations below 14 msec.

6.6.2.3 Visual temporal order judgements in subjects with unilateral head injury

Two studies using visual stimuli other than ASL address the prediction that the left hemisphere should be able to abstract more information than the right about sequences of events in stimuli presented for optimal durations. Both deal with patients with lateralized damage to the cerebral hemispheres. Swisher and Hirsh (1972) presented two different-coloured stimuli in centre field and subjects had to tell which order they came in. The time difference needed to tell accurately the order was longer for fluent dysphasics (with presumed left temporal lobe damage) than for non-fluent dysphasics (with presumed left frontal damage), normal controls or right lesioned patients. The selective disadvantage was not seen for lateralized visual stimuli. Their results were not presented in a statistical fashion, but the impairment could be quite severe: Control subjects could distinguish the order when the two stimuli were separated in time by about 40 msec, right hemisphere patients required 66 msec, non-fluent dysphasics required 116 msec, and fluent dysphasics 170 msec. The impairment in the fluent dysphasics was so great that one may guess that the mechanism subserving temporal order judgement in these patients was of a quite different nature from that in the intact left hemisphere.

Carmon (1978), presented subjects with visual stimuli consisting of 2–6 components displayed in sequence. Each component was displayed for 1 msec, with 20–70 msec between components. Subjects were required to report the sequence of components. Subjects were normal controls, or right- or left- brain-damaged patients. For 2-component sequences, patients with left hemisphere damage were impaired, in that longer temporal separations (about 65 msec) were required to recognise the sequence than in controls (about 40 msec). Those with right hemisphere damage were not significantly different from controls. For more complex sequences of up to 4 steps, both left- and right-lesioned patients were impaired, but the right-lesioned patients were now significantly more impaired than left-lesioned ones. Carmon's interpretation of these results was that, provided the visual pattern is not very complex, the left hemisphere is better at identifying temporal structure in a visual stimulus than the right. However, this difference is reversed for patterns that are spatially complex. This issue is mentioned again in section 10.2.3.3.

One more study should be mentioned: Carmon and Nachshon (1971) used as stimuli sequences of three to five components, which could be visual (light flashes of three different colours) or auditory (sounds of different quality) in any combination. Subjects were required to identify the sequence (by pointing to the various sources). Normal controls and right-hemisphere patients performed equally well, and could report the sequence when individual components had a duration of 200–350 msec. Left hemisphere patients required significantly longer components, of duration 550–800 msec.

6.6.3 Experiments using somatosensory stimuli

Two studies report on lateral differences in recognising temporal patterns in the somatosensory modality. Bakker and van der Kleij (1978), using children aged seven and eleven, tested the ability to recognise the sequence of touching of two fingers in left or right hand. Each "touch" had a duration of 500 msec. At both ages there was a significant right hand superiority in this task. Hammond (1981) used, as a "baseline", a stimulus consisting of a sequence of 5, regularly-sequenced touches, whose inter-stimulus intervals could be from 100 to 500 msec. The sixth touch in the sequence could continue the regular pattern, or could be separated by a different interval. The threshold interval by which the sixth stimulus had to diverge from preceding regular stimuli, in order for the irregularity to be detected was determined by ascending and descending series of intervals. The preferred hand (i.e. the right hand in five out of six subjects) had the better temporal acuity. For instance, with the 100 msec regular interval, the threshold acuity of the right *vs.* the left hand was 25 *vs* 45 msec on the descending series and 65 *vs* 75 msec on the ascending one. While this task is not a temporal order judgement it *is* an assessment of accuracy of time perception, as in other studies mentioned for the auditory modality.

6.6.4 Summary about laterality of temporal order judgements and the amount of temporal information which can be abstracted from brief stimuli

The information reviewed in this section permits a broad generalisation to be made across the three sensory modalities considered. Stimuli presented to the right sensory surfaces of right handed people (and therefore having better access to the left hemisphere) can convey more temporal information (compressed within intervals of 50 msec to several hundred milliseconds) than can stimuli presented to the left sensory surfaces. These findings are in accord with the predictions of section 5.2.8.

In the auditory mode, this conclusion is supported overwhelmingly from studies using speech (or speech-like) sounds, as well as that using non-speech sequences such as Morse code. The left hemisphere shows its superiority over the right for temporally-complex stimuli lasting roughly 50 msec. The left hemisphere advantage for consonant or consonant-like stimuli declines when the rate of acoustic change is decreased, as in liquids or vowels. Overall, the differences between hemispheres shown in studies of unilateral head injury is more dramatic than those shown in persons with normal brains. The left hemisphere advantage for temporally complex sounds is particularly striking in contrast with the results of Sidtis (1980) and Sidtis and Volpe (1988), discussed in section 6.5.1, where tones that were "spatially" complex (i.e. containing many harmonics, but without rapid transients) showed a right hemisphere advantage. From several studies, using both speech-like and non-speech-like stimuli, there are indications that the left hemisphere dominance also applies to organised temporal patterns substantially longer than the time of temporal integration of perception. This is also discussed in Chapter 10 under the heading of short term memory.

There is very little evidence that a right hemisphere superiority can appear in temporal order tasks with stimuli shorter than 50 msec. What there is, is for vowel sounds (rather than consonants), and these have little temporal structure, but may be *spectrally* complex, which would confer a right hemisphere advantage for other reasons. The absence of reports of right hemisphere advantage in temporal order judgements, except when spectrally complex sounds are used, is strong confirmation of the model discussed in Chapter 5.

In the visual modality, studies of brain-injured patients suggest that there is a comparable left hemisphere superiority for abstracting temporal information, provided the spatial configuration simultaneously defined is not too complex, in which case right hemisphere superiority emerges. In subjects with normal brains most of the available evidence is about the perception of American sign language. The spatio-temporal structure of this language is different from spoken language, being a code where spatial complexity is more important than temporal complexity, at least in the time span of 50–150 msec. American sign language is not strongly lateralized to either hemisphere. This difference from spoken language is predicted, from the central hypothesis of this book, and the difference of spatio-temporal structure between ASL and spoken language.

Most of the experiments reviewed in section 6.6 used non-linguistic stimuli. Therefore the strong conclusion that the left hemisphere is superior in abstraction of temporal information indicates a basic perceptual difference between hemispheres, rather than a derivative of linguistic specialisation of the left hemisphere. There is however some suggestion that abstraction of temporal information from longer sequences may reverse this left hemisphere dominance: Right hemisphere advantages have been seen for extended stimulus sequences using Morse code (see above). Moreover, studies of temporal processing of short melodic fragments by head-injured patients also indicate a role for the *right* hemisphere (see Chapter 10). This may reflect "holistic" aspects of temporal sequences, captured by short-term memory mechanisms, whether in speech or non-speech input. This is discussed further in Chapter 10. Despite this qualification, the evidence reviewed in section 6.6 as a whole gives considerable support to the predictions proposed in section 5.2.8, and therefore also supports the anatomical hypothesis about differences in axonal conduction delays between hemispheres, from which those predictions were derived.

6.7 OVERALL CONCLUSIONS ABOUT DIFFERENCES BETWEEN LEFT AND RIGHT HEMISPHERES IN PROCESSING SENSORY INPUT

In Chapter 5, various predictions were made about the influence of a number of variables on the relative perceptibility of sensory stimuli directed to left or right hemispheres. The present chapter has reviewed a large body of evidence related to these predictions.

Overall, this body of data provides impressive support for the predictions made in Chapter 5. The interactions between hemispheric advantage and stimulus variables such as stimulus duration, stimulus intensity, definition, contrast, retinal eccentricity, size and complexity are all derivatives of the same relationships described in section 5.2.2. Evidence on all these matters supports the predictions, sometimes very strongly. There are a few studies which do not fit with the predictions. Bearing in mind the large number of interacting variables, it is not surprising that a few discrepant results arise, because all relevant variables cannot be controlled in all studies.

One of the most specific predictions was that, for reaction time measurements, or timing responses, there should be a right hemisphere advantage for short duration or low intensity stimuli, which reverses to a left hemisphere advantage as the stimulus duration is lengthened, or as higher intensity stimuli are used. The same reversal of lateralization was also predicted for measures of accuracy of report, provided that information about temporal ordering was not required. Table 10 summarises the more precise data reviewed above in which hemispheric advantage for stimuli of different durations is assessed. Out of 45 comparisons (in 30 studies), 42 use visual stimuli, the rest auditory. The various results are arranged in order of increasing stimulus duration.

Arranged in this way, these 45 comparisons fall into three groups, according to stimulus duration. For *eighteen experiments with durations less than 100 msec*, all but four find a right hemisphere advantage, whatever stimulus type is used. One of the exceptions (Sergent, 1984), with a duration of 50 msec, used words, a stimulus for which a strong left hemisphere bias may be expected. Another (McKeever and Suberi, 1974) found a right hemisphere advantage for durations of 20 msec and below, with left hemisphere advantage for longer durations. The stimuli were letters, again favouring the left hemisphere, and this might have led to the critical duration at which reversal occurred being lower than in average studies. A third (Nicholls and Cooper 1991) was prevented from showing a right hemisphere advantage at short duration by a "floor" effect. The fourth exception (Nicholls and Arkinson, 1993), showing a progressively increasing left hemisphere advantage as duration of a non-linguistic shape is decreased, is difficult to explain, and is probably the single study most seriously in conflict with predictions in the whole of Chapter 6. For *sixteen comparisons with a duration of 100 msec,* eight gave right hemisphere advantages, five gave left hemisphere advantages and three gave equality between hemispheres. Most of the studies where a left hemisphere advantage emerged used the stimulus types where (apart from duration) it would be expected that they do so (e.g. low eccentricity stimuli (Marzi *et al.*, 1979); local letters (Martin 1979); words (Pring, 1981); letters rather than faces (Rizzolatti *et al.*, 1971)). For *eleven comparisons using stimuli longer than 100 msec,* seven give a left hemisphere advantage, three give hemispheric equality. Only one study finds a right hemisphere advantage, and that used global letters, a stimulus for which a bias towards the right hemisphere is expected. Thus, this meta-analytic comparison gives an almost complete separation of left from right hemisphere advantage on the basis of duration alone. Most of the exceptions can be explained with references to the hemispheric advantages to be expected from other aspects of the stimuli used. Admittedly there are weaknesses in trying to make exact comparisons of data from different types of experiment. However, the reversal from right to left hemispheric advantage, as duration is increased, has also been seen several times *within* some of the individual studies reviewed.

The data in Table 10 permit one to make a rough quantitative estimate of the range of durations over which temporal integration of perception occurs in each hemisphere. It would

Table 10 Cerebral laterality versus stimulus duration

Authors	*Sensory Modality*[¶]	*Method*	*Dependent Measure*	*Stimulus Duration*	*Hemispheric Advantage*
Jeeves & Dixon (1970)	V	Bright dots	Simple RT to onset	2 msec	R hem
McKeever & Suberi (1974)	V	backward masking	Completeness of report	0–20 msec*	R hem
Sergent (1983)	V	Letters	Compound RT	20 msec	R hem
Rizzolatti and Buchtel (1977)	V	Faces	Compound RT	20 msec	R hem
Fodor and Bever, (1965), Bever *et al.* (1969a,b)	A	Clicks in one ear sentences in other	Relative time	<25 msec	R hem
Bertelson & Tissegre (1972)	A	Clicks in one ear sentences in other	Relative time	<25 msec	R hem
Gregory (1978)	A	Clicks in one ear, music in the other	Relative time	<25 msec	R hem
Davidoff (1977)	V	Dots	Temporal detection threshold	~30 msec	R hem
McKeever & Suberi (1974)	V	Backward masking	Completeness of report	30–110 msec*	L hem
Sergent (1982a)	V	Faces, male or female	Compound RT	40 msec	R hem
Nicholls & Cooper (1991)	V	Lines of differing length	Accuracy of report	40–80 msec	L hem[†]
Gardner & Branski (1976)	V	Gap detection	Accuracy of report	50 msec	R hem
Wilkins & Stewart (1978)	V	Letters, physical identity	Accuracy of report	50	R hem
Katz (1980)	V	Numbers, numerical judgement	Compound RT	50 msec	R hem
Pring (1981)	V	Four-letter words to each field	Compound RT	50 msec	R hem
Sergent (1984)	V	Words	Compound RT	50 msec	L hem
Sergent (1982c)	V	Local & global letters	Compound RT	62 msec	R hem for global letters
Nicholls & Atkinson (1993)	V	Lines of differing length	Accuracy of report	65 msec or less	L hem
Curtis & Foster, (1917);	V	Dots	Brightness judgement	100 msec	R hem
Dallenbach, (1923);	V	Dots	Brightness judgement	100 msec	R hem
Davidoff, (1975)	V	Rectangles	Brightness judgement	100 msec	R hem
Rizzolatti and Buchtel (1977)	V	Faces	Compound RT	100 msec	R hem
Sergent (1982a)	V	Faces, male or female	Compound RT	100 msec	R = L
Glass *et al.* (1985)	V	Faces, Degraded (expt 3)	Compound RT	100 msec	R hem
Glass *et al.* (1985)	V	Faces, Clear (expt 3)	Error score	100 msec	R hem
Kitterle & Kaye (1985)	V	Gratings	Accuracy of report	100 msec	R = L
Martin (1979)	V	Local & global letters	Compound RT	100 msec	L hem for local letters R = L for global letters
Marzi *et al.* (1979)	V	Letter arrays	Completeness of report	100 msec	L hem at low eccentricity R hem at high eccentricity
Rizzolatti *et al.* (1971)	V	Faces	Compound RT	100 msec	R hem
		Letters	Compound RT	100 msec	L hem

Table 10 *continued*

Authors	*Sensory Modality*¶	*Method*	*Dependent Measure*	*Stimulus Duration*	*Hemispheric Advantage*
Pring (1981)	V	Four-letter words to each field	Compound RT	100 msec	L hem
Previc (1982)	V	Gratings	RT	100 msec	L hem
Cohen (1976)	V	Backward masking	Completeness of report	120 msec*	L hem
Nicholls & Atkinson (1993)	V	Lines of differing length	Accuracy of report	140 msec	R = L
Sergent (1982c)	V	Local & global letters	Compound RT	150 msec	L hem for local letters R hem for global letters
Sergent (1983)	V	Letters	compound RT	150 msec	L hem
Pring (1981)	V	Four-letter words to	Compound RT	150 msec	L hem
Besner *et al.* 1979	V	Numbers, numerical judgement	Compound RT	150 msec	L hem
Glass *et al.* (1985)	V	Faces, degraded (expt 3)	Compound RT	190 msec	R = L
Glass *et al.* (1985)	V	Faces, clear (expt 3)	Error score	190 msec	R = L
Sergent (1982a)	V	Faces, male or female	Compound RT	200 msec	L hem
Wilkins & Stewart (1978)	V	Letters, physical identity	Accuracy of report	990 msec	L hem

* Duration = duration of stimulus plus delay before masking
† Emergence of a right hemisphere advantage may have been obscured by a "floor" effect.
¶ V = visual; A = auditory

seem likely that the right hemisphere performs all or most of its integration in less than 100 msec, while the left hemisphere can perform a significant amount of further integration after 100 msec of a maintained stimulus. However, since the evidence on time-intensity integration performed on both hemispheres together (i.e. without regard to laterality) indicates that complete integration occurs by 150 msec, one can assume that the left hemisphere on its own also completes its integration roughly within this time. As a result one can use the data from section 6.2.1 to develop a further implication from section 5.2.2: The left hemisphere appears to have a greater absolute auditory threshold for stimuli of 200 msec duration or greater. This has the tentative implication that the left hemisphere has a greater total number of cortico-cortical axons (and probably of such cell bodies) than the right.

One study (Sergent 1982b) finds a reversal of laterality from right to left hemisphere as *intensity* is increased, other parameters being held constant. While this is just a single result, it is a critical test of the various models described in section 5.2.2 and illustrated in Figures 5.5 and 5.6. According to the arguments presented there, it suggests that, whether or not the total *duration* over which perceptual integration occurs differs between hemispheres, there is also a more rapid *gradient* of integration in the left than the right hemisphere. The model described by Figure 5.5 (e) and (f) is thus preferred to that of 5.5(b).

In section 5.2.8 another very specific prediction was made, that there be a left hemisphere advantage for temporal order judgements, with no reversal as duration was shortened.

The results described in section 6.6 support this prediction overwhelmingly: There is a left hemisphere advantage in the vast majority of cases reported about temporal order judgements. All the exceptions can be explained either on the basis that complex stimuli are used (for which a right hemisphere advantage is expected) or because short term memory may be involved (which is considered in Chapter 10). It has been claimed that the left hemisphere is specialised for language as such, rather than for temporal analysis of a sensory stimulus regardless of whether it is linguistic. This claim has been made especially with regard to an apparent left hemispheric advantage for American sign language. However, the evidence for this is not strong. ASL appears to be less strongly left-lateralized than spoken language, and this is to be expected from the spatio-temporal structure of the signs used, independent of whether they form part of a language.

Many researchers have suggested in the past that the left hemisphere is specialised for temporal analysis, the right hemisphere for spatial or "quasi-spatial" analysis, without deriving this generalisation from more fundamental structural bases of the differences between the hemispheres. In this work however, the theory is derived from a simple but global morphological premise. This premise allows predictions to be made about psychophysical differences between the hemispheres which are more detailed and specific than in previous formulations.

Overall these results give strong support to the model described in section 5.2.2. The psychophysical data support three specific morphological assumptions of this model:

(i) The right hemisphere has more cortico-cortical axons than the left whose temporal delays are of the same order as the integration time of single neurones.
(ii) The left hemisphere has more cortico-cortical axons than the right whose temporal delays exceed by far the integration time of single neurones.
(iii) Even when one considers durations of stimuli where perceptual integration occurs in both hemisphere (notably the time span from 50 to 100 msec), if there is to be a steeper gradient of integration in the left hemisphere, there must be a larger absolute number of axons which could mediate such perceptual integration in that hemisphere than in the right.

It is pertinent at this point to summarise the form of the arguments which allowed these specific predictions to be made. The reversal of hemispheric advantage in perceptual integration from right to left as stimulus duration was increased depends on the relationship between the range of durations of conduction time in each hemisphere compared with the neuronal integration time. If a predominance of cortico-cortical connections have a delay no longer than the neuronal integration time, the hemisphere with these properties will thereby be more effective in "spatial" ("simultaneous") integration. If there is a predominance of connections with delays exceeding by several fold the duration of the neuronal integration time, the hemisphere with these properties will display superiority in certain perceptual tasks especially for stimuli above a critical duration or intensity; and it will be specialised also for abstracting information about temporal order. The fact that these relationships are actually found thus provides considerable support for the basic model, and suggests that indeed a higher proportion of cortico-cortical axons in the right than the left hemisphere have conduction delays under 20 msec, whereas the left hemisphere has a preponderance compared with the right of axons with delays lying between 20 and 150 msec.

7 Lateralization of the Contents of Memory

7.1 INTRODUCTION

Chapter 6 dealt with lateralization of basic perceptual processes, such as could be studied mainly with stimuli lasting a small fraction of a second. Considerable support was found for the idea that perception achieved by the left hemisphere incorporates temporal structure contained in the input in a superior manner to that achieved by the right. The right hemisphere on the other hand is capable of synthesising Gestalts, which are spatially complex, though temporally unstructured, better than the left hemisphere. Both of these generalisations can be derived from the central hypothesis put forward in section 3.6, according to the detailed predictions derived in Chapter 5.

Perception is a momentary process, but it is also the gateway to longer terms psychological structures known as memory or "knowledge". The transition from perception to memory involves the laying down, in lasting form, of myriads of associations. The process of synaptic modification described earlier (section 3.1) appears to be the physical basis for acquiring these associations. The central hypothesis of this book implies that the associations between converging active afferents to a single neurone in the left hemisphere can span intervals of time extending up to 100–200 msec, while the right hemisphere specialises in laying down effectively simultaneous associations. Proceeding from such implications, one might expect that the differences in basic processes of perception between the hemispheres might lead to differences in the content of memory or "knowledge" in the two hemisphere. The present chapter deals with such lateralization of the content of memory.

Much of the content of memory in humans is connected with language. Experimentally the easiest aspect of knowledge to test is also that associated with language, especially using single words as probes in experiments. If the left hemisphere is specialised for "language" as such, one would expected *all* aspects of language to be lateralized to the left hemisphere. However, the policy adopted here is to regard language in terms of its physical attributes as sensory or motor signals, rather than as a form of signal metaphysically different from other sensory and motor signals. In these terms, there are aspects of language where a right hemisphere dominance might occur. Such instances might be particularly important because they would refute the notion that language *per se* is the function for which the dominant hemisphere is specialised. They would support the idea that a more basic, and more precisely-definable functional dimension distinguishes right from left hemisphere.

We have already seen hints that language might not be centred on the left hemisphere in a unitary fashion, when discussing American sign language. However, a role for the right hemisphere in some aspects of language has been known since the 1960s: Initially it became clear, from split-brain and hemispherectomy studies that the right hemisphere mediates important aspects of the visual recognition of words (Smith, 1966; Gazzaniga and Sperry, 1967; Gott, 1973; Levy and Trevarthan, 1977; Bergego *et al.*, 1993). Such a function might be expected from what we have already discussed in terms of perception of complex stimuli. Subsequently, interest has grown in the differential lateralization of semantic as opposed to phonetic attributes of words. This also came initially from study of split-brain patients. For instance, Gazzaniga and Sperry (1967) found that the isolated right hemisphere was able to recognise individual words, both written or spoken (e.g. by pointing to a corresponding picture or object). However, occasionally substitutions were made between the correct word and a synonym (as though the meaning had been correctly identified rather than the word itself).

Since then specific interest has grown in the differential representation of words designating abstract or concrete entities, and in the ease of "imaging" a word. This development has arisen partly from study of patients suffering from "deep dyslexia" (otherwise known as "phonemic dyslexia"). This syndrome is described by Marshall and Newcombe (1973), Shallice and Warrington (1975), Saffran and Marin (1977) and Patterson and Marcel (1977). Deep dyslexic patients retain a large reading vocabulary, and can adequately read words designating "concrete" objects. They fail however in reading abstract or function words. In reading, they may, like the right hemisphere of split-brain patients, substitute a word by its synonym. The difficulty such patients have in converting written words to spoken form is seen particularly in reading nonsense words, when there are no cues from familiarity of the visual form of the word. These patients are also deficient in recognising rhymes or homophones, or in recognising a misspelt word.

Evidence on the lateralization of phonetic and semantic memory for words is reviewed in the next two subsections (7.2 and 7.3), followed by a more detail dissection (7.4) of semantic representation, dichotomised in terms of the abstract *vs.* concrete meanings of words. The chapter finishes with a discussion (7.5) of this evidence in relation to the central hypothesis of this book.

7.2 PHONETIC REPRESENTATION OF WORDS

A number of studies report on the different losses of word representation in patients with right and left unilateral brain injury. Lesser (1974) assessed the comprehension of auditory words by requiring subjects to select a picture appropriate to each word. Right brain-damaged patients were not impaired when the alternatives were phonologically related, while left brain-damaged patients were. Gainotti *et al.,* (1981) also found that right hemisphere patients were not impaired compared with normal controls in phonetic discrimination. This applied to words presented either visually or auditorily. Chiarello and Church (1986) asked patients to make similarity judgements on pairs of words presented simultaneously. The words might be similar on the basis of rhyme. Left hemisphere patients were impaired on rhyme judgements, while right hemisphere patients were minimally impaired.

Studies such as these suggest that the left hemisphere has an overriding superiority in phonemic representation. However, this conclusion needs to be qualified by emphasising that the differences between the hemispheres are not sharp categorical ones, but differences of degree. From split-brain and hemisperectomy studies, it is clear that the isolated right hemisphere has some ability to represent acoustic aspects of words. Zaidel (1976; 1978) showed that in split-brain or hemispherectomy patients the intact right hemisphere has an important capacity to understand spoken language, and has an auditory lexicon richer than its visual one, only slightly less than that of the isolated left hemisphere. This is also seen in patients with left hemispherectomy following brain tumours (Smith, 1966; Gott, 1973): The isolated right hemisphere has a considerable power of comprehension of spoken language, far greater than might be expected from its poor expressive capacity. Dennis and Whitaker (1976) describe three children aged 9 or 10 years, one with left hemispherectomy, the others with right hemispherectomy. Right hemisphere removal led to a marginally impaired performance on auditory language, and left hemisphere removal led to a marginal impairment of visual language. However, overall semantic judgements were similar in all three children.

More detailed study showed that there are some qualitative differences in acoustic representation in the right hemisphere compared with the left. Specifically, phonetic perception of a word by the right hemisphere is impaired compared with the left hemisphere. Springer and Gazzaniga (1975) tested callosotomy patients in the CV-syllable recognition test, originally used by Shankweiler and Studdert-Kennedy. If the CV syllables were presented dichotically, syllables could not be recognised in the left ear. In another explicit test of consonant identification, the isolated right hemisphere was found to be severely impaired in a dichotic listening test (Zaidel, 1978). The right hemisphere of split brain patients can recognise (by pointing) common words spoken to them, but may make errors by substituting an unvoiced for a voiced consonant (Zaidel, 1978). Zaidel concludes that the right hemisphere perception of spoken words is acoustic rather than phonetic, i.e., the right hemisphere somehow forms a representation of an "auditory Gestalt" which does not contain its sequential phonetic components. However, the right hemisphere probably also has some limited capacity for phonemic analysis, even though it is usually overridden by the left hemisphere: In Springer and Gazzaniga's study, CV syllables could be identified by the right hemisphere provided they were presented monotically. (The capacity of the left ear for recognising phonetic structure shown here is unlikely to be due to uncrossed pathways to the left hemisphere, because it could be shown with left hand (right hemisphere) responses).

7.3 SEMANTIC REPRESENTATIONS OF WORDS

Many studies show an important role for the right hemisphere in representation of word meanings. Some of these involve the study of unilaterally brain-damaged patients. For instance Gainotti *et al.* (1981) found that right hemisphere patients were impaired compared with normal controls in semantic discrimination of words (presented either visually or auditorily). Bergego *et al.* (1993) carried out a test in which patients were asked to name an object they were presented with. Right hemisphere patients made many mistakes in such a task by accessing the right category but the wrong instance of it (e.g. cat instead of tiger).

Although these studies of right brain-damaged patients show a role for the right hemisphere in representing word meaning, explicit comparison of left and right hemispheres, in several studies of brain-damaged patients in semantic tasks, show that left and right hemispheres are equally involved. In Lesser's study of 1974, where subjects were required to select a picture appropriate to each word, right brain-damaged patients were found to be impaired when the alternatives were semantically related to the correct word, and left brain-damaged patients were impaired to an equal degree. Chiarello and Church (1986) asked patients to make similarity judgements on pairs of words presented simultaneously. The words might be similar on the basis of rhyme, visual appearance or meaning. Right hemisphere patients were minimally impaired except on meaning judgements, and then to a similar degree to left hemisphere patients. From such results it seems that semantic representation requires both the right as well as the left hemisphere: Damage to either can impair selection of the exact meaning, though superordinate categories or semantically related words can still be selected.

Two studies in normal brains support this view. Day (1977) found that improvement of reaction time by previous presentation of a semantically related word was equal for either visual field. Martin (1978) visually presented four pairs of words (one pair every 500 msec). From each list of eight words, subjects were asked to report all of a particular physical category (e.g. small or long words) or all of a particular semantic category (e.g. animal or plant). There was a right field advantage for physically cued items, and no field advantage for semantically cued ones.

A few studies however show a distinct right hemisphere advantage for semantic representation. In a study of brain-injured patients Gardner *et al.* (1978) required subjects to produce the antonym on presentation of either words or pictures. Left hemisphere patients preserved their ability to select an antonym (though the Wernicke's aphasic patients were impaired on the verbal version of the test). Right hemisphere patients however were markedly impaired, and often chose synonyms rather than antonyms, especially on the verbal task, but not only on this. The authors believed that this "bespeaks a general looseness of conceptualisation". Another paper also indicates that accurate access of semantic attributes of a word requires an intact right hemisphere. Glosser and Goodglass (1991) carried out a word-association test on normal controls and right brain-damaged patients. On most measures there was no difference between groups, but some of patient groups (especially those with frontal lesions) tended to produce more idiosyncratic or unrelated associates than controls. Rausch (1981) used a method for accessing the full range of associations of a given word. Lists of 60 words were presented in sequence, visually or auditorily, and subjects were required to recognise any words which were repeats of words earlier in the list. The distracters in the list could be similar in category, in common associations, in rhyme (for auditory presentation), or were homonyms (different spelling, same sound, in visual presentation). In the auditory task, patients with left temporal removals made more false recognition errors than controls or right temporal patients, especially with respect to erroneous categories ($p < 0.02$), or common associates ($p < 0.01$). Errors based on rhymes showed a weaker non-significant trend. The right temporal patients were not different from controls. On the visual task left temporal patients made more false recognitions, cued by categories ($p < 0.001$), associates ($p < 0.02$) and homonyms ($p < 0.02$) than controls. Presumably these errors reflect the functions of the intact right hemisphere. Right temporal patients made *less* false recognition errors compared with controls, especially of homonyms. These results suggest that the right hemisphere has a wider associative network for words than the left, especially for semantic associates. This is revealed when the left hemisphere is damaged and phonetic memory, as an alternative strategy for solving the task, is thus impaired. When the right hemisphere is damaged, the phonetic superiority of the left hemisphere undisturbed by semantic associations, can give a better performance based on phonetic memory than in control subjects.

Three papers in normal subjects also suggest that the right hemisphere has a richer semantic network for words than does the left. Chiarello (1985) studied "semantic priming". Subjects had to make a rapid decision whether a letter string was a word or a non-word, and each string was preceded 500 msec earlier by a briefly presented "priming word" which might be somehow related to the target word. Priming is demonstrated when reaction time or accuracy of report improves as a result of the prime. Under some circumstances, priming by semantically related words was better for prime/target pairs presented to the left than the right visual field. Rodel *et al.* (1992) presented subjects tachistoscopically with pairs of words, which differed in the closeness of their semantic relatedness, and required subjects to state whether or not they were related. When the pairs were presented to the left fields, a semantic relation was detected for more remote semantic associations than for pairs presented to the right fields. Lambert and Voot (1993) used semantic relatedness to *interfere* with performance rather than to improve it. A target letter string was presented in centre field for 150 msec, and another (unattended) string was presented to one or other side for 33 msec starting at the same time as the target. Subjects again had to respond to the attended words with a word/non-word decision. If the unattended word was semantically related to the attended one, it slowed the reaction time (an effect which has been described previously). This effect was however found only when it was presented in left visual fields, not in the right ones.

The implication of these several papers is that the right hemisphere has a wider network of semantic associations than the left, and can represent subtleties of meaning more accurately. This is a different conclusion from the first group of five studies considered in this subsection, showing that semantic judgements were equally impaired in left and right brain-damaged subjects.

It is possible that there is a real superiority in the right hemisphere compared with the left in accessing semantic associates of a particular word. However, the issue is not clear cut, for another reason: All the above studies of semantic judgements of words, including those which show a right hemisphere superiority, use mainly words designating concrete objects. As we will see in the following subsections, such words give a right hemisphere bias for reasons different from semantic processing *per se*. The finding of a better semantic network in the right hemisphere may therefore arise from this fact alone, and may thus be more an apparent than a real superiority.

7.4 IMAGEABILITY AND THE ABSTRACT/CONCRETE DISTINCTION

As mentioned in section 7.1 (above) study of deep dyslexic patients suggests that there may be a separate representation of concrete, imageable words, compared with abstract ones. Independently, scales have been created rating words (especially nouns) on these dimensions (e.g. Paivo *et al.*, 1968). Most of the words used in the studies of semantic lateralization described in the previous section would be classed as concrete or highly imageable. As a result of the clue from deep dyslexia, a number of studies have been carried out on the interaction between these variables (concreteness, imageability and abstractness) and laterality in normal persons. Table 11 gives details of studies in the visual modality using tachistoscopic methods. In this table imageability and concreteness are considered together. Although they are not identical concepts, rating of many words on these two dimensions are similar, and similar laterality effects occur for each. The discussion which follows in section 7.5, suggests a better way of formulating this laterality effect than either of these concepts.

Table 11 deals only with effects for nouns. (Different effects are described for verbs, adjectives and function words). The table contains some quite unexplained inconsistencies. For instance some attempts at exact replication (e.g. Ellis and Shepherd (1974) *vs.* Orenstein and Meighan (1976)) give quite different results. Even attempts at replication within the same paper may be inconsistent (Jackman, 1985). Nevertheless a consensus view can be seen in this Table: Out of twelve comparisons, eight provide support for the idea that the right hemisphere is favoured in dealing with concrete or highly imageable nouns, in comparison with abstract or low imageable ones. Of these, two studies find an advantage for concrete over abstract nouns in the left fields, and an advantage for abstract over concrete ones in the right fields. The majority (five studies) find an advantage for concrete over abstract nouns in left fields and no difference between the two in right ones. One study (Shannon 1979) finds no difference between word classes in either field, yet nevertheless finds that the overall left field advantage was greater for concrete than abstract. Apart from these eight studies there are three reports (tabulated) of tachistoscopic experiments showing no differential effect between fields for abstract *vs.* concrete nouns, including one (Lambert and Beaumont, 1983) in which a variety of modifications to the experimental design were tried in order to reveal the effect described by others, without success. There is also one study (Schmuller and Goodman, 1979) in which the effect is reversed (increasing right hemisphere dominance for concrete compared with abstract nouns).

Lambert and Beaumont (1983) discuss the field thoroughly. They consider that the order of report may introduce a bias, due to habitual left-to-right habits associated with reading. However, this potential confound applies to the bilateral studies only, yet the consensus effect is also found in some unilateral studies. In some studies stimulus duration is individually adjusted to give a standard percent of correct responses. This may also complicate interpretation, in ways which are difficult to predict. Nevertheless the majority of studies indicate that concrete or highly imageable words are dealt with better by the right hemisphere than are abstract or low imageable ones, and this difference is less in the left hemisphere. It should however be noted that there is still, in almost all cases, an absolute right field advantage for words overall.

Studies using methods in the visual sense other than tachistoscopic presentation mainly give support to the consensus view. McFarland *et al.* (1978) employed a "running

Table 11 Laterality and the dimensions of imageability or concreteness for visually presented words

Authors	*Duration msec*	*Unilateral or Bilateral*	*Mode of Response*	*RVF*	*LVF*	*Comments*
Ellis & Shepherd 1974	150	bilat	AOR	A = C	C > A●	overall RVFA
Hines 1976	20	unilat and bilat	AOR	A > C●●●†	C > A●●●†	RVFA > for A than C●●●
Orenstein & Meighan 1975	150	bilat	AOR	A = C	A = C	overall LVFA
Day 1977 expt 1	100	unilat	RT	A > C	C > A	C: R = L A: RVFA●●●
Bradshaw & Gates 1978 expt 3	150	unilat	RT	A = C*	C > A*	RVFA > for A than C●
Day 1979	125	unilat	RT	A = C	C>A●●	C: R = L A: RVFA●●
Shannon 1979	150	unilat	RT	A = C	A = C	LVFA > for C than Ao
Schmuller & Goodman 1979	IA<60	bilat	AOR	C > A	C > A	C: RVFA●●● A: R = L
Marcel & Patterson 1978	IA (70%) of mask delay	unilat	AOR	A = C	C > A	C: R = L A: RVFA
Lambert & Beaumont 1983¶	IA(50%)	unilat	AOR	A = C	A = C	overall RVFA
Jackman 1985 Expt 2	<150 (IA)	unilat	AOR	A = C	C > A●	C: R = L A: RVFA●
Jackman 1985	<150 (IA)	unilat	AOR	A = C	A = C	C: R = L Expt 3 A: R = L

C = Concrete or highly imageable words; A = abstract or low imageable words
AOR = accuracy of recall; RT = Reaction time; IA = exposure time individually adjusted to give criterion of stated % correct.
* Comparison of low frequency words
† Comparison of familiar words
o = Trend (non-significant); ● = p < 0.05; ●● = p < 0.01; ●●● = P < 0.005
¶ = similar result in four different experiments

memory" task, in which subjects were presented with a long sequence of words presented monaurally, and had to say when a word was repeated. When the task was made especially difficult by interference with speech in the other ear, there was a right ear advantage when the target words were abstract, but not when they were concrete. Emmorey and Corina (1993) found in deaf subjects that ASL signs for abstract ideas were perceived better by the left hemisphere, but the advantage switched to the right when concrete objects were signified. Jones-Gotman and Milner (1978) presented a sequence of word-pairs, with subjects required subsequently to name the word previously associated with a given word in the list. When subjects with right brain damage were instructed to mediate this "paired associate learning" by constructing a mental image, they were impaired mildly, but when abstract words were used which were to be linked by an imagined sentence, they were not impaired. Villardita *et al.* (1988) used right brain-damaged subjects, and normal controls, and required subjects to learn lists of twelve words, which could be either abstract or concrete. Right brain-damaged subjects were impaired in learning the concrete words, but not the abstract ones. Against these three studies there is however that of Farah *et al.* (1985) who compared the two hemispheres of a commissurotomy patient in imagery of letters (a decision of the physical form of a lower case letter, when the subject was presented with its upper case equivalent.) Only the *left* hemisphere could perform the task. This result will be mentioned again later, in the context of the manipulations of information possible within short-term (working) memory (see section 10.2)

Two studies in the auditory modality should also be mentioned. These involved recall or perception of word pairs presented dichotically, and failed to show a right hemisphere ascendancy for concrete versus abstract words (Kelly and Orton, 1979; Young and Ellis, 1980).

The question of the relative richness of semantic networks for concrete versus abstract nouns in the two hemispheres has also been addressed in two studies. Chiarello *et al.* (1987) used the "semantic priming" technique, described in section 7.3. When the priming situation occurred unexpectedly (i.e. relatively rarely) it was found that the priming effect was equivalent in each visual field for both abstract and concrete primes. When priming was expected (that is, it occurred on most trials), abstract primes were less effective in the left field than the right, while concrete primes were similar in either field. This suggests that under these circumstances, the right hemisphere cannot use the meaning of abstract words as well as that of concrete words. Villardita *et al.* (1988) in the study involving learning word lists mentioned above, studied "semantic clustering". This is the tendency for recall of one word to follow that of a semantically related word. Such semantic clustering was markedly impaired in the right brain-injured patients, for concrete words, but not for abstract ones. These two studies suggest that the tendency for the right hemisphere to have a wider semantic network than the left derives from its superiority with concrete words, rather than an inherently wider semantic network. Although such a widened network of associations in the right hemisphere might be expected on theoretical grounds, it has not yet been conclusively demonstrated.

7.5 DISCUSSION OF THE EVIDENCE

From section 7.2 it is clear that the left hemisphere has a more precise representation of phonemic structure of words. This is exactly what is predicted from the central hypothesis of this book: The left hemisphere has an ability superior to that of the right in analysing temporal structure in perceptual tasks, and this is reflected in the way it forms an acoustic representation of words. The right hemisphere appears to represent acoustic structure of words, but with a greater degree of temporal blurring of the acoustic image, telescoping adjacent phonemes together in time. It may thus be capable of synthesising all the acoustic events which make up a syllable (or even a word) as an "auditory Gestalt". This form of representation of speech may still permit some word identification, especially for familiar words, and when supplemented by a residual right hemisphere capacity for representing temporal structure. Provided the spoken word can actually be identified by such a right hemisphere mechanism, the meaning of the word can of course be identified as well as it could be with the left hemisphere.

The right hemisphere also has an important role in representation of word meanings, though it does not amount to a right hemisphere superiority. However, the evidence reviewed in section 7.3 and 7.4 shows that this capacity of the right hemisphere is not for word meaning as such, but better for particular categories of meaning, specifically, the meaning of "concrete" or "imageable" words. It is difficult to take this evidence at face value, without enquiring further into the meanings of terms like "abstract", "concrete" and "imagery". The operational definition of words along these dimensions is given in the scale published by Paivo *et al.* (1968). Examining this scale shows that words that are highly "concrete" or "imageable" are those physical objects which are perceived *visually*, and these are also often of high frequency use. Amongst the words rated in the list of Paivo *et al.* there are almost none which are perceived with the auditory sense. This is understandable because we have very few names in English for specific sounds. (Perhaps the best examples of such categories are the naming of a species of bird on hearing its song. For many species, the bird-song is a more familiar identifier than the visual image of the bird.) Nevertheless we do have a considerable capacity for recognition on the basis of sound quality. For instance the timbre of different musical instruments or the quality of different person's voices are instantly recognisable and readily imageable. Spoken language depends on this faculty: We easily recognise different speech sounds, though we seldom need to name each speech sound in the way we name visually-perceived objects.

Thus it is clear that "imageability" is an inaccurate designation of the class of words for which there is a right hemisphere ascendancy. There is however an important spatio-temporal difference between visual objects and acoustic stimuli which are equally discriminable and imageable, which gives the clue to the real distinction. The visual objects are perceived as spatially complex but temporally simple, whereas the acoustic stimuli are temporally complex, even if their temporal structure is at a scale so fine as to defy conscious awareness and reporting of that structure. When this distinction is born in mind the perceptual differences between the hemispheres described in the previous chapter can be employed to explain hemispheric specialisations for abstract and concrete words. "Imageable" and "concrete" words (as rated in Paivo *et al.*'s scale) are formed by association between (a) the image of a word (initially acoustic, with its visual form grafted on) and, (b) a complex visual perception of the object to which the word refers. The first of these two associates would be more easily represented by the left hemisphere, the second by the right. Abstract words, on the other hand, would have no clear visual percept as referent. They would be formed by association of the audio-visual image of the word, a network of other words associated with the target words, and a variety of complex situations (not necessarily *perceptually* complex) which are examples of the abstraction. In this case neither the word image itself, nor any of these associates would be expected to have a strong right hemisphere dominance. Thus the considerable right hemisphere capacity for representing meaning of "imageable" and "concrete" words, and the left hemisphere dominance for "abstract" words turn out to be easily predictable from the basic sensory associations needed to acquire the meaning of each class of words. Once these factors are accounted for, there is no evidence that meanings of words *per se* are lateralized to the right hemisphere. The lateralization effects for concrete or abstract words are thus what would be expected from the central hypothesis of this book. Imageability itself is not lateralized: In contrast to static visual images, many auditory images (such as those mentioned above) are likely to be lateralized to the left hemisphere. Even for visual images, there is evidence that active construction or "manipulation" of the image is performed better with the left than the right hemisphere (see section 10.2)

The dimension of concreteness (imageability) versus abstractness is not independent of word frequency, word familiarity or age of acquisition of the word. Generally speaking, concrete and easily imageable words would be familiar, of high frequency, and acquired early in life. This is significant in the context of laterality studies, because words acquired early in life are likely to be bilaterally represented, since lateralization of function develops later than language itself. However, in Table 11 the right hemisphere ascendancy for concrete or imageable words is not restricted to high frequency words (e.g. Bradshaw and Gates, 1978).

It is as yet not proven that the right hemisphere has a richer semantic network for word meanings than the left. To prove this it would be necessary to explore semantic networks using abstract as well as concrete words as probes, explicitly comparing left and right hemispheres for words of equal associability. Two studies described above, one in normals and one in right brain-injured patients suggest that left and right hemispheres do not differ in their semantic networks for abstract words. However, more rigorous investigation of the issue is needed. This is because the central hypothesis of this book actually makes the prediction that the right hemisphere has a richer repertoire of semantic associations for each word than does the left: A network of connections rich in delay lines (such as hypothetically exists in the left hemisphere) will be particularly suited for representing one-dimensional strings of information. In other words, given a sequence of nodes, A,B,C … etc., it will be impossible to form an association between A and C without B as an intermediary stage. A network rich in *simultaneous* links (such as hypothetically exists in the right hemisphere) will not be constrained in this way. It can represent the relationships possible in a two- (or more-) dimensional space. In other words, the potentiality for linkage between different nodes will not be constrained by strict specification of sequence.

This conclusion is just a restatement of what is implicit in Figure 5.7. Its obvious relevance is to the matters discussed in Chapter 6, namely the difference between representation of temporal and "spatial" structure. However, there is no reason in principle why the same difference should not distinguish other classes of one- *vs.* two-dimensional fields of information. Here, it implies that a semantic node for a particular word in the right hemisphere can have a richer range of associations with related words than would such a node in the left hemisphere.

Overall, if one bears in mind the significant capacity of the right hemisphere for acoustic representation of a word, and its ability to represent meanings of concrete or visually imageable words, one can understand the capacity of the right hemisphere for understanding spoken language, which is substantial, though less than that of the left hemisphere.

8 Motor Aspects of Lateralization: Theory and Predictions

8.1 INTRODUCTION

The evidence reviewed in the last two chapters dealt with lateralization of the processing of information derived immediately from sensory input, or stored in the form of long-term memory as a result of that input. However, one of the most obvious facts about functional laterality in humans is that *motor* performance as well as sensory processing is lateralized. The most commonplace example of this is that the majority of persons are right-handed. Bradshaw and Nettleton (1983) in fact conclude that *comprehension* of linguistic input is the *least* lateralized of language functions, while the motor function of articulation is most highly lateralized. Therefore, in attempting to formulate a hypothesis of the biological basis of functional lateralization it is necessary to take motor lateralization at least as seriously as sensory (perceptual) lateralization. The structural basis proposed above for the lateralization of perceptual function was that there is a greater temporal dispersion of signals via cortico-cortical delay lines in the left hemisphere compared with the right. If lateralization of motor function is to be explained parsimoniously on the same basis as that of perceptual function, new arguments must be presented demonstrating how a greater repertoire of long delay lines in the left hemisphere might contribute to better motor performance in manual and articulatory tasks by this hemisphere than by the right. This is not straightforward. To develop these new arguments, it is necessary to summarise another body of theory, about the role of the forebrain in the acquisition and representation of motor programs as such. Only then is one adequately prepared to analyse the *lateralization* of such motor functions. Before attempting this difficult task however, it is necessary to explore in more detail the claim of Bradshaw and Nettleton (1983) that it is motor rather than perceptual lateralization which is more fundamental (section 8.2). From this, the new ideas can be derived about how the central hypothesis of this book can be adapted to explain such strong left hemisphere dominance for spoken language and hand use (section 8.3). Anatomical implications of these new ideas (which are really extensions of the central hypothesis, but not something completely new) are included with this discussion. After this, in Chapter 9, evidence relevant to these new hypotheses will be reviewed.

8.2 LATERALITY OF SPOKEN LANGUAGE AND HAND USE

A variety of evidence suggests that lateralization of motor function (especially of the hands), and of linguistic expression, is stronger and more robust than that of sensory perception and speech comprehension:

(i) In commissurotomy patients, the right hemisphere has some language comprehension, as discussed in the previous chapter. For instance the right hemisphere can obey instructions given verbally (i.e. responding with the left hand). However, the isolated right hemisphere in such patients has almost no *expressive* ability (Gazzaniga and Sperry, 1967).

(ii) Lateralization of manual praxis or spoken language to the left hemisphere applies in a substantially larger proportion of the population than does lateralization of language perception or of other perceptual tasks: Right handedness has been found in about 85% of nine to eleven year old children (Roberts, 1969; Fennell *et al.*, 1983), and similar proportions are reported in adults (Subirana, 1969). Only about 10% of adults show strong left-hand preference (Subirana, 1969). According to Roberts (1969) aphasia occurs in 60–70% of patients with left hemisphere damage (an underestimate of the proportion with left hemisphere language dominance, because some left hemisphere lesions might not affect the language zones). Aphasia is however quite rare in cases of right brain damage (according to Roberts [1969] in 2.6% of right handers, and 31% of left handers). Using the amobarbitone test (Rasmussen and Milner, 1977) or unilateral electroconvulsive therapy (McManus, 1979, cited by Bradshaw and Nettleton, 1983 [p. 196]) it is found that between 90 and 95% of right handers have left hemisphere dominance for spoken language, while in left handers the corresponding figure is about 70%. In contrast to these figures for linguistic expression, estimates of the percentage of right handed subjects with left hemisphere dominance in a dichotic listening task ranges from 63.4% (Fennell *et al.*, 1983, in eleven year old children to 72% (Lake and Bryden, 1976), 80–90% (Zurif and Bryden, 1969) to 87% (Roberts, 1969) of right handed eleven year old children, while Shankweiler and Studddert Kennedy (1975) give an estimate of 90%. Bradshaw and Nettleton (1983, p. 101) sum these figures up as indicating that 70–80% have a right ear advantage in this test, far lower than the incidence of left hemisphere dominance for speech. Zurif and Bryden (1969) also report a 75–80% incidence of right field advantage amongst right handers, in tachistoscopic tests where right field advantage is to be expected.

(iii) Practice or choice of strategy can readily reverse lateralization of language comprehension or of performance in perceptual tasks (see Chapter 10). This occurs much less easily for speech production laterality or for handedness. Admittedly the disadvantage of the left hand in right handers can sometimes be overcome by extended experience or systematic training (for instance in expert typists, as mentioned in section 9.3.4). However, these adaptations are acquired much more slowly and with more difficulty than the attentional shifts of language comprehension or perceptual discrimination discussed in Chapter 10.

(iv) Apraxia, a syndrome where some aspects of learned purposeful motor control are lost, is much more common after left than right hemisphere lesions. In addition when apraxia occurs after left hemisphere lesions, it commonly affects control in both hands (Poeck, 1986). A more detailed discussion of apraxia is given in section 9.4.1 below.

This sort of evidence has lead to the idea that motor lateralization is *the* fundamental form of lateralization in humans, and that perceptual and language comprehension laterality is somehow secondary. Bradshaw and Nettleton (1983) recognise a major issue here. They ask (p. 174) whether:

> "... the left hemisphere [is] adapted not for symbolic functions *per se*, but rather for the execution of some categories of motor activity that happen to lend themselves readily to communication. Alternatively, are *all* the apparent left hemisphere language superiorities *really* just manifestations of the left hemisphere's processing analytically and sequentially?"

Previous chapters have advocated the view that hemispheric specialisation is nothing to do with language *per se*, but rather with more basic information processing strategies, ultimately deriving from the hypothetical difference between hemispheres in repertoire of cortico-cortical delay lines. However, that position has so far been developed with reference only to processing of sensory input. Motor performance presents new difficulties. In such functions one is apparently not dealing with the brain's ability to abstract temporal structure from a stimulus external to the brain, but with the brain's ability to *construct* a temporal structure, a plan for ensuing action as it were, without immediate instruction from the environment. This seems to raise theoretical problems categorically different from those underlying analysis of sensory input. Indeed the solution seems very difficult indeed if it is true that the construction of a temporal structure is truly without instruction from the environment. However, in the sections below a schema will be developed in which there is such instruction, albeit by more complex and less direct means than for registration of sensory patterns.

It may seem difficult to resolve the issue raised by the above four points, which appear to suggest that motor lateralization is *the* fundamental lateralization. However, there are some simple-minded points that can be used to lessen the force of these arguments. For instance, with reference to point (i) above (about spoken language in the right hemisphere of commissurotomy patients), it should be mentioned that the isolated right hemisphere *does* have a facility in spoken language, but this is seen better in left hemispherectomy cases which are carried out *early* in life (and is seen only occasionally in adults) (Smith, 1966; Gott, 1973; Dennis and Whitaker, 1976). Likewise Kinsbourne (1971) studied three right handed persons who became aphasic after left hemisphere lesions, and found, with the amobarbital test, that the remaining speech, though aphasic, was actually served by the *right*, not the left hemisphere.

A more fundamental consideration, however, is that language expression, unlike language comprehension, is (technically) an *instrumental* function. Instrumental functions are modifiable according to the effects they produce, and whether these effects have motivationally favourable or unfavourable consequence. Generally speaking, the processes of reward and punishment, acting in the forebrain, ensure that only favourable programs of motor performance and behaviour become represented in the brain. Certainly this is likely to apply to both language production and skilled manual praxis: A capacity for spoken language, or skilled hand use would not be acquired in childhood if it did not serve the vital interests of the child.

It is likely that the fact that language expression and skilled hand use are such instrumental functions has a major bearing on the strength of its lateralization. Assume that initially there is a mild degree of left hemisphere dominance for spoken language or hand use, comparable to that found in language comprehension or in some perceptual tasks. What the biological basis of this dominance is need not be specified, though, on grounds of parsimony, it would be convenient if it were similar to that for lateralization in language comprehension and in perceptual tasks. Regardless of what its basis is, continual use of expressed language is likely to strengthen its representation in the neural network where its representation is initially strongest, that is, for most people, in the left hemisphere. Likewise continual use of the most skilled hand is likely to increase its superiority over the other hand. Although the right hemisphere may not be totally incapable of spoken language or skilled manipulation via the contralateral hand, under normal circumstances it will never realise its potential, because the left hemisphere will always do better. Therefore the left hemisphere representation of spoken language will receive differential reinforcement when it is used effectively, and this will confound any such differential reinforcement for motor performance attempted by the right hemisphere. This line of reasoning may also explain why behaving experimental animals show consistent tendencies to use one or other forepaw for skilled tasks, indicating that motor representation for instrumental tasks is concentrated in one hemisphere or the other, rather than being diffusely represented in both (see Bianki, 1988, pp. 23–28). The only circumstance where the human right hemisphere's potentiality for spoken language or skilled manual control is realised is in cases of early left hemispherectomy. In these cases, since performance depends entirely on the right hemisphere, differential reinforcement of successful speech *can* ensure that the resources of that hemisphere for control of speech or manual skill will be maximally utilised. In view of these arguments, it is proposed that the pronounced left hemisphere dominance for spoken language and manual functions may well still derive from a similar basic premise as for the receptive aspects of language. However, to substantiate this proposal a significant and quite complex extension of the original hypothesis is also required, described in the next section.

8.3 MOTOR-SENSORY CORTICO-CORTICAL FEEDBACK IN TIME-STRUCTURED MODELLING OF LATERALIZED MOTOR PROGRAMMES IN HUMANS

8.3.1 The basic model for reward mediated learning and cortico-striatal interaction and their application to motor control

The extension of the original hypothesis for lateralization, concerning side-to-side differences in axonal conduction delays, requires that it be integrated with another body of psychobiological theory. In psychological terms, this is the theory underlying acquisition and performance of instrumental behavioural tasks. The corresponding neurobiology concerns the striatum, its dopaminergic innervation from the brainstem, and its interplay with the cerebral cortex. The present subsection deals with processes not specific to humans, but part of mammalian motor learning generally. However, it is an essential prelude to section 8.3.2, dealing with the way in which interaction between cortex and striatum may be employed in a more specialised way in the specifically human aspects of motor control.

The psychobiology of instrumental behaviour has been discussed in a number of previous publications (Miller, 1981; Miller, 1988; Wickens, 1991; Miller and Wickens, 1991). The start of this work derived from consideration of reward-mediated learning. The fundamental premise usually made about learning as a whole is that it is realised by change in efficacy of selected synaptic junctions. That change may be either functional strengthening or functional weakening of synapses. Since learning can occur quite rapidly, the corresponding synaptic changes are nowadays usually envisaged as biochemical rather than morphological in nature, at least initially. The process of selection of which synapses should be changed is addressed by the postulate of Hebb (1949) already discussed. Hebb envisaged that synapses were strengthened when there is coincident activation of the pre-synaptic and post-synaptic sides of a junction. Such a selection rule, applied throughout a network, is generally believed to underlie types of learning such as Pavlovian conditioning and pattern learning, as discussed in Chapter 3. In these forms of learning, at least in their idealised form, what is learned depends on contiguities within the sensory (or other) input to a neurone. It does *not* depend on the nature of any subsequent behavioural output, and whether or not that output fulfils the requirements of any system of motives or intentions. In other words it does not depend on the *effect* of a behavioural response.

However, learning dependant on the effects of responses, *does* occur, and is very important in all mammals. It has been studied for a century now, since Thorndike (1898; 1911) formulated his celebrated "law of effect". This type of learning is called instrumental learning, or, almost synonymously, learning mediated by reward or punishment. The words reward and punishment are however a little confusing. They may refer to the fact that the consequences of a behavioural response fulfil some favourable motivational or intentional requirement (for reward) or fulfil a negative requirement (in the case of avoiding or escaping punishment). In a different sense however, these words are used to refer to the way in which behaviour which fulfils these conditions is also *modified* by these same occurrences, leading to the increase of the likelihood of a portion of behaviour (if it fulfils positive requirements) or the decrease of it (for fulfilment of negative requirements). The word *reinforcement* is also commonly used to refer to the strengthening of behaviour consequent on occurrence of reward.

8.3.1.1 The "three factor" rule

If learning is to be interpreted as a change of selected synapses, there is one very important reason why the Hebbian rule cannot underlie instrumental learning. In Hebbian synaptic change all the conditions needed for operation of the rule are microscopic, local to the neurone whose synapses are to be changed. In instrumental conditioning there is an essential step at the macroscopic level, the level of behaviour of the whole organism, and the assessment by the brain as to whether that behaviour (as a whole) does or does not fulfil motivational or intentional requirements*.

* The argument used in the previous section to explain why the left hemisphere might achieve greater dominance over the right in instrumental tasks, compared to perceptual ones was based precisely on this fact. The behaviour of the organism *as a whole* depends on the part of the brain which offers the best programs of instrumental action. Other parts of the brain, whose programs are less satisfactory may not benefit from contingencies of reward and punishment, because these functions are aspects of behaviour as a whole rather than assigned to specified structures in the brain. These other parts can however achieve Hebbian (local) learning without impediment.

There is another important aspect to instrumental learning which helps to define the rule for selection of synapses for change. Amongst learning theorists this is known as the *problem of retroflex action* (Hilgard and Marquis, 1964). As a problem in learning theory, this arises because the behaviour which is to be reinforced cannot be specified at the time the behaviour is emitted, but only a short time later when its effects become apparent. Expressing this in terms of synapses, when rewarding effects occur, the reward process must somehow act retroactively on the synapses which were active a short time previously, though they need not be currently active.

A solution to this problem was suggested (Miller, 1981, pp. 80–83), that the process of strengthening selected synapses (which for Hebbian learning is achieved in a single stage) becomes subdivided into two subprocesses for instrumental learning. *Initially* there is a *selection* process: Synapses which are active and which activate post-synaptic neurones are envisaged to generate a short-lived trace, which identifies these synapses as candidates for change. This trace was called a "state of readiness" (Miller, 1981). The *second process* is the actual *strengthening* of the synapses. It occurs a short time after establishment of the state of readiness, and requires interaction between a "reward" signal and the state of readiness. The "reward" signal is envisaged to be activated, in the simplest model, by the consequences of behaviour emitted following the original activation of synapses. According to this scheme, reward mediated learning should not be possible if the reward signal is delayed to a time when the state of readiness has declined to the point where it has died out. Since the establishment of the state of readiness requires two initial factors (pre-synaptic and post-synaptic activation) and the reward process (usually occurring a little later) is also essential, this paradigm for synaptic change has sometimes been called the "three factor rule". Barto *et al.* (1981), using very similar reasoning, postulated a comparable scheme to enable simulated neural networks to solve control problems similar to instrumental learning in animals. They referred to an "eligibility trace" which corresponds exactly to the "state of readiness".

8.3.1.2 Physical embodiment of the three factor rule

We have made a number of suggestions about the physical embodiment of this abstract scheme. There is a great deal of psychopharmacological evidence that dopamine is a major component of the reward system (see Beninger, 1983; Miller *et al.*, 1990). The highest concentration of dopamine in the brain is in the striatum (caudate, putamen, as well as limbic striatal structures such as the nucleus accumbens). Dopamine is also present in the cerebral cortex, but in much lower concentrations. Therefore the striatum is likely to be the most important site where synaptic changes for instrumental conditioning occur. Consequently, the population of synapses which is susceptible to change in such conditioning is likely to be a major element of striatal synaptology. The most numerous cellular components of the striatum are the medium spiny neurones, making up about 90% of nerve cells there. Their dendrites have spines, upon which are excitatory synapses derived widely from all over the cerebral cortex. There are several reasons for thinking that these synapses are the population affected by the dopaminergic reward signal:

(i) They are axo-spinous synapses, and there are now good reasons for thinking that synapses on dendritic spines are modifiable ones (e.g. Wickens, 1988).
(ii) This population of synapses receive signals from the cerebral cortex, and projects back to it (by way of pallidum and thalamus). They can thus form a modifiable

component in the relay between sensory input and motor output in instrumental behaviour.

(iii) Many of the dendritic spines upon which these synapses are formed are closely related to release sites of dopaminergic axons in their vicinity (Freund *et al.*, 1984; Kubota *et al.*, 1987).

As for the hypothetical state of readiness, Wickens (1990) has made the suggestion that it is embodied by a transient elevation of calcium ion concentration in (and localised to) the spines heads to which cortico-striatal synapses are apposed. The exact duration of the state of readiness is a matter of conjecture at present. Behavioural evidence suggests that it is of the order of one second or less. Physiological estimations of it are needed.

Preliminary physiological evidence has been presented by Begg and Wickens (1993), working with a cortico-striatal slice preparation, to show that synaptic strengthening can occur provided all the three designated factors coincide (conjunction of pre- and post-synaptic activation together with a pulse of dopaminergic stimulation).

8.3.1.3 What triggers natural dopamine release?

This scheme for synaptic modification has the merit that it translates into synaptic terms the basic information structure of instrumental conditioning. However, it leaves important problems unanswered. Suggestions can be made about some of these unanswered questions. For others, no solution is yet available. One elaboration of the basic scheme depends on asking about the circumstances under which dopamine is naturally released from terminals in the striatum. In other words, what normally activates the midbrain dopamine neurones? An answer to this question will help to define the contingencies under which reinforcement occurs, and so specifies exactly what can be learned by the above paradigm for synaptic modification.

In the earliest formulations of striatal function it was envisaged that midbrain dopamine cells were activated by consequences of behaviour which are motivationally favourable with respect to basic visceral motivational needs (such as obtaining food when an animal is hungry, or water when it is thirsty, or for moving to an environmental temperature which is more favourable). There is some evidence for this sort of control of dopamine release, by food (Heffner *et al.*, 1981; Blackburn *et al.*, 1986; Radhadishun *et al.*, 1988), or sexual activity (Pleim *et al.*, 1990; Pfaus *et al.*, 1990; Wenkstern *et al.*, 1993). However, evidence from Schultz (1986; see also Schultz *et al.*, 1992, 1993) suggests that the dopamine cells are usually controlled not by such primary rewarding stimuli, but by any stimuli available which allow *prediction* of such a primary type of reward. Thus, the midbrain dopamine neurones are not controlled directly by such primary reward unless it is *un*predictable (Mirenowicz and Schultz, 1994).

Beyond the concept of primary reward however are a number of secondary, or indirect kinds of reward, some of which are initiated internally rather than as a result of external consequences of action. These are discussed next.

8.3.1.4 Reward mediated learning and acquisition of motor control

Germane to the arguments about motor or speech lateralization, there is some evidence that the rubric of reward mediated learning is also relevant to the acquisition of motor control. First, it is very clear that in humans this control is acquired *very slowly* in the years after birth, which probably means there is an element of learning in the acquisition. Second, the co-ordinated control of motor actions may be used for many purposes without direct fulfilment of the above basic visceral motives. Therefore it is unlikely that those primary motives play a direct part in acquiring such co-ordinated control. Third, there is nevertheless a body of theoretical work which directly compares Thorndike's law of effect (which refers to behaviour producing "satisfaction" to the animal) with the processes leading to acquisition of control (where the effect of a response produces "knowledge of results": Adams, 1971). Fourth, in young animals and humans a prominent aspect of behaviour is *play*. This seems to be a "pleasurable" activity, repeated over and over again with endless variations. This suggests that it is maintained by some form of reward, and is perfecting some aspect of behaviour, on a trial and error basis, with reinforcement of the most successful versions. This play behaviour may be carried out despite conflicting demands of other basic visceral motives, and so must be fulfilling a different requirement. It is plausible to suggest that play is important in acquiring and maintaining motor skills by experimentation with many versions of the basic motor programs for control of limbs (or other combinations of effectors). Each successive improvement on old programs may then be "stamped in" as a replacement of the older ones, by a reward/reinforcement process. Finally, there is a little evidence that in patients with Parkinson's disease — characterised by a deficit in striatal dopamine — the motor impairment applies not only to motor *performance*, but also to *acquisition of new patterns* of motor performance (Frith *et al.*, 1986; Elias *et al.*, 1988; Pillai, 1989).

In such motor learning, acquisition of control amounts to acquiring a set of relationships between each motor program and its consequences. The motor act thus becomes an *intention*, and the consequences become *fulfilment of the intention*. In the more obvious sense, this could refer to the consequences produced on an object in the outer environment by a motor act. However, more fundamentally, and at an earlier stage of an animal's or child's development, exactly the same processes presumably apply to producing consequences on the organism's own limbs. In terms of motor control theory it matters little whether one refers to control over an arm or a leg, or over a tennis racket or a hockey stick. The only major difference is that, in the case of acquiring control over one's limbs, the results about which one gains knowledge (in order to mediate the learning) will be visual or direct kinesthetic information about the limb, rather than visual information or indirect kinesthetic information about the object one is holding.

Consider reaching or aiming movements in more detail. Imagine a particular object of interest to the organism, at a particular position in extra-corporeal space. Reaching or aiming movements can be made in a variety of different limb orientations, hand positions, and at a variety of different speeds. The range and variety of these motor acts is large but it is not unlimited. It is envisaged that for each such object and position, extensive practice leads to formation of the complete set of motor programs which can efficiently aim at or reach for the object, from any specified starting position, and at any specified speed. Play, involving learning by knowledge of results, is envisaged to mediate this process. Some of the earliest stages of such learning can be seen in human infants, when they repeat with evident interest and satisfaction, the movement of their own fingers and toes with intense visual monitoring. More advanced play develops later, involving more complex behaviour of the whole body. As a result of such play, a repertoire of approach responses for every accessible objects will be acquired, based on reinforcement of those movements or movement sequences which have previously succeeded in aiming, reaching (or in other ways interacting) with the object. Given the presence of an object in a particular place, and the ancillary conditions leading the organism to interact with it, the optimal motor programs might then be made available. However, a major aspect of this process, which requires further thought, is the formation of the *correctly-timed sequences* within the motor programs, both for individual movements and for longer co-ordinated sequences. This is particularly important for *rapid* aimed responses. Before this can be explicitly discussed however, we need to expand on the theory of instrumental learning, in so far as it involves interaction of the striatum with the cerebral cortex.

8.3.1.5 The cerebral cortex as the site of storage of motor or behavioural programs

It is not envisaged that all motor programs are stored in the striatum itself. The cerebral cortex, which is in close relation with the striatum, by both input and output pathways, is a much larger structure. It is likely to be the main store for memory, including memory for motor programs acquired with the help of the striatum. Our best concept for representation of such memory is the cell assembly (section 3.1), and some version of this concept is likely to be involved in representing motor and behavioural programs.

Previous papers (Miller, 1981, 1988; Miller and Wickens, 1991) have developed ideas about how motor and behavioural strategies, acquired by reinforcement mechanisms in the striatum can lead to permanent memories being laid down in the network of the cerebral cortex. The most detailed account of the interaction between cortex and striatum (Miller and Wickens, 1991) advanced the concept of *cortico-striatal cell assemblies*. This concept is based on anatomical evidence (Alexander *et al.*, 1986) of re-entrant loops of connections from cerebral cortex, to striatum, to pallidum, to motor thalamus, and then back to the cerebral cortex. To explain this concept, imagine an active neural assembly which is at first confined to the cerebral cortex. This will activate striatal cells via cortico-striatal afferents, and will thus set synapses there into the hypothetical "state of readiness". If the active assembly includes components in a motor region, and produces a motor or behavioural output, there is a possibility that, a short time later, a rewarding consequence of that output will be produced and detected by the organism. At this stage, the striatal synapses which are in a state of readiness will be definitively strengthened. Therefore, the next time that same cell assembly is active, signals will be transmitted more effectively from the assembly through the striatum to pallidum, motor thalamus and back to the cortex. There is then the possibility that thalamo-cortical synapses will be activated upon cells in the cortical assembly which are already active. According to Hebb's rule, such thalamo-cortical synapses will become strengthened. At this stage, a self-activating assembly will have been formed, incorporating neurones in striatum, pallidum and thalamus, as well as in cortex.

Such a configuration, once formed, is capable of maintaining activity during a delay between salient events. Take, for instance, a typical delayed response task. In this, a cue is first presented which provides information about which response, from a set of alternatives, will ultimately deliver reward. However, the response is capable of delivering a reward only some time delay after the presentation of the cue. The time at which reward becomes available is indicated by another stimulus (the "imperative stimulus"). Animals are trained in such a task by starting with negligible delays, and then extending the delay between cue and reward availability. Thus, the initial training is a conventional instrumental training procedure. According to our model of striatal function, the sequence "cue — imperative stimulus — response — reward" occurs sufficiently quickly that the "state of readiness" established in striatal synapses by the cue is still present when the reward comes, and thus the link from cue to response is strengthened. As the delay between cue and imperative stimulus is lengthened, sufficient conditions for the response to occur are a signal from the cue about *which* response to produce, combined with the indication from the imperative stimulus about *when* to produce it. The striatal synapses which are strengthened in the initial part of training now serve to activate a reverberatory cortico-striatal loop, representing the response which should be produced at the end of the delay period. The role of such a reverberatory circuit to extend the time after an informative cue stimulus when an appropriate motor response can be produced is illustrated in Figure 8.1.

It should also be noted that, when the structure of strengthened cortical and striatal synapses is initially formed, reward processes play an essential part; but in subsequent performance of the behavioural output so acquired, reward is not needed. Indeed, once the structure is robust, performance becomes a routine matter, and may even be dealt with by the cortex alone, with the striatal link itself becoming mainly redundant. This

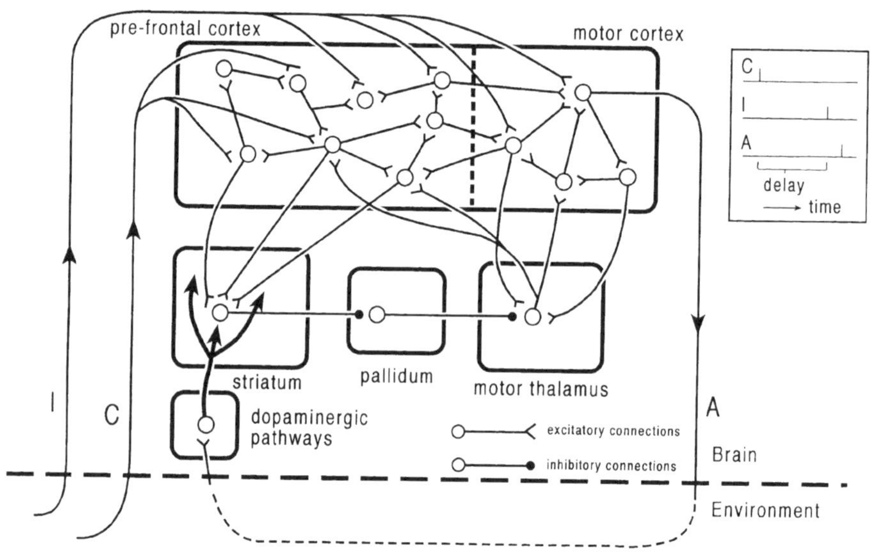

Figure 8.1 The Role of Cortico-Striatal Circuits in Extending the Time After a Stimulus when an Appropriate Response can be Produced. An initial cue stimulus (C) provides information about which response is required to obtain reward. This activates a cell assembly (depicted as a connected set of neurones in prefrontal and motor cortices). By itself, activity in this assembly would quickly die out. However, this assembly can provide input to the striatum, and via pallidum and motor thalamus, can disinhibit the neurones in the motor thalamus. Provided such neurones have another tonic source of activation (shown here as originating from cortico-thalamic projections), this circuit can lead to re-activation of the cell assembly, which thus has the potential to maintain activity over a delay. Given that activity still persists in the assembly at the end of the delay, an imperative stimulus (I), occurring after the delay, can act as a trigger for the appropriate action (A). The critical synapses for this self-amplifying loop are excitatory ones from cortex to striatum. They are envisaged to become strengthened during the initial stages of conditioning, under the influence of a dopaminergic reinforcement signal (neurone and its connections shown in bold line, and arrow heads), itself controlled by behavioural output.

idea was first expressed in 1981 (Miller, 1981, p. 104–113), and still applies with the more complex version using the concept of "cortico-striatal cell assemblies"

8.3.1.6 Cortico-striatal interplay and synfire chains

This initial insight about cortico-striatal interaction allows activity in a cortical cell assembly to be maintained or even amplified during a delay. However, the cortical cell assembly in this case is envisaged to be a static one, without organised temporal structure in its firing, and in which conduction delays between sequentially activated neurones have no necessary part. In acquisition of motor control however, individual acts may need to be produced in a co-ordinated and accurately-timed sequence. There is a real problem in using reinforcement mechanisms to lay down such sequences. This arises because, during the state of readiness (which may last about 1 sec), a succession of several independent motor acts can be emitted. By the time the reward signal is generated, the state of readiness will have decayed more completely for the first-emitted compared with the last-emitted act. Therefore, the reward will be directed most strongly to the *last* rather than the first of these. In other words the problem of retroflex action (see section 8.3.1.1) still applies.

This circumstance obtains when the relation between items of motor outflow to each other are *independent* of each other, occurring in sequence but without exactly measured time between them. What happens to cortico-striatal interplay when the cortical cell assemblies are time-structured synfire chains of neuronal activation? If the sequence could be emitted as a whole, with accurate timing linking the various elements of it, one has the possibility of overcoming the problem mentioned in the previous paragraph: Because of the organisation of the sequence of cortical activations, the later parts could never occur unless preceded by the first parts. Hence the sequence, if emitted at all, would be emitted as a whole, and therefore would be reinforced as a whole.

Consider this possibility in more detail. Suppose that a synfire chain exists in the cortex, which, once set into activity, produces a progression of accurately-timed activations of specific neurones. The termination of the sequence is envisaged to be a rewarding consequence, which may be either a primary reward, or some other salient stimulus which activates the dopamine neurones, as discussed in section 8.3.1.3. The members of the synfire chain sequence will not only activate the next in the *cortical* chain. They will also activate *striatal* neurones, via cortico-striatal projections, and therefore will set synapses on these neurones into a state of readiness. At the end of the chain of activation, reward is envisaged to occur. Provided that the chain is not of longer duration than the state of readiness, these synapses will be definitively strengthened. The next time the synfire chain is activated, the first neurones in the cortical chain will, as before, tend to activate the following ones; but in addition there will be an increased tendency for specific neurones in the striatal loop to be activated, which in turn will feed back disinhibiting signals to the thalamus, thus activating the cortex. A short transmission period, fixed by conduction time and synaptic delays, is likely to be introduced via this striatal loop. Since the synfire chain itself is a fixed temporal sequence, there will be regularly-repeated coincidences of the activation of certain later neurones in the chain by cortico-cortical links and by thalamo-cortical ones, released from inhibition by the striatum. The net effect of such a scheme is that the striatal loop is involved in enhancing the security of transmission in the temporally organised chain. This role is to be compared with that mentioned above (Figure 8.1) for a conventional Hebbian assembly, where the striatal loop serves merely to maintain or amplifying activity, in a self-exciting assembly which has no measured temporal structure.

We can readily apply this scheme to the cascade of events leading to a motor action. The most relevant motor action is the utterance of a speech sound by a child learning to

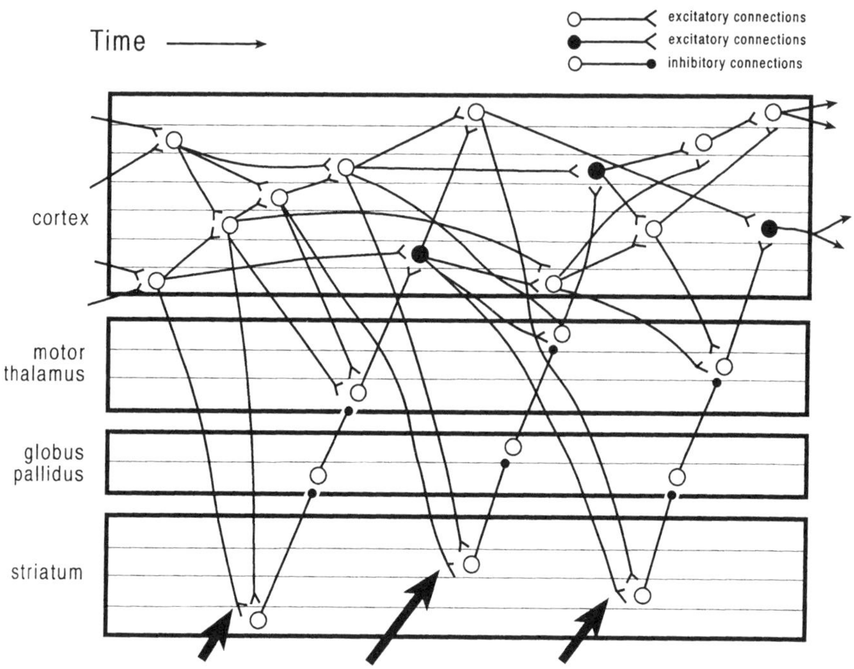

Figure 8.2 Role of Striato-Pallido-Thalamic Loops in Feeding Forward Amplifying Activity to Increase the Security of Transmission in a Temporal Chain of Cortical Activity. Like Figure 5.7 (but in contrast to Figure 8.1.), the horizontal axis of this figure represents time, and the different rows represent different neurones. Circles in a particular row represent *activity* in the corresponding neurone, and at the corresponding time. The upper "stave" of neurones represents the cortex, while lower "staves" indicate motor thalamus, globus pallidus and striatum. In the cortical "stave" is depicted a temporal ("synfire") chain of activity. A requirement for two-fold convergence for secure transmission is shown. For most neuronal activations, this is provided by two *cortico-cortical* synapses. However, three of the cortical activations (filled circles) have only a single active synapse derived from within the cortex. For these, thalamic afferents provide additional synaptic activations, so maintaining the security of transmission in the cortical chain. These thalamic neurones are tonically activated by cortico-thalamic excitation, but this activity may be inhibited from tonically active pallidal neurones. However, when striatal neurones projecting to these pallidal neurones are activated, by cortical afferents, the pallidal neurones are inhibited, and the thalamic neurones, in turn, are disinhibited. Under these conditions, secure transmission of the cortical chain becomes possible. A critical part of this role of the striatum is the existence of strengthened cortico-striatal synapses. These are envisaged to have been strengthened previously by the dopaminergic reward signal (shown as bold black arrows). Since there are conduction delays in both the striato-pallido-thalamic loops and cortico-thalamic connections the influence of the subcortical circuitry on the cortex will be directed at cortical activations temporally "downstream" from the cortical activity which initiated such influences. Given that both cortico-cortical links and the subcortical links operate via relatively *fixed* conduction delays, there will be regularly-repeated coincidences of activity in cortical neurones from both sources. Therefore the cooperative aspects of the Hebbian mechanism will ensure the strengthening of synapses participating in such coincident activation (e.g. synapses on the neurones indicated by filled circles).

talk. This requires a temporally-structured motor program, with a tendency to be lateralized to the left hemisphere. (However, the application is not restricted to production of speech sounds) It is known that in the hundreds of milliseconds leading up to the initiation of such a motor act, there is a build-up of neural activity in the cortex, especially in the prefrontal and premotor areas. This is known as the "readiness potential" (Kornhuber and Deecke, 1965; Deecke *et al.*, 1969). For a motor act which is itself a sequence, it is likely that there be detailed temporal structure in this build-up of activity, as well as in the actual final motor program. Thus, the pattern of neuronal firing in the run-up to utterance of the speech sound is a good example of a synfire chain in action.

The preliminary activity in the premotor cortex may be envisaged to set synapses on striatal neurones into a state of readiness. However, one may envisage that the sequence of activations within the cortex is not very secure: These early, preparatory parts of the chain of cortical activity may sometimes occur without leading to the production of the actual motor program. However, if the speech sound *is* uttered, the perceived speech sounds may match memories of speech sounds previously heard (e.g. from the child's parents). It may thus in some way be perceived as "salient" and activate the reward processes. This would strengthen the striatal synapses activated by the premotor neurones. With repetitive activation of the synfire chain the initial (preparatory) part of the synfire chain would be able to recruit pathways through the striatum as well as via cortico-cortical links, so producing a stronger tendency than before to activate the actual executive part of the motor program.

This schema allows sequences of motor programs to be acquired as "chunks" — temporally organised segments, forming parts of longer time-structured chains of neural activity. There are however constraints on the temporal length and complexity of the chunks which can be reinforced in this way. As already mentioned the sequences of events which are reinforced could never be longer than the hypothetical "state of readiness". Therefore sequences of behaviour longer than the state of readiness must be "chunked" into segments no longer than this, at least during the phase of acquisition. There is however, an additional, more severe constraint which probably means that the chunks in a motor sequence may be substantially shorter than the duration of the state of readiness: Since each chunk is a combination of parts in a timed sequence, the likelihood of a chunk occurring by chance, with correct timing and sequence is much greater for simple sequences than for complex ones. Therefore the chances of a particular sequence (which also has predictable rewarding consequences) being repeated often enough for progressive reinforcement to occur declines steeply for the more complex sequences. This means that if a motor sequence has to have a temporal structure on a fine time scale, only quite short segments are by chance likely to occur often enough to become strengthened by the reward process.

This is a major constraint for organised chains of motor performance such as used in human speech. The above scheme may permit the acquisition of a relatively short motor program, such as that for producing the accurately timed sequence of acoustic transients defining a single stop consonant. This consists of a particular timed sequence of perhaps three or four recognisable acoustic transients, and a similar number of essential motor features, all within a time of about 100 msec. Together these control the production of a sound such as the child may have heard from its parents. The motor program leading to emission of such a sound may not occur very often. However, if it does occur a few times, its similarity to the parents' speech sounds will be recognised, and the motor program leading to its utterance will be reinforced, not element-by-element, but as an organised whole. If however, one considers a longer sequence, with (say) six or eight acoustic transients and motor features, the chances of a program ever being emitted leading to such a more complex sequence will be more remote. Chunks of speech sounds as long as this are unlikely to be acquired directly by the mechanism suggested above.

Nevertheless, building on this mechanism, construction of more complex schemata may be possible, allowing the chaining of such elementary speech sounds to form longer and more complex utterances, such as uni- and polysyllabic words and longer phrases. To explain this chaining we need to start with a proposal, which was put forward some years ago (see Miller, 1981, Chapter 7) for the modelling of the likely consequences of a set of actions, carried out in advance of the actual production of that set.

8.3.2 Motor-sensory modelling and lateralized motor functions in humans

8.3.2.1 "Open-loop" and "closed-loop" motor control

Motor control theorists often draw a distinction between "open-loop" and "closed-loop" control. In open-loop control, a motor program is generated centrally, and executed by central edict, without moment-by-moment sensory guidance. In closed-loop control, exteroceptive sensory guidance is used continually throughout a movement, to ensure that the movement actually performed approximates to an external target. Many sequences of motor output, including those of walking and running, contain a temporal structure too fast to be continously guided by sensory feedback from the external environment, and so are subject to some form of open-loop control. Closed-loop control typically occurs in slower movements.

Examples of this distinction can be found for very many types of motor action. Rapid reaching or aiming for a target may be mainly open-loop control, especially if vision of the target is prevented once the movement has been initiated. On the other hand, visuomotor tracking of a target moving irregularly or unpredictably must involve closed-loop control. Eye movements include saccades — rapid "jumps" of eye position which bring the visual image of an object to the "target", that is the central fovea of the retina. Here, as with rapid reaching, the trajectory of the movement is planned in advance, and is therefore open-loop control. On the other hand, eye movements may be used for tracking an irregularly moving object, under continuous visual guidance. Such "smooth pursuit" movements correspond to closed-loop control. In principle the distinction could also apply to the motor control involved in production of sounds. Open-loop control would consist of producing a sound or sequence of sounds on the basis of a pre-existing internal program. Closed-loop control would involve moment-by-moment adjustment of the motor apparatus, in response to changes in actual sounds produced. (For instance this can occur in driving a car where there are rapid adjustments to the foot controls, in response to the tone of the engine.) Lashley (1951) discussed this issue, and reached the specific conclusion that spoken language consisted of such a rapid sequence of events that it too must have an open-loop component in its control. Kim (1978) extended these arguments, pointing out an additional reason why open-loop control is likely to apply to speech: Conduction time from the forebrain to different speech articulators may vary by as much as 30 msec, which would create more severe control problems for closed than open-loop control. When a person makes a "false start" in speech, he cannot correct himself before completion of at least the first syllable. Moreover, stutterers stutter, and infants babble in *syllables*, not in the individual speech sounds (phonemes). This all suggests that closed-loop control, phoneme by phoneme, does not occur, but rather complete syllables are preprogrammed in advance.

Both of these initial concepts of control require considerable elaboration before they say much about the real mechanisms of motor control in mammals. In so-called closed-loop control sensory guidance becomes complicated, especially for rapidly moving targets, because of the delays involved in relay of signals. Special predictive computations may therefore still be needed in order derive the motor program which correctly tracks the target. Open-loop control also may not be strictly "open-loop": Although there may be no feedback from exteroceptors (such as visual information about hand position), there may be kinesthetic feedback from proprioceptors. Moreover, if a rapid sequence of actions can be broken into segments, it would be possible for programming of each segment to be open-loop, but with closed-loop sensory "checking" of the accuracy of each segment as a whole. Above all, the concept of open-loop control begs the crucial question of *how* the central program for action is acquired in the first place. In fact, under some circumstances, kinesthetic feedback can provide the basis for the construction of central programs, although they may not be required for their execution, as explained below.

8.3.2.2 Components of the primate motor apparatus permitting accurate temporal relationships in motor-sensory associations

The mammalian motor apparatus is, in principle, capable of producing a varied and complex output. This apparatus is controlled by parts of the brain (notably the cerebellum) which are phylogenetically quite old. Probably these parts of the brain are specialised to perform biomechanical computations in which variables such as position, linear and angular velocity and acceleration, momentum, inertia and such like, are important. However, in addition to such control mechanisms common to all vertebrates, humans have a *unique* ability in two motor functions: manual dexterity and expressed language. In the paragraphs below these are considered together. The former of these functions is a development of co-ordination (limited to the hominid primates) of a rather specialised motor organ — the hand. This organ is similar in many primates, including both apes and monkeys (Hill, 1970), but is capable of such sophisticated control only in humans. Expressed language depends on co-ordination of a series of effectors, including lips, tongue and soft palate, but especially of a very complex motor organ — the larynx — whose complexity is again not unique to humans, but is found in similar form in many primates (Hill, 1970). However, the sophisticated co-ordination of function of these speech effectors, as of the hand, *is* unique to humans.

Both these special motor capabilities are closely linked with the function of the left hemisphere, and the properties of this part of the brain are undoubtedly a necessary condition for the development of this phylogenetically new form of co-ordination. Probably a principle of co-ordination is implicated which is quite different from those more typical of vertebrate animals, which rely on the cerebellum and other brain structures. However, it is unlikely that left hemisphere specialisation is a sufficient condition for this new function to emerge, in view of the fact that it is mainly these two complex effectors, rather than the whole human locomotor apparatus, which are so closely related to the activity of the left hemisphere. Distinguishing features of these special effector organs may thus be a contributory condition for the emergence of lateralized motor functions.

Take the hands and arms first. When the hands are used for delicate manipulation, they usually act against quite small resisting forces. In reaching or aiming movements of the arm, the load against which the arm moves is exactly the same on successive occasions. Thus, any program of movement for these effectors will be associated, uniquely and unconditionally, with a specific pattern of kinesthetic feedback. If the movement program has a precise temporal structure, the kinesthetic feedback will have another temporal structure with unique correspondence.

The larynx and other vocal effectors do not act against a significant mechanical load. The component parts of these effectors are not bulky, and so have little inertia. In any case, the load of these effectors, when used in their most specialised manner, is likely to be constant, which is not the case for the prime movers in locomotion, for instance. Sensory feedback from these effectors will therefore have the same precise and unconditional spatio-temporal relationship to the motor program as for unloaded movements of hand and arm. Kinesthetic feedback may be part of this. However, the most important sensory feedback is of course in the auditory sense, the relay back to the brain of the vocal sounds produced by each vocal gesture. Here of course there are no additional loads or other conditions which might modify the timing of that feedback, except in the circumstances of some special experiments, reviewed below. Again there will be a unique correspondence between the temporal structure of the motor program and that of the sensory (auditory) consequences it produces.

The movements of the eye can also be considered here, although lateralization of their motor control has not been studied as extensively as for manual praxis or speech production. Eye movements dictate any changes in visual input in the same unconditional and precisely-timed manner as movements of the unloaded hand and arm dictate kinesthetic feedback, or as vocal gestures dictate auditory feedback.

For these effectors it is possible for there to be very regular and precise temporal relationships between a motor command emanating from the motor parts of the cerebral cortex, and the kinesthetic or other sensory feedback of the effects produced by those commands on the effectors. It is argued below that this can permit the acquisition of a new, and more precise form of motor control, which, when acquired, appears to be a centrally programmed sequence, not requiring moment-by-moment sensory guidance. However the programs for this class of "open-loop" motor act place particular demands on the temporal delays within the cortex. *This*, it is suggested below, is the reason why these motor functions are lateralized to the dominant hemisphere of humans.

8.3.2.3 Motor-sensory association neurones in the cortex

This new and more precise form of motor control, for which the dominant hemisphere of humans is so important, involves the psychological function of *modelling* an action in advance of its actual performance. The fundamental concept needed for such modelling is a class of neurones first described by Mountcastle *et al.* (1975) and called by them "command neurones". The distinctive feature of such neurones is that they represent the association between a motor act and its sensory consequences. In other words, they are motor-sensory association neurones. For instance neurones have been found in the parietal cortex which fire if and only if a reaching movement is made under visual control, or with tactile or kinesthetic feedback. Similarly there are neurones which respond to the combination of a particular eye movement and the particular change of visual stimulation such a movement is likely to produce. Evidence suggests that the cortex contains a very large number of neurones with such response properties, representing a very wide repertoire of motor-sensory associations.

It is not difficult to see how such motor-sensory association neurones acquire their response properties. Evidence is available from monkeys for axonal projections from arm and face premotor areas to corresponding parts of area 5 and 7 in the parietal cortex (e.g. Cavada and Goldman-Rakic, 1989; Neal *et al.*, 1990; Andersen *et al.*, 1990). There are also comparable projections from the frontal eye fields to area 7 (Pandya and Kuypers, 1969; Pandya and Vignolo, 1971; Pandya *et al.*, 1971; Cavada and Goldman-Rakic, 1989; Neal *et al.*, 1990). Areas 5 and 7 also receive projections from the primary sensory regions for vision and somatic sensation. These regions are the ones in which Mountcastle *et al.*'s command neurones were found. Assume that the synapses of such cortico-cortical axons are subject to Hebbian modification (as we assume for other cortico-cortical synaptic terminals). Each action of the organism is likely to elicit its own characteristic consequences. The correlation between the motor program for the action and the sensory consequences of that action can than be registered by Hebbian strengthening in the terminal synapses of the motor-sensory cortico-cortical axons, co-operating in the process with those of projections from primary sensory areas. This is depicted in Figure 8.3.

These considerations apply initially to motor-sensory associations linking visual, or kinesthetic consequences to programs for movement of limbs or eyes. In principle, there might also exist motor-sensory association neurones representing the correlation between a motor program for utterance of a vocal sound, and the actual emitted sound. If such neurones exist, it is likely that they would be found in an auditory association area, perhaps equivalent to Wernicke's area in humans. Such neurones have actually been sought in one study, in squirrel monkeys (Muller-Preuss and Ploog, 1981). Single unit recordings were made in a wide area of the superior temporal gyrus during spontaneous phonation by the monkeys or during play-back of the sounds produced by such phonation. In about half of the responsive neurones the response pattern was similar during spontaneous phonation and during play-back of the same sounds. The remainder of responsive neurones responded only during play-back, the response to the sound somehow being suppressed during its spontaneous phonation. *No neurones were found which responded only during the spontaneous phonation,* or more vigorously during this than during play-back. Thus, in monkeys, the evidence available suggests that there are *no* motor-sensory association neurones in the vocal/acoustic modality.

The failure to find such neurones in monkeys may be because the cortico-cortical connections necessary for their acquisition do not actually exist. The appropriate connections would originate from a premotor or motor area concerned with movements of the speech apparatus (namely part of the inferior frontal lobe, a region similar to the human Broca's area) and would project to the auditory association area in the superior temporal

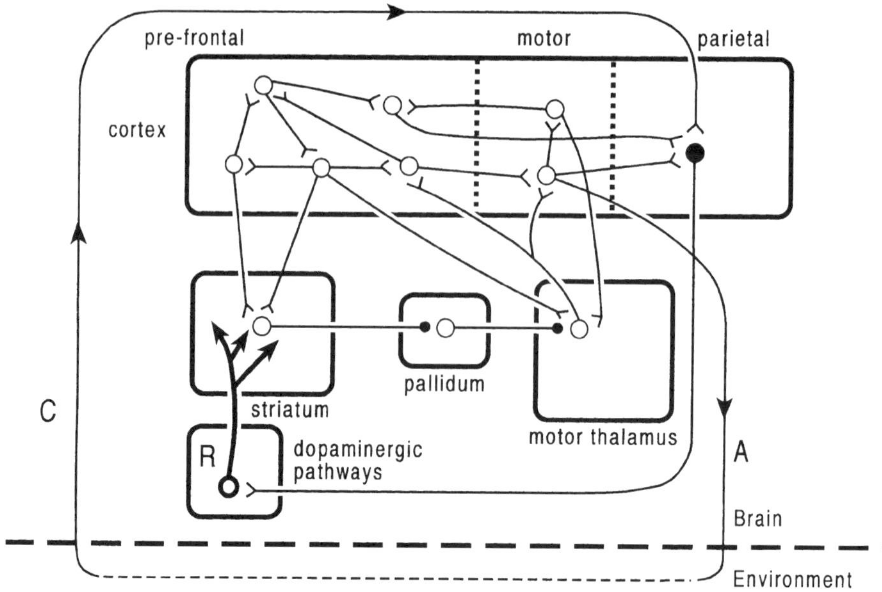

Figure 8.3 Motor-sensory Association Neurones (Mountcastle's "command" neurones) Representing the Consequences of Action. A cortical cell assembly is depicted as located in prefrontal and motor cortical regions. The parts of it in the prefrontal cortex are envisaged to become active during the preparation for a motor act, and when activity spreads to the motor cortical parts, the act (A) is executed. The consequences of that act (C) are detected by sensory systems and relayed back to the cortex. In the parietal cortex, neurones receive convergent activation from (i) the preparatory components of the cell assembly, located in the prefrontal cortex, (ii) the executive parts of the cell assembly, located in the motor cortex and (iii) the sensory consequences following execution of the act. Hebbian strengthening occurs as a result of this convergent activation. Thus neurones (such as that depicted as a filled circle, in the parietal cortex) can come to represent the specific consequences of a specific motor act.

gyrus and planum temporale/insula region (similar to Wernicke's area). Anatomical evidence for the existence of such a projection in monkeys is very slim. Admittedly, several studies have traced projections from prefrontal cortex to temporal lobe (Pandya and Kuypers, 1969; Jones and Powell, 1970; Pandya and Vignolo, 1971; Pandya *et al.*, 1971; Markowitsch *et al.*, 1985; Pandya and Yeterian, 1985; Shiwa, 1987; Ban *et al.*, 1991; Deacon, 1992). However, few of them specifically trace pathways originating in the inferior frontal gyrus (Pandya and Kuypers, 1969 lesion 4D; Pandya and Vignolo, 1971 lesion 4B; Pandya and Yeterian, 1985 fig 23C). None of these identifies a region of termination in the superior surface of the posterior temporal lobe (equivalent to planum temporale) (though Pandya and Yeterian [1985] trace projections from inferior frontal gyrus to the insula). One study (Leichnetz, 1989) traces cortico-cortical connections specifically from area 45, but these do not terminate in the relevant region.

In what follows however, *it is assumed that such projections actually do exist* ***in humans***, and then more richly in the left than the right hemisphere. (This is discussed in more detail below where morphological aspects of laterality of motor functions is dealt with.) This assumption would then be part of the explanation that humans have the facility of speech, whereas monkeys do not. How could the existence of a projection from Broca's to Wernicke's area in humans play such a decisive role in spoken language?

8.3.2.4 *Conduction time in motor-sensory cortico-cortical axons and the registration of accurate temporal relationships in motor-sensory associations*

In the previous section evidence was mentioned that cortico-cortical connections exist allowing an association between manual gestures and their kinesthetic consequences, or between eye movements and their visual consequences. We predicted also (in the absence of direct evidence) that in humans (perhaps uniquely) cortico-cortical connections also exist allowing association (probably in Wernicke's area) between an articulatory gesture and its auditory consequences. The special characteristics of arm, hand, eye and vocal apparatus can lend themselves to special forms of motor control, *providing these cortico-cortical links representing motor-sensory associations also have special properties.* In humans the cortico-cortical connections on which motor-sensory associations depend would be quite long ones (spatially): The length of prefronto-parietal connections would be much longer than the direct distance, because they would have a trajectory passing deep to the central sulcus. The arcuate fasciculus, passing from Broca's to Wernicke's area is likewise a long, even circuitous route. The range of conduction times of such axons is entirely a matter of conjecture, but there is no reason to believe it would be different from that of other cortico-cortical projections of equivalent length. Therefore one may envisage a repertoire of delay lines in these pathways up to several hundred milliseconds, for reasons suggested in an earlier chapter.

When a time-structured sequence of neuronal firing is initiated in the premotor and motor areas, a characteristic time-structured speech sound is uttered and perceived by the auditory association cortex, following its sensory relay in the ascending auditory pathways. *It may be envisaged that cortico-cortical relays will become strengthened which form a link which is exactly matched, at its origin, to the temporal structure of the neural sequence for preparation and execution of the motor act, and exactly matched, at its termination, to the sequence of sensory activations received in the auditory association cortex.* The same is likely to be true for motor programs governing manual reaching or aiming, or eye movements. This is depicted in Figure 8.4. It should be noted especially that both the origin and the termination of this cortico-cortical link are spatio-temporal patterns characteristic of the speech sound. Thus, the cortico-cortical links which make possible the motor-sensory association in the auditory sense would be a *spatio-temporal template of the exact sound to be expected from the motor program activated.* This may also be true for other examples of motor-sensory association neurones; but in the case of human speech mechanisms the formation of such an accurate spatio-temporal template may turn out to be pivotal.

8.3.2.5 *Longer chaining of motor sequences: the basic modelling function*

As just explained, motor-sensory association neurones may be activated when the preparatory and executive parts of a motor program sequence are activated and the motor act is performed. However, as suggested in section 8.3.1.6, the preparatory part of the sequence may come into activity without following through to the full motor program and the action. Nevertheless such an "abortive" activation, limited to the preparatory parts of the program sequence would tend also to activate the same representation of the consequences of that sequence, via the motor-sensory links described in the previous section. As a result of this the brain may have the possibility of *testing the combined*

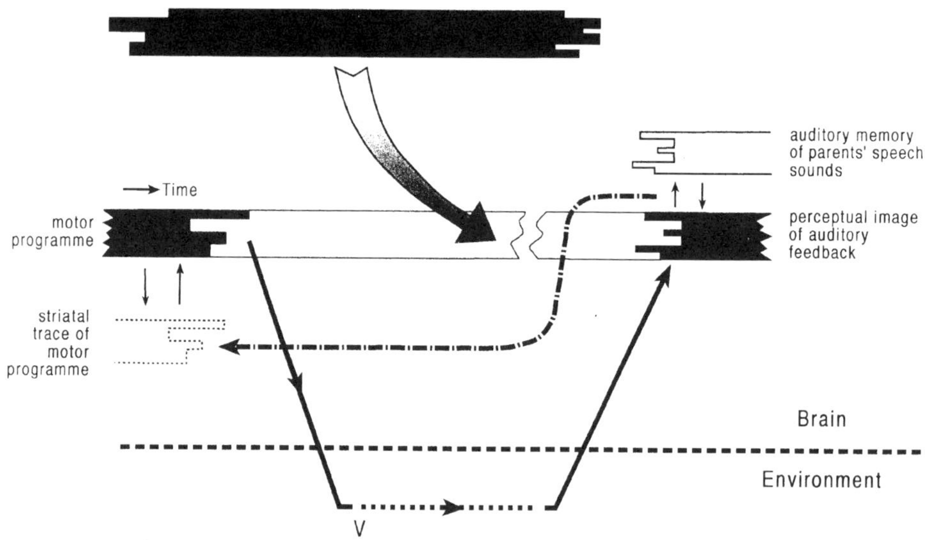

Figure 8.4 Motor-Sensory Cortico-Cortical Links as Spatio-Temporal Templates of the Sound to be Expected from a Specific Speech Motor Program. In the central "stave" of this diagram is shown time structures of neural activations, representing the pre-motor and motor program for a speech sound (left), and the perceptual (acoustic) image produced when that sound is uttered (right). For simplicity of drawing, these are shown in blocked form, but could equally well be shown as "synfire chains" as in Figure 5.7. Note however, that the stave on the left does not depict the activations of the same neurones as that on the right: The former represents activity in motor and/or prefrontal cortex, the latter, that in auditory association cortex. Suppose a child emits a motor (speech) act with the temporal structure of that shown on the left. Initially this occurs by chance. However, it will set up a trace of the same temporal pattern in the striatum (the "state of readiness"). The speech act will be executed by the vocal apparatus (V) and, after a short delay, will produce the perceptual pattern shown on the right. If this randomly-emitted speech sound happens to match the time-structured memory of speech sounds heard from the child's parents and others (depicted upper right, open blocks), the activation of the perceptual/mnemonic image will be much greater than normal, and is recognised as specially "salient". This is envisaged to activate the reward signal (arrow shown as — • —), and so will reinforce the striatal synapses holding the trace of the initial motor program. Interplay between striatum and cortex (see Figure 8.1, shown here as a two-way vertical arrows, on left) can then form a strong representation of the temporal sequence which leads to production of this speech sound. This is similar in principle to that shown in Figure 8.2. The particular motor program emitted (rather than others, which do not produce a familiar sound) will subsequently be produced more often than expected by chance. As this speech sound is subsequently repeated many times, there will be coincident activation of association cortex neurones by sensory stimuli (acoustic consequences of the emitted speech) and selected cortico-cortical pathways. Hebbian strengthening of terminal synapses of the latter pathways will then occur. As a result, a new synfire chain of cortico-cortical connections is elaborated (upper left) which fits exactly the temporal structure of the original motor program for the speech sound (both its preparatory and its executive parts), the temporal structure of the perceptual (acoustic) image of the sound, and the delay between the two. This is the spatio-temporal template of the speech sound, a representation of exactly what sounds are to be expected from a specific sequence of vocal gestures.

consequences of various different combinations of motor programs, without actually emitting any of them. This is the faculty known variously as "mental planning", "modelling", or "thought". In fact, thought becomes possible only when the preliminary, preparatory part of a motor sequence can be split off from the executive part, allowing modelling of a planned action without actual trial and error of the act itself.

Two varieties of modelling may be distinguished. The basic variety involves modelling the consequences of action as a *sequence* from action to consequences, without the sequence being measured precisely in time. Given this basic process, it may be developed into an elaborated version, in which "*time-structured modelling*" is possible. Here the complex motor sequence and its consequences are both given a precise temporal metric, and (most importantly) so also is the relation between the two. This elaborated version is hypothetically implicated in the motor functions of the left hemisphere. In the present subsection the basic variety is dealt with, while time-structured modelling, more relevant for understanding spoken language, is dealt with in the next subsection.

The archetypal example of the basic variety of modelling, discussed in an earlier work (Miller, 1981) is the chimpanzee described by Köhler (1927). This animal sees a bunch of bananas out-of-reach, and there are also accessible in the chimpanzee's environment the various means by which the bananas can be reached, such as boxes to pile on one another, and a stick with which to reach the inaccessible object. Chimpanzees can solve such problems, and human infants acquire this ability before they learn to speak (Piaget and Inhelder, 1969).

It was suggested previously (Miller, 1981) that this is done by modelling the consequences of a combination of motor acts, by the appropriate combination of motor-sensory association neurones (Mountcastle *et al.*'s "command neurones"). Assuming that preparatory activity can be divorced from the executive part of the motor program, one may envisage a variety of patterns of neuronal activation in the parts of the frontal lobe representing preparation for a motor act, with associated neuronal activations in parietal or temporal lobes representing the likely consequences. If the act can be considered a deliberate *intention*, activation of a representation of the consequences will also activate an image of fulfilment of the intention. This may activate the reward process, just as does actual fulfilment of intentions (see section 8.3.1.4). Over time the organism will acquire a rich representation of such programs for fulfilment of the intentions.

At their origin, the motor-sensory links are likely to be parts of cell assemblies for motor programs, and at their destination they are likely to activate cell assemblies for sensory events (consequences). One of the advantages of cell assemblies, is that they are capable of versatile and numerous recombination and regrouping. Thus, when a new combination of premotor assemblies is activated, it is likely to set into activity a new combination of several cell assemblies each representing the consequence of one component of the combined act. In order for modelling (in advance of actual action) to be possible, it is required that *when the cell assemblies for the consequences of the individual motor programs are combined, they will represent in approximate form the combined consequence of that unrealised, but specific combination of motor acts.* In other words, it is required that if assemblies "a", "b" and "c" represents events A, B and C, then the combination of the three assemblies — "abc" — accurately represents the combination of the three events, even if that combination includes quite new emerging subevents. This requirement specifies an elementary, non-symbolic form of *deduction*.

Consider this with respect to the chimpanzee's problem. Three cell assemblies for events (consequences of action) can be specified: "the consequence of piling boxes on one another", "the effect of the chimpanzee standing on the pile", and "the effect of reaching upwards from that position with a stick, towards the inaccessible bananas". It is suggested that this combined representation will have important aspects of neural representation in common with a fourth event for which a cell assembly is likely to be already in existence. This fourth event is "reaching with the unaided arm to obtain an accessible, and desired object", an event which would previously have become strongly linked with activation of the reward system. If the combined activation of the first three cell assemblies also activates the fourth cell assembly sufficiently to "ignite" it, that ignition is also likely activate the reward system strongly, and reinforce the representation of the complete sequence of the four events.

Thus, the organism can acquire a representation of the expected fulfilment of intentions by simple *sequences* of acts. In this process, there may be moments of "insight" when the consequence of a sequence of actions is foreseen in advance of actually emitting the sequence, as Köhler described in chimpanzees solving a puzzle. This general scheme is depicted in Figure 8.5.

It should be noted that, in the scheme depicted in Figure 8.5, the cell assemblies representing components of action and their consequence do not contain temporal structure. That is, the preparatory and executive sequences and the sequence of consequences, all acquired by the modelling process described so far, are sequences of "snapshots" arranged one after the other, without defined temporal relationships. Referring to an earlier chapter, these chains of snapshots are each represented by a conventional Hebbian assembly, not by synfire chains (that is, they are more like the networks shown in Figure 5.7C than those in Figure 5.7A). Such sequencing is achieved by the processes involving the striatum, and the hypothetical "state of readiness", for recognising the patterns of information which have been precursors to each occasion when the reward signal is activated. Not only is no precise temporal metric involved in each preparatory part of the motor assembly or the consequent perceptual cell assembly, but above all, no temporal measure is built into the relation between premotor and perceptual cell assemblies.

Admittedly fixed time delays may become part of the sequential firing of assemblies, encoded by axonal delay lines. However, the time delays involved *within* the modelled sequence of actions or *within* the modelled sequence of consequences are likely to be less extended than those *linking between* the actions and the consequences. Thus, it is conceivable that delay lines could be long enough to represent the delays within the preparatory program for a sequence of actions, or within the program of its consequences, while still being insufficient to represent the temporal relationships between the two. In this case, an extended motor program with precise intrinsic temporal structure could be gradually constructed by linking smaller time-structured "chunks" of neural activation. In this case, each chunk would have to have been reinforced individually following instances of actual execution of the chunk (as shown in Figure 8.2), before the extended chain could be constructed. However, without cortico-cortical delay lines which are long enough to link a preparatory part of the motor program with its likely consequences, there can be no active time-structured modelling (i.e. in advance of action) of the extended chain taken as a whole*.

* A point also made in the 1981 account of modelling is that for such modelling to be accurate it is necessary for the correct *context* of neural activity to be evoked, as well as the representations of the information items directly involved in the modelling. This need arises, in a general way, because modelling of untried actions is a nonverbal form of *deduction*. As first suggested by Immanuel Kant, deduction requires use of the appropriate context as well as the appropriate premises for the deduction. Therefore it is likely that motor-sensory modelling requires activation of the correct context representation. This goes well beyond the scope of the present book, but has been discussed in more detail elsewhere (Miller, 1989; 1991; 1994). The neural realisation of the interaction of context and individual items is envisaged to involve interplay between hippocampus and cortex. Such interplay is envisaged to be necessary for modelling, whether of problems like the chimpanzee's or in planning a linguistic utterance.

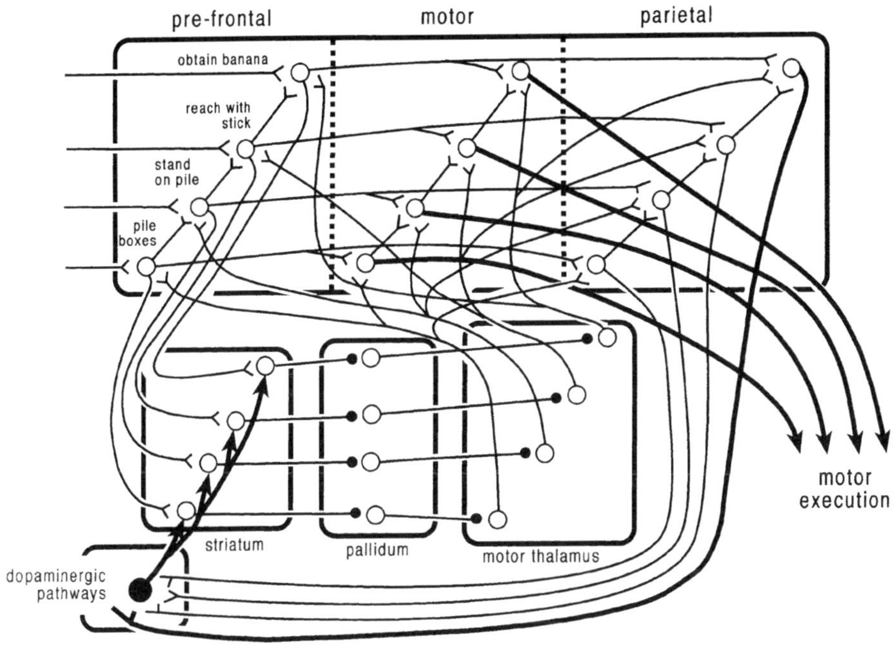

Figure 8.5 Motor-Sensory Modelling of the Combined Consequences of a Complex Motor Program. In the prefrontal cortex (upper left) are depicted four neurones (circles). These are components of different cell assemblies (not shown) representing preparatory activity for different parts of a complex motor program (Köhler's chimpanzee obtaining a banana). As a result of previous experience, each of these has been connected with a neurone in the parietal cortex, each of the latter also part of a cell assembly (not shown) representing the consequences of the corresponding component of the program (circles, upper right). The arrangement of these sets of neurones in their "boxes" (and their interconnections) is meant to depict relationships of sequence but not of timing. Each prefrontal cortical neurone, when active, also sets into a "state of readiness" synapses in the striatum. Activation of the corresponding parietal cortical assembly amounts to activation of the representation of fulfilment of an intention, and therefore activates the dopaminergic ("reward") neurones (shown in bold lines, lower left). This will strengthen the striatal synapses in the "state of readiness", and, on subsequent rehearsals of the model, increased activity will be directed, via the striatal loop, back towards the cortex. By a succession of such fragments of modelling it may be possible for the chimpanzee to model the combined consequences of a sequence of several acts ("piling boxes", "standing on the pile", "reaching with a stick", and ultimately "obtaining the banana"). The final event will exert the strongest rewarding effect. As this sequence of events is rehearsed many times, two effects will make the sequence become more fluent: (i) The prefrontal cell assemblies representing each preparatory motor component may become sequentially linked by cortico-cortical connections. (ii) The reward signal may reinforce the whole sequence, if it does not precede the final rewarding event by a longer duration than the "state of readiness". All this development is achieved as a result of activation of *preparatory* components of the motor program, rather than actual emission of the combined act. Hence the consequences of the combined act will be known in approximate form before it is performed. The chimpanzee may thus be able to work out a strategy for obtaining the banana. However, each prefrontal cell assembly is likely to have links with a corresponding executive assembly in the motor cortex. Hence the increase in activity directed back from striatum to cortex when the model becomes robust may cause an activation of the actual executive components of the modelled sequence, provided the right triggering stimuli also occur. Note that there are no delay lines in this scheme of modelling, so that none of the cell assemblies contains a temporal structure. Without such temporal structure in the cell assemblies in prefrontal and parietal cortices, and in the relation between the two, a consistent *temporal* chain of behaviour cannot be organised.

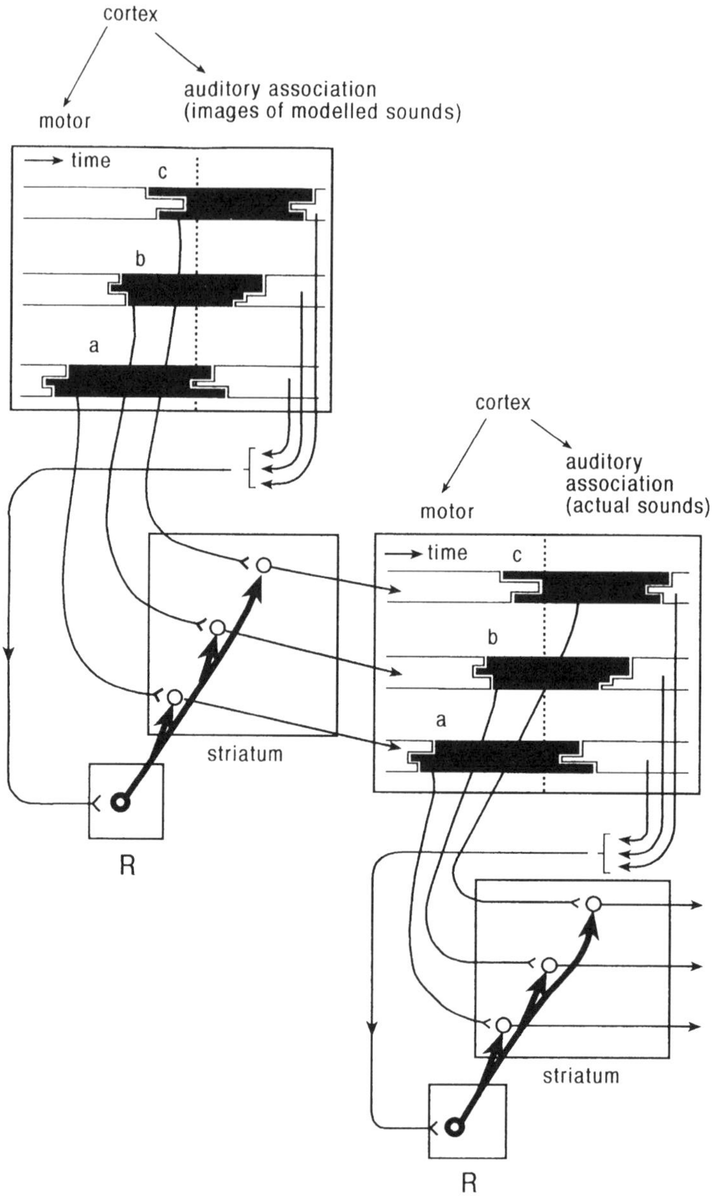

Figure 8.6 Time-Structured Modelling, for Instance of a Complex Sequence of Speech Sounds. It is envisaged that the complex speech sound consists of a timed sequence of three elementary speech sounds (phonemes). In the planning phase (upper left half of diagram), a premotor representation of each phoneme is activated in the prefrontal cortex (open bars to the left of uppermost block of the diagram). Each has its own temporal structure (see complex shape of right-hand end of open bars), and they are activated in the correctly timed sequence (potentially as a single time-structured chain of activity, though in the diagram they are shown as separate elements). This activates the motor-sensory delay lines for each phoneme, represented as the filled bars (a,b,c, upper left). The termination of these delay lines corresponds to the activation of modelled images of the sequence of sounds. These motor-sensory delay lines represent the temporal structure of the plan for action as well as that of the sounds which are to be expected from the plan (see complex shape at left and right hand end of filled bars). In addition, activation of the preparatory motor sequence sets into a state of readiness synapses on a small set of striatal neurones.

8.3.2.6 Longer chaining of motor sequences: time-structured modelling

In Figure 8.5, and in the preceding paragraphs (as in the 1981 schema for modelling, Miller, 1981) no explicit reference was made to motor programs being modelled as *accurately-timed chains* of smaller motor program elements. However, provided that motor-sensory association neurones constitute delay lines long enough to link the neural activity representing preparation for action with the activity representing consequences of action, these neurones can also be employed to model such accurately-timed chains. When this occurs, an altogether more powerful method of modelling becomes possible.

This scheme is depicted in Figure 8.6. It will be noticed that the lower right half of this figure is an almost exact replica of the upper left half, indicating that the same cortico-striatal machinery is involved in various stages of the planning and performance of the timed sequence of actions. In fact, one may easily extend this sequence to produce a long sequence of preliminaries to action, a veritable cascade of cortico-striatal interactions used for advanced planning. The planning of an action to be performed in the *immediate* future may thus be a very small part of human thought. In a wider sense, thought may become nearly completely divorced from immediate action.

The essential aspects of the hypothesis builds upon the principles depicted in Figures 8.2 and 8.5. It amounts to testing the combined consequences of a preparatory version of a motor sequence, the model being constructed along an accurate temporal scale. Clearly the existence of a wide variety of motor-sensory association connections is an absolutely necessary anatomical substrate for such modelling to occur. Using such connections a large set of spatio-temporal templates may be acquired representing all the motor-sensory links defining each phoneme, as described above (Figure 8.4). Combinations of small numbers of these templates can be used for modelling longer utterances, such as mono- and polysyllabic words. The process can be broken down into a series of stages:

(i) Envisage that a novel combination of preparatory programs for phonemes is activated from the prefrontal cortex. The acoustic image of the corresponding combinations of speech sounds will be activated, with an internal temporal structure corresponding exactly to what would be expected from the temporal relations of the preparatory motor program elements. Thus, for the new combination, it will be possible to model, in advance of actual speech, the sounds produced by that combination. The model which is constructed is a time-structured one, the underlying principles of which have already been shown in Figure 8.2 and 8.4.

(ii) If that model matches a memory of sounds previously heard from parents or others, the reward system will be activated, and the link from each motor element to a

---

Figure 8.6 *(continued)* If the modelled image of the speech sound matches a memory of previously perceived speech sounds (e.g. from parents or others) the reward system (R) is activated, which converts synapses in a state of readiness into definitively strengthened ones. This "opens" a set of pathways back to the cortex, directed particularly at program elements downstream from those hitherto activated (as shown in Figure 8.2). These downstream elements may include the program for the definitive performance of the sequence of vocal acts (depicted in lower right half of diagram, another battery of cortico-striatal interactions). Again a set of motor-sensory delay lines are activated in a timed sequence (many in common with those used at the modelling stage). Their termination will synchronise with perception of the sequence of actually emitted phonemes. If these speech sounds match a memory, then, in a fashion similar to the planning phase (shown upper left), striatal links can become strengthened which direct activation at later links in the preparatory or executive parts of the program.

subsequent one will be strengthened via the striatal loop: This will involve strengthening of specific cortico-striatal synapses by the operation of the three factor rule, followed by Hebbian strengthening of thalamo-cortical ones activated in synchrony with elements of the cortico-cortical chain. As previously depicted (Figure 8.2), since both the circuits via the basal ganglia (and thalamus), and the cortico-cortical links impose fixed delays, this process strengthens the security of transmission downstream in the temporal chain.

(iii) The preparatory model of the complex speech sound will have many neurones in common with the program for actual emission of the same speech sound. That is, they are subdivisions of a single larger assembly configuration. When activation of this combined assembly is amplified by the striatal loop, it is likely that the modelled sound will be followed by its actual utterance, by the principles depicted in Figure 8.5. This is the process of first discovery of how to produce a new time-structured speech sequence or other motor sequence.

(iv) With repeated performance of the sequence, further categories of synaptic change consolidate its representation, and make performance more fluent and secure. Whenever reward processes are activated, any striatal synapse activated within the duration of the preceding state of readiness will be strengthened; and, via striato-cortical pathways, that will lead to enhancement of the security of transmission in any aspects of the cortical chain (preparatory, executive, modelled sounds, actual sounds; see Figure 8.2). Moreover, because there is now a tendency for the different elements of the cortical motor program to occur in definite temporal relations to each other, these temporal relations will become more secure as a result of Hebbian strengthening of cortico-cortical delay lines from the initial to the later sections of the sequence.

(v) In each performance of an extended chain of speech acts, the activation of the template of consequences of the model will have a precise temporal relation with the activation of parts of the definitive program for actual performance. Apart from cortico-cortical links from prefrontal or motor areas of cortex to the sensory association area, there may also be strengthening of links from sensory association area back to prefrontal area, provided the part of the preparatory plan which is the destination of these links represents a later fragment of the uttered speech than the acoustic image which is their origin. These would represent the consistent relations between a fragment of the chain of consequences to a fragment of the plan for an motor act later in the chain.

(vi) Given that such robust temporal chains of activity can be formed, representing the various aspects of word production, each chain can then become an instrument for goal fulfilment, in the specific circumstances in which it leads to favourable consequences. The rubric of instrumental conditioning therefore can be applied, with the complete length of each chain forming the instrumental program. However, it was noted in section 8.3.1.5 that with extensive training, instrumental tasks can become "routine", probably requiring cortical pathways alone, without the subcortical reward effects and the striatal loop. Thus with further extensive rehearsal these subcortical events become no longer an essential part of the speech production. Speech may come to be emitted routinely, using cortical pathways alone.

Thus the rapid implementation of a well established program whose preparatory elements have just been modelled will ultimately depend not on amplification by the reward effect, but mainly on cortico-cortical delay lines. Some of these will represent the actual temporal structure of the motor program, some will represent the modelled structure of the to-be-emitted speech sound. Others will represent necessary temporal relations between the motor program and ensuing speech sounds (motor-sensory association links); and yet others will relate the representation of emitted speech sounds early in the sequence to later elements in the motor program. The final representation of the chain of motor acts and their consequences will have a rich and secure set of links via delay lines between all elements, whether they represent preparation for the act, the modelled consequences of the act, the execution of the act, or the actual speech sounds resulting from it.

In such a scheme there is an obvious role for cortico-cortical connections from Broca's to Wernicke's area, and intrinsic to each area, but there is also a role for connections in the reverse direction, from Wernicke's to Broca's area. The full range of cortico-cortical links used in this way is depicted in Figure 8.7.

An important feature of such modelling of a chain of motor acts is that the number of elements which can be sequentially combined in any modelling step is quite small, possibly no more than three or four (as depicted in Figures 8.6 and 8.7). The reason for this is similar to that explained in section 8.3.1.6. Each sequence of motor elements to be used in modelling (like each "chunk" of a motor program for a phoneme acquired more directly by conditioning) is a *combination* of parts in a timed sequence. The likelihood of a sequence being arranged and timed correctly in the chance activity of the "naive" neural network is much greater for simple sequences than for complex ones. Therefore the chances of a particular sequence of phonemes being repeated often enough in modelling for progressive reinforcement to occur declines steeply for the more complex sequences.

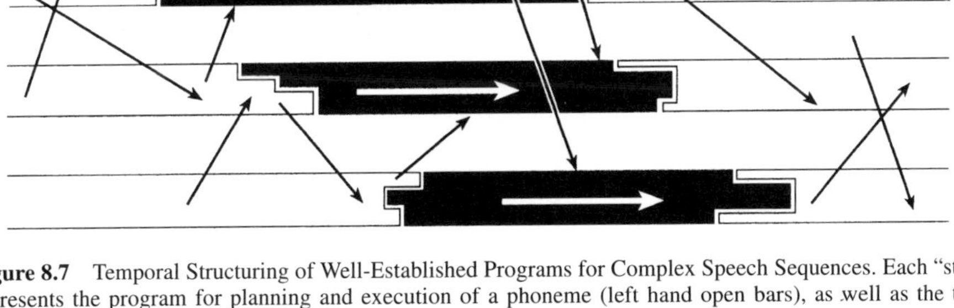

Figure 8.7 Temporal Structuring of Well-Established Programs for Complex Speech Sequences. Each "stave" represents the program for planning and execution of a phoneme (left hand open bars), as well as the time-modelled images of the expectations of speech sounds (central filled bars), and the template for the actually emitted sounds (right hand open bars). The temporal structure of each component, indicated by the complex shape at the end of the bars, is encoded by cortico-cortical delay lines. However, other such delay lines will exist in the well-established representation, linking different "staves" (oblique arrows). These would link different parts of the programs (preparatory, executive, expected sounds, actual sounds) for different phonemes, which, in the complex speech utterance, occur regularly in accurately-timed association. Note that some of these connections pass from perceptual representations (right side bars of one stave) to later parts of the motor plan (left side of another a stave).

This means that if a modelled motor sequence has to have the correct temporal structure, only combinations of small numbers of motor components are by chance likely to occur often enough to become strengthened. In section 8.3.1.6, phonemes were considered to be the longest combination whose representation could be directly acquired by the mechanisms depicted in Figure 8.2. In comparable manner, there is a limit to the complexity of arrangements of phonemes whose representation can be acquired by the modelling process depicted in Figure 8.6. This may correspond to the larger building blocks of speech (single syllables or simple words). More complex speech sequences could not be acquired directly, and so would need to be broken down into units of such size.

Thus, at some stage in acquisition of spoken language, the program for each word is envisaged to be constructed by the above process of time-structured modelling. This would be a laborious and painstaking process, but the slowness of speech acquisition is completely hidden in the fluency of connected chains of speech sounds eventually achieved. The same painstaking process presumably applies to extemporised music. The facility by which "playing by ear" is acquired is likely to be acquired mainly by lengthy internal modelling and externalised practice of a sequence of sounds in advance of their production in fully fluent manner.

8.3.2.7 Anatomical and functional predictions about lateralization of motor functions

According to the hypothesis advanced above, the process of modelling of words and phrases is not possible without a rich population of cortico-cortical motor-sensory links to form spatio-temporal templates of each elementary speech sound. Thus, it becomes plausible to think that the presence of such connections from Broca's to Wernicke's area in humans may be a decisive specialisation permitting *spoken* language to develop. Indeed, in mammals generally, it is hard to see any need to register the association between the details of temporal structure of a motor program and the details of temporal structure of the noises made when that program is carried out. The only circumstance when such a set of associations would be needed is in communication by acoustic signals.

We have seen in preceding chapters that the lateral specialisation for receptive aspects of language can plausibly be accounted for in terms of a global difference between hemispheres in the spectrum of cortico-cortical axon calibres, and that such a difference can give rise to a wide range of other perceptual and cognitive differences between the hemispheres. Given the basic hypothesis that the repertoire of cortico-cortical connections in the left hemisphere has a longer range of delays than in the right hemisphere, the extension of this hypothesis to explain the lateralization of motor performance is obvious: *The projection from speech-controlling motor and premotor areas to auditory association areas should not only be richer on the left than the right side, but also the connections in these pathway should give a richer repertoire of long delay lines in the left than the right hemisphere.*

As we have seen, in experimental primates, there is no substantial evidence for the existence of the required set of connections allowing auditory motor-sensory association neurones to form; and physiological experiments have also failed to find such neurones. It is possible that the projections needed for such auditory motor-sensory association neurones to form really do exist in monkeys, but have not been discovered by lesion or tracer placements yet studied. In humans it is well known from gross dissection studies that there is a large fascicle of fibres connecting frontal to parietal and temporal lobes. In particular, the arcuate fasciculus can be defined, linking the inferior part of the frontal lobe to the temporal lobe. It is plausible to suggest that this contains axons passing from Broca's area to Wernicke's area, which would suffice for the motor-sensory pathways subserving spoken language. However, such large scale morphology tells nothing about the precise areas of origin and termination of these fibres, nor does it even indicate which direction the axons within the arcuate fasciculus travel. There are no data available about the details of cortico-cortical projections, mainly because of the ethical limitations on performing the necessary experiments in humans. However, although there are no data about the details of cortico-cortical projections in human brains, cytoarchitectural data *are* available about the relevant regions of cortex. These data also provide evidence of differences between the sides, and so are relevant to the left-hemisphere specialisation for expression of language.

It will be recalled from section 4.2.5.1 that in the *anterior speech area* laminae I-III of the cortex are thicker than these laminae in the corresponding part of the inferior frontal lobe on the right side. In addition lamina III contains some large pyramidal cells in this area; and the larger of these pyramidal cells are larger on the left side than the right side. In fact, there are hints that these distinctive features on the left side may be found more widely in the prefrontal area than Broca's area alone. Since laminae II and III are the predominant site of location of cell bodies of cortico-cortical cells (and contain few other projection neurones), one may conclude that these regions, especially that on the left side are specialised as the source of such projections. Hayes and Lewis (1993) regard the projection target of the large lamina III pyramidal cells in the left side in area 45 of humans as unknown. They speculate that this projection may be unique to humans.

Large size of neuronal cell bodies (such as in those of lamina III of Broca's area) is usually taken to imply that those cell bodies have to support a large amount of axoplasm, either because the axons are long, or because their calibre is large. A third alternative, which is compatible with the axons having a small calibre, is that large cell bodies have to support a large number of collaterals, somewhere along their trajectory. There is no data to say which of these possibilities applies to the large lamina III pyramidal cells in the anterior speech area.

In the *posterior language zone*, columns of cells are further apart on the left side, implying a larger ratio of neuropil to cell bodies, and dendritic trees are, by implication more widespread than on the right (see section 4.2.5.2). These features are compatible with a specialisation of the posterior language area for reception of a large number of afferent connections.

Thus the combined cytological specialisations of anterior and posterior language areas are compatible with there being a rich projection starting in the anterior area and terminating in the posterior one. This possibility remains to be proven by more direct evidence.

It should also be mentioned that the anterior and posterior language areas are larger than the corresponding regions of the right hemisphere. If it is true that there exist specialised cortico-cortical relays from the anterior to the posterior area, the larger size of the areas on the left would make this link of greater potential importance for the motor-sensory associations needed in speech.

Given that the decisive condition needed for spoken language is the existence of a set of connections with not only substantial conduction delays, but also between rather circumscribed areas of cortex, this would seem to be a more specific specialisation than the global one supposed to underlie the receptive aspects of language. Thus, hemispheric specialisations may be of two types: There may be a global difference giving the left hemisphere a richer repertoire of cortico-cortical conduction delays than the right, and leading to a left hemisphere specialisation in receptive aspects of language, and to many other perceptual and cognitive differences between hemispheres. In addition, a particular difference between the hemispheres is in the pathway from Broca's area (and perhaps its surroundings) to Wernicke's areas, for which pathway both the richness of this projection, and its range of conduction times may be greater in the left hemisphere. This more specific lateral difference is likely to be particularly related to the human facility for *spoken* language. If it is true that the anatomical specialisation of the left hemisphere permitting spoken language is different from the global one supposed to underlie receptive aspects of language, and other perceptual aspects of lateralization, one would expect that the left hemisphere advantage for spoken language need not correlate with that for perceptual tasks or the receptive aspects of language. This is discussed in Chapter 11.

An important point should be made clear about the above proposal for the role of cortico-cortical connections in production of speech: In the past, it has usually been assumed that these connections are sensory-motor ones, that is they project from Wernicke's to Broca's area. The reason for this assumption is that lesions which damage this pathway give rise to a syndrome known as *conduction aphasia* (see section 9.4.3) in which patients lose the ability to repeat sentences they have just heard. The apparent reasoning is thus that to repeat a spoken sentence involves sensory perception in the auditory regions of the cortex, and the representations so formed somehow directly *instruct* the motor regions in the corresponding performance. This begs the question of how a motor program for extended speech can be formed. As the preceding chapter has shown, it is far more complex than simple one-to-one instruction of the motor act needed for each perceived speech sound. An entirely different proposal has been offered here, which incorporates a scheme for *construction* of programs for speech sounds. The facility for repetition of speech, seemingly so easy, relies on the fact that there has already been years of construction and elaboration of the programs for speech. If it were not so, one could repeat a sentence in a foreign language one had never heard before, just as easily as one could those in one's native language. In fact however, for this lengthy acquisition phase, connections from Broca's to Wernicke's area appear to be more important than in the reverse direction, though both are involved.

The previous subsections have concentrated on the acquisition of spatio-temporal templates for modelling speech, and the neuroanatomical requirements for this. To complete this section on predicted anatomical substrates for motor lateralization, the argument should be extended beyond lateralization of speech production. It is well known that humans have a pronounced tendency for manual praxis to be lateralized to the left cerebral hemisphere (giving right hand superiority for many tasks). If a lateral difference exists in the repertoire of delay lines from vocal premotor to auditory association cortex, it is plausible to extend this lateral difference to other motor-sensory association fibre systems. In experimental primates, the motor-sensory connections needed to register the association between an arm or hand movement and its consequences *do* exist, unlike their counterpart in the vocal/auditory modes. It is likely that they also exist in humans. However, if in humans it is also true that these connections have a richer repertoire of delay lines in the left hemisphere than the right, one could explain a lateralization of manual praxis, with superior control for sequences of action by the right hand and arm than the left. The left hemisphere would then represent a more accurate kinesthetic image of the movement to be expected from a particular sequence of commands than the right. In the next chapter we will discuss evidence that the right hand superiority in most people corresponds to just such a specialisation of motor programs. In this section however, the prediction can be made that there be a richer repertoire of long delay lines in the left than the right hemisphere for prefronto-parietal links as well as for those from Broca's to Wernicke's area.

8.3.2.8 The functional interplay between premotor and sensory association cortical areas with regard to language

Wernicke's area, being a sensory association area, is linked much more directly to auditory sensory input than to linguistic output. However, in the scheme outlined above, Wernicke's area is envisaged to be a site at which the sensory image of speech sounds can become associated with precise time-structured templates for production of the same speech sounds. The functions of neurones in this sensory association area would thus become modified by their input from regions involved in motor planning. In this scheme, the most important cortico-cortical connections are from premotor to sensory association area, with only a relatively minor role for connections in the reverse direction (see Figure 8.7). Suppose however we envisage that connections exist equally in *both* directions between premotor and sensory association areas (as indeed they do exist between prefrontal and parietal cortex). If such two-way connections do exist, we may then envisage a *reciprocal* functional influence between the functions of speech motor planning and speech perception. This is depicted in the two halves of Figure 8.8.

Figure 8.8A is a version of the model already shown in Figure 8.5, using "neurones" as the construction element rather than time-structured "blocks" of neural activity. In Broca's area a time-structured motor program is activated, leading to articulation of a speech sound, and also to activation of motor-sensory delay lines. In Wernicke's area neural activation representing the perceptual aspects of the time-structured sound patterns are set up, and neurones in which the convergence of such activation with representative connections from the motor program will form to represent the overall speech sound. One of the consequences would then be that, when speech sounds are perceived, they are categorised not simply on acoustic criteria, but according to the manner in which each sound is produced. In other words the perceptual analysis of speech sounds is accomplished in part by a speech motor command system. This is a viewpoint advocated by Liberman *et al.* (1967), on the basis of a variety of psycholinguistic evidence: Consider two acoustically similar speech sounds, which nevertheless are produced by quite different effectors (e.g. /p/a and /k/a). One would expect that the motor-sensory association fibres terminating in Wernicke's area and linked with the production of one of these sounds, would have no overlap (no links in common) with those for the other sound, because they arise from quite separate speech production cell assemblies. Thus the potential continuity between one sound and another as perceptual or acoustic events would be broken by the fact that their production mechanisms are sharply discontinuous. The structure of the perceptual image of speech sounds would be biased by the structure of the corresponding speech production mechanisms.

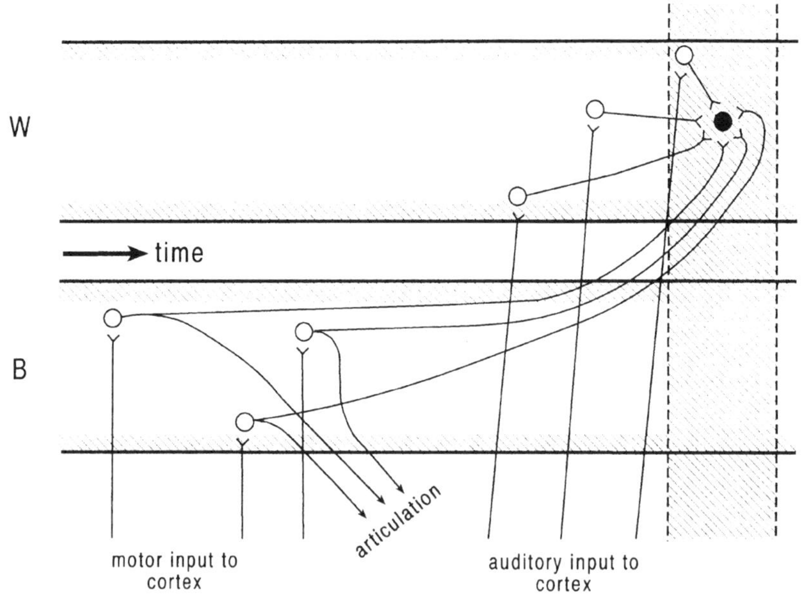

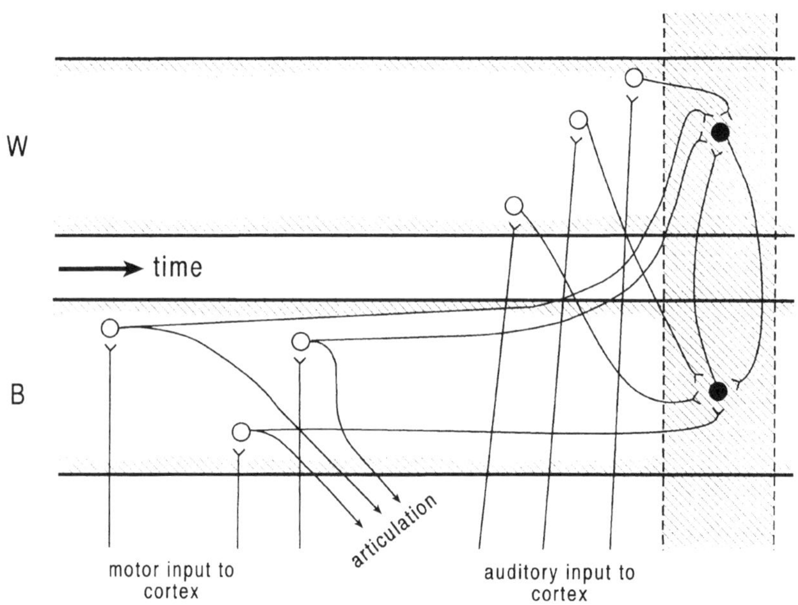

Figure 8.8 Two Versions of the Interaction Between Broca's and Wernicke's Areas in Representation of Production and Perception of Speech Sounds. In each diagram time flows from left to right, and Broca's (B) and Wernicke's (W) areas are depicted. In each diagram there is a configuration of active neurones representing the motor sequence needed for articulation of a speech sound, and another representing the sensory sequence needed for its perception. In the bands on the right of each diagram neurones are shown in which convergence occurs accurately enough to represent the speech sound with respect to both its motor planning aspects and its perceptual characteristics. In **A** the only connections between the two areas pass from Broca's to Wernicke's area, and therefore Wernicke's area is the only one where the combined motor-sensory representation can be formed.

Suppose, on the other hand, that connections between Broca's and Wernicke's area, are reciprocal. This is entirely plausible, because most cortico-cortical links, including those between premotor and parietal cortex, are known to be reciprocal. In this case another equally likely scenario is possible, as illustrated in Figure 8.8B. To the right of this figure are shown a pair of neurones, one in each area, reciprocally connected by relatively fast-conducting (i.e. approximately vertical, according to the conventions of such space-time diagrams) connections. If, as is quite plausible, such linked pairs of connections exist, such neurones in either cortical region could become the focus for convergent activation of delay lines representing either the motor or the sensory parts of the sequence. Thus, in the diagram, a neurone in Wernicke's area is depicted receiving coincident activation by two input lines from the motor program, and one from the acoustic trace; while a neurone in Broca's area receives two inputs from the acoustic trace and one from the motor program. The reciprocally-linked pair of neurones, functioning together, then represent the speech sound as an abstraction in which motor program and acoustic image are equally important. Figure 8.8B is perhaps *more* plausible than 8.8A, since in the former case the speech sound is represented by an elementary cell assembly, rather than a "grandmother cell".

A consequence of this would be that neurones would be present in both the premotor area and in the sensory association area with very similar feature-detection properties. These properties would be defined as much by the time-structured combinations of motor events needed to produce each speech sound as by its acoustic properties. Thus, in this case a perceived speech sound could be phonetically identified as easily by neurones in the premotor area as by those to which it is more directly relayed in the sensory association area. Indeed because of the limitations of the speech production effectors, the phonetic representation of speech sounds might be expected to have less tolerance of bias than that relying on principles of acoustic perception. In other words, the premotor area, which is directly involved in phonetic production, might contain a more "authoritative" representation of speech sounds (being phonetically-classified) than one based on acoustic properties. Thus one might expect that the premotor areas play an important role in perception of speech sounds, especially when fine discrimination on phonetic principles is required. It thus becomes possible not only for the motor planning functions of the premotor areas to influence the functioning of the sensory association areas, but also for the premotor areas themselves to take on some role in speech perception.

This reciprocal influence of premotor and sensory association area can be taken even further. The semantic associates of a perceived word, according to the simplest version of neurolinguistic theory, are likely to be registered as links of the acoustic image of that word. It might thus be expected that representation of the semantic attributes of a word are laid down in regions close to the primary or secondary auditory regions representing that acoustic image. (The angular gyrus has been suggested as a locus where there is sufficient multimodal convergence to represent all the associations an acoustic image of a word might make to its semantic referents (Geschwind, 1965)). However, if perception of

---

Figure 8.8 (*continued*) In **B** connections can pass in either direction between the two areas. A pair of reciprocally-linked neurones (one in each area) can thus act as a functional unit, with either of its members receiving convergent input along delay lines from either the motor or the sensory structure corresponding to the speech sound. The linked pair of neurones, rather than a single neurone becomes the representation of the motor-sensory abstraction for the speech sound.

words is to be achieved in part by identifying the means of their production, and as much by anterior as by posterior language zones, one might anticipate that associations within the frontal lobe would also play a part in representing the semantic attributes of words.

In section 9.2.2, a number of experimental studies are described which show unexpected involvement of premotor cortex in language perception, and semantic representation, and of auditory association cortex in language production. The preceding paragraphs provide a theoretical framework which makes these somewhat paradoxical findings easier to understand.

8.3.2.9 Motor specialisation of the right hemisphere

Our hypothesis for motor lateralization is that there is a richer repertoire of motor-sensory delay lines from premotor to sensory association areas in the left hemisphere than in the right. This morphological difference could give rise to a superiority in open-loop motor control, involving the construction of time-structured models of movement in advance of action. The question is raised whether there be any compensatory motor superiority in the right hemisphere. The present subsection argues, from premises related to the above hypothesis, that there could be a right hemisphere superiority in *closed-loop* motor control, involving moment-by-moment sensory guidance. The detailed anatomical requirement for this to occur is another special case of the central hypothesis proposed in section 3.6, namely *that associational pathways from sensory association regions to premotor regions in the right hemisphere should have a faster conduction time than their counterparts in the left hemisphere.* The derivation of this prediction follows:

In closed-loop motor control, the signals controlling motor performance need to be guided by sensory messages received about the consequences of immediately preceding motor performance. For such guidance to be possible, cortico-cortical association pathways are needed projecting from the recipient regions of sensory cortex (or sensory association cortex) to the region of cortex controlling emission of motor signals. The faster conducting are such association pathways, the faster can be the sensory guidance they mediate, and therefore the more accurate can be the sensory guidance. If there is an overall tendency for cortico-cortical connections to be faster in the right hemisphere than in the left, one might expect sensory guidance co-ordinated in that hemisphere to be faster and more accurate than that co-ordinated in the left hemisphere. Evidence relating to this prediction is reviewed in Chapter 9 (sections 9.3.4 and 9.4.4).

Apart from closed-loop control it is likely that the right hemisphere has its own special role in motor modelling. Because of the probable difference between the right and the left hemisphere in temporal representation, the models constructed in the right hemisphere should be different from those constructed in the left, in that they should not be so accurately structured in time. (The left hemisphere will always dominate in this function.) Instead, the right hemisphere is likely to specialise in constructing models of the slower components of motor performance and linguistic expression. Moreover, *the motor models constructed by the right hemisphere should co-ordinate and synchronise a larger number of effectors into a single fluent synergy* than can those constructed in the left hemisphere. This prediction is an exact counterpart of the right hemisphere specialisation for dealing with complexity in perceptual Gestalts. Both of these right hemisphere specialisations correspond to the chains of neural activation such as those depicted in Figure 5.7(C) rather than those in 5.7(A). As a result, during linguistic expression, the motor programs the right hemisphere should represent would include those for prosodic expression, and manual and body gesture accompanying the verbal content. While these programs would not in themselves be accurately structured in time, it is likely that they be well co-ordinated in time to fit the concomitant stream of verbal expression. However, it would be the latter stream which determines the timing of elements of prosody and gesture, presumably by left-to-right transmission of signals through the corpus callosum.

Prosody is a subject which has not yet been considered. As a pattern in the sensory input it is temporally extended too far to have been included in Chapter 6, which dealt only with *immediate* perceptual analysis. Instead, it falls within the domain of short-term memory, and will be dealt with in Chapter 10. In that chapter we will also deal with longer term aspects of motor planning, which includes the planning of streams of prosodic and gestural expression.

8.4 SUMMARY OF THEORETICAL ARGUMENTS ABOUT LATERALITY OF MOTOR FUNCTIONS, WITH PREDICTIONS

The arguments developed in section 8.3 are quite complex. They involve a number of separate theoretical suggestions. The earlier of these have been presented in detailed form in previous papers. However, these ideas have been extended in the present chapter to provide some new combinations of concepts, essential for the present purposes. Before moving on to the discussion of evidence to evaluate these hypotheses, let us summarise the structure of the argument in this chapter.

Individually we can list the main premises of the preceding sections: First we were reminded of the logical structure of instrumental conditioning. A three-factor rule for synaptic modification was suggested as the fundamental embodiment of this logical structure. In terms of real brain anatomy and physiology it was suggested that the various components of the three factor rule correspond to important aspects of structure and function in the mammalian striatum. Although the striatum plays a crucial part in the initial learning of instrumental behaviour, it is not itself likely to be the final repository of the memories so acquired: Because of the rich connections between striatum and the cerebral cortex, it is likely that those memories are transferred to the latter (much larger) structure.

In earlier parts of this book it has been explained how neural networks in the cortex can form chains of activation, with precise temporal structure, and the central hypothesis for cerebral lateralization has been offered which predicts that such temporally structured chains of activation can be formed more effectively in the left than the right hemisphere.

Other components from which the schema for motor lateralization was constructed concern neurones which can represent the association between motor programs (or just the preparatory stages of such programs) and their consequences. It was assumed, on the basis of suggestive cytoarchitectural evidence, but without direct proof, that, in humans, such motor-sensory links exist for vocal/auditory function.

In addition it has previously been suggested that combinations of activation of such motor-sensory links can allow the modelling of possible consequences of combinations of actions, in advance of their actually being emitted. Thus it becomes possible to prepare motor programs that have never before been emitted, without first subjecting them to trial and error in the real environment.

The final premise is that certain motor effectors (the primate arm and hand under many circumstances, and the vocal apparatus) have mechanical properties which allow themselves to be used in a manner where each motor command dictates unconditionally the sensory consequences (in the auditory or kinesthetic modalities), as well as their exact timing in relation to the motor commands.

Given these premises individually, they may be combined in ways which permit radically new forms of motor co-ordination to emerge. When temporally structured chains of cortical activation are "ignited", activity can be transmitted with much greater security if interaction of the cortex with the striatum occurs than without striatal involvement. Such a facilitatory process is not envisaged to be unique to humans. However, when motor-sensory links also exist which have a rich enough repertoire of long delay lines, the temporally structured chains representing motor programs can become linked with the time-structured representations of their consequences in a manner which exactly encodes the temporal relations between the two. This process *is* envisaged to be unique to humans, and can improve the motor programs for those special effectors where unconditional temporal correlations between motor output and sensory consequences are possible.

This critical fact then makes it possible that the consequences of long chains of motor output can be modelled in advance of action, with encoding of the precise temporal structure between motor outflow and consequences, even in long chains. This is the necessary condition for spoken language, and the same condition makes possible special forms of control in manual praxis.

Lateralization of such modelling derives from an asymmetry hypothesised in either the number or the conduction time (or both) of the motor-sensory links required for such modelling. Thus, the essential aspect of the hypothesis to be evaluated is that in spoken language and in manual praxis the left hemisphere advantage over the right lies in more accurate representation of the temporal relations between motor outflow and the unconditional sensory feedback (auditory or kinesthetic) associated with the use of the effectors for which cerebral lateralization applies. This is the factor which permits the accurate time-structured modelling of a movement in advance of its performance.

Modelling as such has been previously suggested by others as the basis for the distinctively human variety of motor control of hands and speech apparatus (for instance by Luria, 1973). However, the distinctive contribution offered here is that such modelling is *time-structured*, and that it relies on motor-sensory delay lines, for which there is envisaged to be a better repertoire in the left than the right hemisphere.

This account, emphasising the role of motor-sensory association pathways, is also rather different from the traditional one in which sensory-motor association pathways are given greater significance. However, at the end of section 8.3.2.5 a role was found for the latter class of connection, such as those from Wernicke's to Broca's area, partly in strengthening the temporal structure of the chain of activation involved in actual speech production, but also for forming abstract representations of speech sounds defined jointly by the motor program and its sensory consequences. In the latter process, it is envisaged that there be complex interplay between premotor and sensory association areas involved in language which blurs the functional distinction between the two. Furthermore, sensory-motor association links may play an important role in some aspects of right hemisphere-mediated sensory guidance.

These hypotheses permit a number of more explicit predictions to be made:

(i) Construction of a full repertoire of time-structured motor models, for language as well as for limb movements is likely to be a very slow process. For lateralized motor functions there should therefore be a very long stage of learning, not typical of most examples of mammalian motor learning. This learning should permit the acquisition of representations of time-structured motor-sensory links and combinations of such links. This should apply to both motor-kinesthetic links in arm/hand reaching movements and for vocal-auditory links related to speech production.

(ii) Since a *reciprocal* functional influence between the motor planning and speech perception is envisaged (see section 8.3.2.8) it is predicted that there be some degree of interchange of functions between premotor and sensory association areas. For lateralized motor functions, fluent performance of the motor task should be accompanied by activation of the corresponding sensory association area, in which convergent activation by both the actual consequences and the template of the consequences (modelled by cortico-cortical axonal links) should occur. Another likely prediction is that premotor region involved in speech production should also be actively involved in perception of fine phonetic discriminations. These predictions are capable of being evaluated by positron emission tomography, as well as by other methods.

(iii) For left hemisphere-dominant motor functions in the adult, the dominance should consist of superior preparatory modelling of an action to be emitted. The circumstances where such preparatory modelling can occur are in carrying out manual or vocal gestures where the gesture is generated as an internal volition, or is specified entirely by instructions (e.g. to verbal command, or in response to imitation) given before the initiation of the movement. In such cases one might expect to find evidence of superior planning ability for the left hemisphere compared with the right. In addition to evidence about the planning process, the execution of such preplanned gestures should be better with right than left hand.

(iv) The alternative type of motor task is one involving guidance by an external sensory stimulus, whose movements are not specified by a prior plan of the movement or gesture. A typical example of this is visuomotor tracking of an object moving in a random or unpredictable manner. Such tasks, or closely related ones, can be devised using various combinations of effectors for performance, and sensory modalities for the guidance signal. In such tasks the guidance is dependent on signals passing from sensory to motor regions of cortex, rather than on motor-sensory messages. In such tasks, a left hemisphere advantage would not be expected. Whether or not there should actually be a right hemisphere advantage cannot be definitely predicted. However, if there was a richer set of rapidly conducting sensory-motor cortico-cortical connections in the right hemisphere than in the left, a right hemisphere advantage *would* be predicted. However, at present there is no morphological evidence indicating whether this condition might apply. Behavioural evidence reviewed in the next chapter helps to address this question.

(v) It is also predicted that motor control by the right hemisphere should co-ordinate a larger number of effectors into a single fluent synergy than that achieved by the left hemisphere.

(vi) Although preplanned actions do not rely on external sensory guidance, the acquisition of the programs did at one stage depend on unconditional sensory input, informing about consequences of each segment of action. When fully-acquired preplanned actions are performed, the same kinesthetic or auditory feedback is still present, and the exact match between it and the template of the segment of action represented centrally will still occur and act as a check that the action has been carried out correctly. In some experiments it is possible to change artificially the timing of this otherwise unconditional sensory feedback, and disruptive effects on performance are produced. This is the "delayed auditory feedback" paradigm, of which a number of variants have been tried. It is predicted that such disruptions should be stronger within the left hemisphere than in the right, because in the right there will be a weaker template to form a mismatch when timing of the relay of information about consequences is altered.

Most of these predictions can be studied in normal persons, though important evidence may also be obtained from clinical material. In the case of language functions, the effectors themselves are not divided into left and right (as in the hands). For these functions, it is more difficult to study the processes of cerebral lateralization in normal subjects. Patients with unilateral hemispheric injury (usually on the left) are thus of greater importance for understanding language lateralization than for lateralized manual control. The predictions which can be made follow closely from those made above. Both speech and lateralized manual ("dextrous") functions should be impaired more by left than right hemisphere lesions. With appropriate selective lesions, it may be possible to demonstrate selective loss of the modelling function without loss of either the basic motor components, or of the perceptual discrimination of external stimuli. In such cases, patterns of errors and inaccuracies should also betray disturbances of a motor-sensory modelling function. In principle, there could also be a different type of selective motor loss after right hemisphere damage. Specifically, in right hemisphere patients, it is important to ask whether there are deficits in closed-loop motor control (involving sensory guidance, in visuomotor and other tracking tasks), or in co-ordination of complex synergies of many effectors.

9 Motor Aspects of Lateralization: Evidence for Evaluation of the Hypotheses of Chapter 8

This chapter reviews the evidence relevant to each of the predictions mentioned at the end of the last chapter. Section 9.1 discusses (in relation to prediction (i) above) the evidence that there is a lengthy period of learning for the lateralized motor functions of the left hemisphere. Section 9.2 deals with evidence about the interchange of functions between the premotor areas primarily involved in motor planning and the parieto-temporal association areas primarily involved in sensory analysis. The functions of each region may initially be separate, but come to interact, so that both areas to some extent take on functions attributed to the other as part of a larger more unitary function. This section thus evaluates predictions made above (item (ii)) that the left sensory association cortex be involved in tasks requiring motor modelling, and that the premotor regions be involved in some aspects of language perception. Sections 9.3 and 9.4 deal with the a complex body of evidence relating to the central prediction made above (item (iii)), that there be superior open-loop control by the left hemisphere compared with the right. The first of these two sections deals with the evidence in normal persons, pertaining to motor control of the hands and eyes. In addition there is a subsection on evidence obtained using the delayed auditory feedback as an indicator of motor lateralization in the vocal apparatus and other effectors (see prediction (vi) above). Section 9.4 deals with similar evidence obtained from clinical material (i.e. patients with unilateral hemispheric injury). Most of the evidence relating to predictions about lateralization of language production functions is clinical evidence, and so this section goes in some detail into this evidence and its interpretation. Although these sections deal mainly with the special motor functions of the left hemisphere, evidence is referred to, for comparative purposes, of motor specialisation of the right hemisphere. Subsections 9.3.4 and 9.4.4 pull together this evidence from normal and brain-damaged subjects, and summarises it in relation to prediction (iv) made above. Little evidence is available about prediction (v), but the issue is raised again in Chapter 12 as one of the aspects of lateralization requiring further study.

9.1 BABBLING AND THE ACQUISITION PHASE FOR MOTOR-SENSORY MODELS

The scheme outlined in Chapter 8 envisages that representations of left hemisphere-lateralized motor function are acquired by a process of learning. This is envisaged to be especially slow, because it involves selection from a wide range of chance combinations of motor fragments those which happen to match prior specifications. Thus a repertoire of modelled motor programs is slowly built up, by a process which is much more protracted than acquisition by direct instruction from the environment, the more usual process of learning. Empirically it is undoubtedly true that these phylogenetically "new" forms of motor co-ordination, using hands, arms and vocal apparatus require long stages of learning: Speech and written language require years for perfection. Other forms of manual dexterity also develop over a period of years. For instance, the percentage of children showing pronounced right handedness increases considerably from age 2 to 8 years (Satz, 1988; Gudmundsson, 1993).

What could be the nature of that learning? According to the scheme just described both instrumental (reward-mediated) and Hebbian components play a part in fashioning the specific combinations and sequences of motor-sensory links. We would expect these to occur in a rather sophisticated fashion, involving aspects of "play" to acquire control, and aspects of "modelling" to construct combinations of the basic elements of each motor program. Observational studies of motor and speech development in infants tell us a lot about the nature of this learning, and provide an important means of evaluating the hypotheses put forward in Chapter 8.

Human infants in the months before they acquire speech display a form of vocal behaviour called "babbling". In this behaviour the infant produces a variety of vocal utterances. These are not words and, initially, are not recognisable phonemes. However, as babbling proceeds, the infant may be heard to be repeating over and over particular consonant-vowel syllables. The repertoire of these gradually increases, and later may develop into bi- and trisyllabic sounds (Stoel-Gammon and Otomo, 1986). At the later stages of babbling the phonetic structure of the sounds produced progressively approximates those of the language to which the infant is exposed (e.g. French or English) (Boysson-Bardies *et al.*, 1981). Clearly the detailed phonetic structure in babbling is influenced by the sounds the infant hears. One thus may suspect that babbling serves a purpose in developing the motor programs which are later used for producing phonemes and words of the real spoken language to be acquired.

This suspicion is confirmed by evidence that babbling in deaf infants is deficient in the production of well-formed syllables. The repetitive, "canonical" babbling is delayed and less frequent (Oller and Eilers, 1988). Consonant repertoire and use of multisyllabic utterances develops more slowly (Stoel-Gammon and Otomo, 1986). From a study of identical twins, only one of whom was deaf, Kent *et al.* (1987) found a deficiency in production of labial and dental consonants in the deaf twin. Eilers and Oller (1994) find these differences so clear cut that they recommend using them for detecting early deafness. However, development of speech programs in babbling requires not only intact hearing, but also requires sound production. In a single case study of an infant tracheostomised between ages 5 mo and 1.5 yr (Locke and Pearson, 1990), it was shown that, on decannulation, repetitive syllable production was severely deficient compared with normal infants of the same age.

What is more, although hearing was unimpaired, the first production of real speech was delayed, compared to infants who had been able to babble normally.

Motor gestures of the *hands* identifiable as babbling also occur in prelingually deaf children when they are first learning sign language (Petito and Martenette, 1991). As with spoken babbling, the configurations of such manual babbling are related to the linguistic signs to which the infant is exposed (in this study, American sign language), and these configurations gradually increase their complexity of syllabic organisation between ages of 10 and 14 mo. As with vocal babbling this manual babbling is produced without meaning or reference.

In sensorimotor terms these examples of babbling also bear some similarity to the process seen in infants with respect to control of fingers and toes: The infant, with apparent intense fascination, studies the movements it can produce in these body parts. In such cases, the relation between motor control signals and sensory feedback is of course not related to any shared system of social communication.

In all these cases, it is hard to avoid the conclusion that the child is engaging in a sort of experimentation. Although one cannot *prove* this on the basis of available evidence, it is plausible to suggest that the object of this experimentation is to establish the correspondences between every possible motor program (directing the actions of vocal apparatus, limbs, fingers or toes) and the sensory feedback resulting from the program in the auditory, visual or kinesthetic senses. Given that all the fragments of motor behaviour so emitted permit precise and unconditional *temporal* correspondence between the program and the sensory feedback, it is likely that what is being established is the time-structured template, laid down as patterns of strengthened cortico-cortical connections, to which the small time-structured chunks of sensory feedback can be matched. In other words elementary fragments of motor program are being laid down such that the temporal structure of each element produces an expectation of a particular temporal structure of sensory consequences.

Babbling (whether vocal or manual), as well as the infant's intense study of its own finger or toe movements all seem to fit the concept of "play". This presumably implies that these behaviours are maintained by a variety of reward. In the case of vocal or manual babbling, this reward is likely to be produced by the matching of babbled sounds (or hand signs) and the sounds (or signs) the infant hears (or sees) from the linguistic gestures of others (e.g. parents). When such a match occurs, the reward process may reinforce the motor programs for the babbled sounds or hand signs. This process seems to be a plausible account of the early stages of babbling. In the later stages, where bi- and trisyllabic words are babbled, the reward would seem to come from matching the sounds constructed from *short sequences* of previously acquired sounds to the sequences of sounds the child hears from others. Likewise when hand signs of greater complexity emerge in manual babbling, they allow matching of combinations of signs the child produces to the combinations the child sees other people producing, the match again being rewarding. This is akin to the process of modelling complex motor sequences proposed in the previous chapter (except that, at this early stage of the child's development, it is achieved by external gestures, rather than by internal modelling).

In section 9.3 data are reviewed suggesting that the right hand superiority in reaching and aiming tasks is based on the presence of better time-structured models of action in the left than the right hemisphere. The existence of manual babbling in deaf infants is one indication of the process of acquisition of those models. However, if the right hand of normal adults has access to better time-structured models of action than the left, one would also expect to find evidence of the process of discovery of models of manual action in normal infants. This discovery process would be related to reaching and aiming with the right hand, rather than to a system of communication. Such a process of discovery is likely to be an extension of finger-toe games of early infancy, elaborated, in the right hand by testing of *combinations* of already-acquired motor schemata. In this case the reward does not come from matching the modelled auditory (or visual) consequences of a gesture to auditory (or visual) images which are part of a social system of communication. Instead, the reward is likely to come from matching the modelled kinesthetic consequences of a motor program (for reaching or aiming) to a kinesthetic memory evoked by sight of an object at a particular position in space. Indeed the same processes may be at work for acquisition of some aspects of oculo-visual co-ordination. The author is unaware of any accounts of the development of infant motor dexterity or eye movement control which fit this description, and therefore this idea is offered as a prediction.

9.2 EVIDENCE OF INTERCHANGE OF FUNCTIONS BETWEEN PREMOTOR AND SENSORY ASSOCIATION CORTICAL AREAS

9.2.1 Role of Wernicke's area in speech production

A central aspect of the scheme proposed in Chapter 8 was that there be cortico-cortical pathways from Broca's to Wernicke's area in humans, and that the latter area should undergo neural activation when the former is active in formulating speech programs. In principle this activation can be detected in normal persons by measurement of regional cerebral blood flow, or glucose utilisation, with positron emission tomography (PET) as the most modern and sophisticated method now in use for making such measurements. Several studies address this prediction. Larsen *et al.* (1978) found bilateral activation of both premotor and superior temporal regions during automatic speech (e.g. reciting the days of the week). There was no evidence of activation of Broca's area, but the temporal lobe appeared to be activated more on the left than the right. Petersen *et al.* (1988) using PET found that passive processing of auditorily presented words activated the primary auditory cortex bilaterally, as well as the left temporo-parietal region, and left anterior superior temporal cortex. Repeating words aloud produced similar temporal lobe activation, and in addition activated the sensorimotor mouth region and premotor structures. Friston *et al.* (1991), and Boivin *et al.* (1992) made PET estimates of cerebral activation, whilst normal subjects performed a verbal fluency task (for instance, "name as quickly as possible all the words you can think of beginning with 'd' as in 'dog' "). In fact the two studies produce discordant results. Friston *et al.* (1991) found that regional blood flow decreased in the superior temporal gyrus during task performance, whilst Boivin *et al.* (1992) found a strong positive correlation between left superior temporal lobe glucose utilisation and performance in the verbal fluency task. The reasons for the discrepancy are not clear, and more studies are required.

In all these studies external words could be heard when the subjects spoke, and so these results provide no decisive evidence for activation of Wernicke's area by modelling of words alone. However, one study does show this clearly. Wise *et al.* (1991) asked their subjects to *silently* generate a series of words (verbs expressing possible uses for a given noun). In this situation, involving verbal modelling but not speech perception, not only were the left premotor and prefrontal areas activated, including Broca's area, but so too was the left posterior superior temporal gyrus (i.e. Wernicke's area). (The control situation with which comparison was made was with subjects making categorical judgements of word pairs. This did not selectively activate Wernicke's area.) One other study supports the view that parietal/temporal regions play an essential role in speech production. Ojemann and Mateer (1979) performed experiments during neurosurgery, in which they charted regions of the left hemisphere at which linguistic functions could be disrupted by application of electrical stimuli. In a few cases, reading aloud was affected by stimulation of the posterior part of the left parietal lobe, near its union with the temporal lobe. Overall sentence form was intact, but there were "jargon" errors in production of single words. These two studies go beyond the inference that the parieto-temporal cortex is passively activated during speaking, and indicate that such regions are necessary for speaking, and, by implication from the first study, for the modelling which is necessary for speech.

9.2.2 The role of left inferior frontal area in phonological perception and in addressing semantic content

In section 8.3.2.8 it was suggested that the speech production areas (Broca's and surrounding areas) might play a part in speech perception, especially when it demanded precise phonetic identification. Liberman *et al.* (1967) review psycholinguistic evidence suggesting that speech sounds are categorised in perception according to the way they would be produced, rather than on a purely acoustic basis. This form of phoneme categorisation provides psychological evidence for the prediction. More precise evidence, with morphological detail, was provided by Ojemann and Mateer (1979) using cerebral stimulation in neurosurgical patients. They found that phoneme discrimination was disrupted by electrical stimulation of the left inferior frontal gyrus (as well as of the temporo-parietal cortex). Modern methods confirm this result: Bartlett *et al.* (1987) found that glucose utilisation in Broca's area becomes correlated with that in Wernicke's area and other left side regions, when subjects are asked to perform a phoneme recognition task. Zatorre *et al.* (1992) find with PET that in making phonetic judgements Broca's area is involved, though it is not so in passive perception of speech.

In addition section 8.3.2.8 raised the possibility that phonetic identification by the frontal lobe could be used for addressing semantic content, giving such regions a role in semantic representation usually attributed to more posterior regions. Petersen *et al.* (1988) found, in a PET study, that cerebral activation occurred in a region of the left frontal lobe (anterior to classical Broca's area), during a word association task. Thus there are suggestions for a role of the inferior prefrontal cortex not only in phonetic perception (as predicted) but also in some aspects of semantic identification. Whether the latter finding is an exact fit to the prediction must await further experimental analysis. Clinical evidence related to these issues is mentioned in section 9.4.3.

9.3 EVIDENCE OF BETTER TIME-STRUCTURED MODELLING OF MOTOR ACTS IN THE LEFT HEMISPHERE OF NORMAL SUBJECTS

In section 8.3.2 a hypothesis was advanced for the left hemisphere superiority in constructing a certain type of motor program. A spatio-temporal template for each small chunk of the program was envisaged to be laid down as a pattern of strengthened synapses (these being the terminations of motor-sensory cortico-cortical connections) and the corresponding axonal delay lines. The temporal aspect of this template depended on selection of axons with appropriate conduction delays from soma (in the premotor areas) to synapse (in the kinesthetic, visual or auditory association areas). The selection process was envisaged to depend on simultaneous activation of neurones in sensory association cortex, by cortico-cortical axons and by sensory (e.g. kinesthetic) feedback. In other words, those cortico-cortical synapses which activate a neurone in parieto-temporal regions at the same time as do sensory feedback signals will be susceptible to Hebbian strengthening. Given a richer set of motor-sensory delay lines in the left than the right hemisphere, the left hemisphere would have superiority over the right in the form of control enabled in this way.

On the other hand, for movements controlled by moment-to-moment sensory guidance, left hemisphere superiority would not be expected, and a right hemisphere advantage might even exist.

Is there any behavioural evidence that the aspect of motor programming for which there is left hemisphere dominance corresponds to a more accurate formation of such spatio-temporal templates in this hemisphere, based on the better matching of this template with kinesthetic or other sensory feedback signals? Conversely, is there a right hemisphere advantage for movements guided by sensory stimuli? The section below reviews evidence from normal subjects relating to these questions. There is a substantial body of evidence about the nature of lateral specialisation of the hands (dealt with in section 9.3.1) and a few papers about lateral specialisation of eye movements (section 9.3.2). The vocal apparatus itself is not lateralized, and so it is not easy to infer separately the capabilities of each hemisphere in vocal expression in normal subjects. (*Clinical* evidence from unilateral head injury about the capacities for vocal expression of each hemisphere will be considered below, in section 9.4) However there is a body of relevant evidence based on the method of delayed auditory feedback, carried out in normal subjects, which is reviewed in section 9.3.3.

9.3.1 Time-structured modelling of dominant hand performance in normal hemispheres

As a start to answering the above questions, there is a group of papers which provide preliminary experimental evidence of the manner in which the right hand is specialised in learned motor functions. Many of these studies involve speeded tapping tasks (such as pioneered by Fitts, 1954), or various aiming tasks. Wyke (1967) compared the left and right hands of normal right handers in aimed reaction time tasks, and in speeded tapping tasks. (The latter also involved aiming, in so far as subjects were required to tap alternately on either side of a line.) The right hand was a little faster than the left in the reaction time task, and gave more taps per second than the left. Steingruber (1975) tested 8–10 year old children on a task involving tapping on objects of size 2 mm, 6 mm or 9 mm. In the most difficult task, the majority of children showed a right hand advantage, but as the task was simplified, a higher proportion showed left hand advantage. Wolff *et al.* (1977) asked subjects to engage in repetitive tapping, aimed at a one-inch key, of a rhythm given in advance by auditory demonstration or by visual symbols. It was found that the right hand was more accurate than the left. Todor and Doane (1978), using a speeded alternate tapping task, found a right hand superiority. Podbros and Wyke (1988) used tasks involving tapping on a light plastic cube. Subjects were instructed to tap (with either hand) as fast as possible, without moving the cube beyond a narrow range. The latter constraint ensured that accuracy of control was equal with either hand. Under this circumstance, the right hand could tap at a faster speed than the left. In a different task, involving speeded moving of a series of pegs into corresponding holes, Annett (1970) found a right hand superiority for all ages of children studied.

Several clues indicate that the superior performance of the right hand is dependent on both *rapid aiming* and *preplanning* being part of the task requirements. In Wolff *et al.*'s (1977) paper, a right hand advantage was not found for a task with similar aiming requirements, but no requirement for preplanning of a rhythm (subjects tapped in time to a metronome). Flowers (1975) used a task involving speeded alternate tapping of strips 1, 2 or 4 cm wide, with difficulty also varied by setting the strips different distances apart. For the easiest task there was no significant difference between hands in speed of tapping, but a right hand advantage emerged as the difficulty was increased. In a task involving only preplanning (a variety of speeded rhythmic tapping tasks, with the rhythm given by verbal instruction) but not involving aiming, there were no lateral differences. Peters (1976) found that with prolonged practice in a tapping task, the left hand could become as fast as the right, but was still significantly less accurate. Roy (1983) used a serial tapping task, in which a series of 1/4 inch holes had to be "dotted". The right hand was faster and more accurate. In addition, the left hand deteriorated in accuracy more sharply as speed of response increased than did the right hand. Todor and Cisneros (1985) used an aiming task involving a stylus, whose trajectory was monitored by an accelerometer. There was little difference in movement trajectory between right and left hands until the later "corrective" stage of the movement. During this stage, which for most conditions took more than 50% of total movement time, the left hand took significantly longer than the right. This was so even when error rates were equated between hands. The right hand advantage is not found in tasks where there is no requirement for speed. For instance, Carson *et al.* (1990a) found no asymmetry in a task where subjects had to reproduce as accurately as possible (but without speed constraints) a pre-selected position in space. This finding is consistent with one of Leonard and Milner (1991): In a study of patients with one-sided brain injury involving reproducing a static position, deficits were found after large right sided (frontal) lesions, but not after temporal lobectomies, left frontal lobe damage or small right frontal lesions. These results fit the concept that the right hemisphere is superior to the left in representing complex Gestalts and body image (as discussed in section 6.5.3). Altogether the evidence reviewed suggests that a right hand advantage appears when the task requires both accurate and rapid aiming of a preplanned movement.

This preliminary conclusion leaves undecided two issues about manual specialisation. *First,* the performance measures in the studies discussed correspond to total response time and do not analyse this into separate components of initiation time and movement time. The hypothesis advanced in section 8.3 applies just to movement time. Different considerations apply to initiation time, which might confound attempts to evaluate the hypothesis. In fact several studies have shown that for initiation time, the left hand is faster than the right (Haaland and Harrington, 1989a,b; Carson *et al.*, 1990b; Bradshaw *et al.*, 1990; Chua *et al.*, 1992). However, Bradshaw *et al.* (1990) found, for a single movement to a target, in strong dextrals, that the right hand completed the response faster despite the fact that the left hand initiated the response earlier.

The *second* issue concerns the exact nature of the superiority of the right hand in aiming tasks. The evidence just presented might be taken to imply that the left hemisphere is superior at controlling tasks involving exteroceptive (usually visual) guidance of movement. If this were the case, performance would be dependent on matching the visual image of the moving hand to the visual image of the target. However, there is another alternative to consider: The role of vision may be simply to identify the target position for a movement, and once this position is registered, a predetermined motor program can then be executed. A common children's game may illustrate the difference, and show that the two strategies of aiming are different. In the game of "knuckle bones", the child is required to perform both tasks together: to throw, track, and subsequently to catch one "bone" which has been projected into the air; and while that "bone" is in the air to perform an aiming task, that is to pick up other "bones" on the floor. Invariably the child makes an initial sighting of the objects on the floor and uses the memory of this for the performance of the first task; and then devotes full visual attention to tracking the object moving through the air.

In the aiming tasks discussed above it is not clear whether visual information was required for moment-by-moment guidance, or merely for provision of information about the target position, with movement trajectory then being governed by a central program. Even a task involving aiming at a target on a moving turntable (Wyke 1968) does not necessarily involve visual guidance, since the target's movement was regular and predictable. In the study of Todor and Cisneros (1985) the existence of an apparent "corrective" phase in the arm movement trajectory is also not necessarily an indication of visual guidance: Jeannerod (1986) has shown that the regularly-observed discontinuities in the velocity curve during the deceleration phase of a reaching movement occur even when no visual guidance is possible. Thus, these discontinuities do not represent corrections based on a mismatch of the visual image of arm position with that of target, though they may be corrections of the *kinesthetic* image of the moving arm with respect to the kinesthetic memory specified by the target position.

The hypothesis advanced in section 8.3.2 predicts that the right hand (left hemisphere) has superior accuracy based on a more accurate open-loop, centrally-planned program, rather than on superior moment-by-moment guidance. This would show a left hemisphere superiority because of the hypothesised richer repertoire of delay lines from premotor to sensory association cortex in the left hemisphere compared with the right. These would permit a more accurate representation to be formed in the left hemisphere compared with the right, of the kinesthetic image of the hand moving to a visual image of the target at a particular spatial location. Over time, and with much practice, contingencies of reinforcement would select a motor program which, given the target position, produces a kinesthetic image for an optimal movement to the target. This image of an optimal movement can therefore come to be more accurately represented in the left hemisphere (for the right hand) than in the right hemisphere (for the left hand). The alternative form of motor control involves continual visual guidance, and for this a left hand advantage might be predicted. To decide between these hypotheses further analysis of the right hand superiority in reaching and aiming tasks is needed.

Several studies indicate that the right hand advantage should actually be attributed to central planning, rather than sensory guidance. Bowen *et al.* (1972) confirmed the right hand advantage for speeded tapping aimed at a single target button. However, in the same subjects, there was no manual asymmetry for a visual tracking task, involving a randomly moving target on screen. Roy and Elliott (1986a,b) varied visual feedback by turning off the lights at initiation of movement in some trials. Subjects were also required to point to a target with a range of movement times (100–400 msec). In confirmation of Roy (1983), with the right hand, accuracy deteriorates less with increased speed than for the left hand. The right hand advantage in movement times could not be attributed to better use of visual feedback, because both hands showed equivalent lights-on benefits. For movements performed in less than 200 msec no effect of illumination was apparent, but the right hand accuracy advantage was still found. O'Boyle and Hoff (1987) required subjects to perform mirror tracing of random forms, a task with clear requirement for visuomotor guidance. Males showed significant advantage to the non-dominant hand and females show a non-significant trend in same direction. Watson and Kimura (1989) found a right hand superiority for throwing (which involves accurately directed and timed motor programs) but not for intercepting a thrown object (which involves exteroceptive feedback).

All these data suggest that the right hand superiority in aiming is based on better timing in the right hand control system for a visually specified target. The one apparent exception to this generalisation is the result of Wyke (1968) who required subjects to use a pointer to touch a moving target (i.e. a spot on a turntable). This task, involving both precision and timing was performed better with the right hand than the left, though it appears to be a case where visual guidance is important. However, since the movement of the turntable was predictable, a role for preplanned motor programs cannot be excluded. Use of prediction to identify target position was explicitly tested in an experiment of Carson *et al.* (1992). They employed a task where subjects were required to extrapolate the position of a complex movement on screen, indicated by 4 points shown in sequence over 150 msec. Responses of the two hands were similar in accuracy, but the right hand movements were shorter in duration, with higher peak velocities. Visual feedback from the moving hand could be eliminated by turning off lights at the time of movement initiation. Vision of the moving hand did lead to superior accuracy of the response, but, as with the experiment of Roy and Elliott (1986a,b) the effect was equal with either hand. Thus right hand superiority was not based on better visual feedback.

Two other relevant studies used the peg-moving task. Annett *et al.* (1979) required subjects to move a row of 10 pegs from one row of holes to another as fast as possible. The dominant hand showed an increasing advantage in accuracy as the tolerance between hole and peg was made stricter. Once the peg had struck the side of hole, however, no difference was found between hands. In other words, when there is an unpredictable component of exteroceptive feedback (in this case somaesthetic), there is no difference between hands. Nevertheless, the right hand placed the peg into the hole with fewer such final corrective movements. This again suggests a more accurate spatio-temporal program in the left hemisphere, for the actual flight phase, i.e. from the time when the target has been identified visually to the time when the target has been located somaesthetically. Bradshaw *et al.* (1988) also used a sequential peg moving task. Both movement times and resting times (after each peg movement) were shorter with the right hand. The difference for resting times was a stronger effect statistically, but it is not clear that this measure is equivalent to initiation times for movement in the studies cited above. (For such a sequential movement the "rest period" could be part of a program for temporal chaining of motor activities, rather than that of the routine for initiation of movement from rest.)

Cerebral lateralization of manual tasks can also be studied by assessing the interfering effect of concurrent tasks known to be lateralized. Lomas and Kimura (1976) tested the interfering effect of speaking on a sequential tapping task (sequence of fingers from index to little finger as fast as possible for 15 sec). Speaking interfered with right but not left hand tapping in right handers. Speaking interfered with both hands in left handers. Non-speech vocalisations had little effect. Lomas (1980) carried out a similar experiment. Speaking interfered with a tapping task in the right hand only when there was no visual guidance possible. McFarland and Ashton (1978) demonstrated that concurrent verbal tasks interfered only with right hand tapping, while non-verbal ones interfered with tapping with either hand. These results indicate that a system for control of movement transitions made with minimal visual guidance interferes with (and, implicitly, is involved in) control of speech, and is left hemisphere lateralized in persons with typical cerebral dominance.

A similar study was carried out in deaf persons fluent in American sign language (Corina *et al.*, 1992a). Subjects were asked to tap with one hand and copy in sign language with the other hand the words presented to them visually or auditorily. Tapping in the right hand was more severely impaired by concurrent signing than that in the left hand. No asymmetry in tapping rate was found when the concurrent task involved production of other symbolic or arbitrary gestures of complexity similar to the sign-language signs. The authors conclude that the left hemisphere is truly specialised for language. This conclusion can be questioned, because the sign-language signs may have been subject to far greater practice, with development of more fluent time-structured models than the other gestures. If this is the case, it is an aspect of motor specialisation, rather than language *per se* which determined the asymmetry in producing ASL.

One recent study provides evidence about the preparatory phase of movement sequences, rather than their actual execution. Balfour *et al.* (1991) used a task in which visual cues were given to either right or left visual fields, which indicated the start of a sequence with one or other hand. Each sequence was from 1–8 "double taps" (i.e. an index finger/middle finger sequence). There were no lateral differences in execution time. However the preparation time showed a linear increase with length of sequence for right field/right hand combinations. For left field/left hand combinations the preparation time increased, more steeply than with right field/right hand combination, up to about 6 double taps, and then decreased. The overall effect was an inverted-U function. For left field/right hand or right field/left hand combinations, the dependency was a mixture of the two patterns. Overall these results show that the left hemisphere is capable of planning longer motor sequences, and can plan a long sequence more rapidly than can the right hemisphere.

A study which has often been cited as showing a *left* hand advantage in a ballistic aimed movement is that of Guiard *et al.* 1983 (see also Diaz and Guiard, 1984). In a task without visual feedback of the moving arm a left hand advantage was seen in "constant error". In other words, the average position reached by the aim showed a smaller systematic error with left than right hand. However, despite their larger error, the right handed responses showed a more consistent position of response (lower variance), though this did not reach statistical significance. Moreover, movement times with the right hand were faster. The larger systematic error in this study for the right hand is difficult to explain and is not found in other studies using accuracy as a measure. However, given that there are also trends to right hand advantage in both consistency and speed it is possible that the right hemisphere (controlling the left hand) can sometimes represent the visual target position more accurately, but the left hemisphere (right hand) can still employ more accurate programs for executing the reach.

Overall, despite some difficulties with the evidence, the results of studies into rapid aiming movements provide overwhelming support for the idea that the right hand has an advantage based on more accurate open-loop programs, rather than on the more accurate visual guidance. Presumably this is because open-loop programs in the left hemisphere have better spatio-temporal structure. Suggestions have already been made about how those open-loop programs are acquired by extensive practice which allows the matching of kinesthetic images of movement to the corresponding motor program and reinforcing those programs, identified by their kinesthetic match, which give optimal performance. The fact that the right hand advantage *is* in open-loop programs supports the assumptions behind this scheme.

It might be argued that the kinesthetic information involved in this scheme is a form of sensory guidance, used in closed-loop control. However, the suggestion offered here is that visual information about target position allows the selection of a corresponding spatio-temporal kinesthetic image for an optimal movement, against which actual kinesthetic feedback is checked. This hypothesis still requires a pre-formed and time-structured model of the movement. There is some evidence from the motor control literature (see section 9.4.1) that the checking against actual sensory feedback does not occur moment-by-moment, but on a segment-by-segment basis. Likewise although speech sounds are subject to auditory checking, they appear to be emitted syllable-by-syllable rather than phoneme-by-phoneme (see section 8.2.3.1). Neither of these really amounts to sensory *guidance*.

9.3.2 Lateralized control of saccadic eye movements in normal subjects

It will be recollected from section 8.3.2.1 that the distinction between preplanned actions and actions subject to sensory guidance corresponds, for eye movements, to the distinction between saccades and smooth-pursuit movements. A few papers studying normal subjects address the laterality aspect of eye movement control in these terms. Significant reports concerned with smooth pursuit will be mentioned in section 9.3.4. The data available in normal subjects for control of saccadic eye movements is less satisfactory. It is known that reaction time for saccades is not symmetrical. The RT for saccades directed to the right is faster than for those directed to the left (Pirozzolo and Rayner, 1980; Hutton and Palet, 1986). This is taken to imply that the left hemisphere is faster at initiating a saccade than the right hemisphere. Without more extensive analysis of this finding, it is unclear whether it represents a left hemisphere superiority in preplanning a movement to a target (which would support the hypotheses discussed above) or a left hemisphere superiority in rapidly initiating eye movements (which would be a lateralization opposite to that for manual initiation time). Bracewell *et al.* (1990) studied accuracy of saccadic movements, made in darkness to a target flashed briefly on a screen. Error was estimated for the total eye movement (i.e. primary saccade followed by the subsequent corrective saccade). Right handed subjects were more accurate in saccadic control for targets flashed in the left than the right visual fields. This suggests a right hemisphere superiority in saccadic control, which contradicts the hypotheses discussed in Chapter 8. However, it was also found that the increase in error as retinal eccentricity was increased was steeper for right field targets than left field ones. This is similar to the effect of retinal eccentricity on visual perception, described in section 6.2.4. It is possible that the laterality effect documented by Bracewell *et al.* (1990) depends on lateralization of perception rather than that of motor control. Overall, the data currently available for lateralization of saccadic control in normals does not evaluate the relevant hypothesis adequately. Further data, from brain-injured subjects is mentioned below and helps to clarify the issue.

9.3.3 Lateralized control of time-structured sound production by speech (and other) effectors

In section 8.4 the "delayed auditory feedback" paradigm, was mentioned (item (vi)). In this paradigm the accurately timed and time-structured consequences of a motor act are artificially delayed and lead to disruption of normal motor performance. It was predicted that such disruption should be stronger within the left hemisphere than in the right, because in the right there will be a weaker template to form a mismatch when the relay of information about consequences is altered. A number of papers compare the effects of the delayed auditory feedback to right and left ears.

Disregarding issues of laterality, the delayed auditory feedback disruption effect is shown most strongly for delays of about 200 msec. This is broadly compatible with the explanation offered in section 8.4 for this effect, because delays of this order would produce the most pronounced mismatch with the cortico-cortical template of expected auditory feedback. Similar delays are used in laterality experiments. Several studies have shown greater disruptive effect when the delayed signal is relayed to the right than the left ear. This has been shown as a lengthening of the time to read a prose passage (Bradshaw *et al.*, 1971; Bradshaw *et al.*, 1972; Elias *et al.*, 1977), or as the greater occurrence of errors in a simple verbal repetition task (Abbs and Smith, 1970; Roberts and Gregory, 1973). In the paper of Abbs and Smith (1970) the greater right ear disruption was not seen for feedback delays of 100 msec, and was seen more strongly for 300 msec than 200 msec delay. Although the effect has been replicated in several studies it is not very robust, and may require special conditions. In Bradshaw *et al.*'s experiments, the laterality effect did not occur unless there is simultaneous competitive stimulation of the other ear (either long-delayed, or instantaneous amplified or attenuated auditory feedback). Even with such competitive stimuli to the other ear, the larger disruption for right-sided delays is not always found (Venkatagiri and Hinkle, 1975). On the other hand, Elias *et al.* (1977) obtained a right ear disruption bias using monaural stimulation.

Another difficulty with this evidence is that a *left*-sided disruption bias can also be obtained, if the subject is engaged in musical performance (piano playing (Bradshaw *et al.*, 1971), whistling (Bradshaw *et al.*, 1972)) or in rhythmic tapping (Bradshaw *et al.*, 1972; Roberts and Gregory, 1973). These findings could be taken as undermining the evidence for a superiority of the left hemisphere in time-structured modelling, because they show that laterality effects are more closely related to the type of auditory material produced by the subject (which each have their own variety of laterality), than to the hemisphere *per se*. Nevertheless it is possible that the different lateral bias seen for speech signals and musical sounds is not actually a function of speech *vs.* music, but of the spatio-temporal differences between the sounds emitted. The piano playing in Bradshaw's experiment had a time-structure in less fine detail than the speech output. In a separate test with binaural delayed feedback, it was found that the delay giving maximal disruption was significantly longer for piano playing than for reading prose (Gates *et al.*, 1974). It is possible that the left-sided bias for disruption in music performance depends on disrupted production of musical "contour" (see Chapter 10) rather than the finer temporal structure involved in spoken language, perception of which gives a left hemisphere superiority.

Closely related to the above work with delayed auditory feedback are experiments carried out by Tsunoda and colleagues (Tsunoda and Oka, 1971; Tsunoda, 1975). Subjects had to use a key (with the preferred index finger) to tap out a rhythm ("xxxx xx xxxx xx etc"). This was relayed synchronously to one ear, and with 200 msec delay to the other. The intensity of the delayed signal was then reduced to find the threshold for the disruption. When tapping produced the vowel /a/ (duration 75 msec) there was always a right ear disruption bias. When it produced a pure tone there was a left ear bias. Tsunoda and Oka (1971) found greater right than left ear disruption for CV syllables in English speakers, and for both CV and V in Japanese speakers. Left ear disruption was greater than right for a 1 kHz tone. For aphasics with right hemiplegia there was a left ear bias for both stimulus types. The authors concluded that the left hemisphere is more easily disrupted than the right by delayed feedback of speech-like sounds. The difference between those who speak Japanese and those who speak English appears to be a cultural rather than a genetic one, perhaps based on the fact the Japanese uses vowel sounds in a different way from Indo-European languages (e.g. a single vowel sound in Japanese may signify a definite meaning such as a concrete noun).

Once again however, this finding is far from robust. Uyehara and Cooper (1980) failed to confirm Tsunoda's result concerning the interaction of ear differences with sound type (vowel vs. pure tone) or subject's language (English vs. Japanese). Elman (1983) repeated the experiment with the added variable of oral as well as manual tapping of the rhythm. The response measure was the time (msec) taken to tap a particular rhythm. A significant ear effect was found for oral tapping, with a significantly greater right than left ear disruption by delayed feedback. No ear effects were seen for manual tapping. The subjects natural language was not a significant factor in the results.

Overall, evidence using the delayed auditory feedback paradigm gives some support to the prediction that matching of an internal and a feedback representation of auditory consequences of an action is important for preplanned performance, especially of speech. However, the effects shown are capricious, and the evidence cannot bear the full burden of proof of the hypotheses at issue in this section.

9.3.4 The right hemisphere and motor control by sensory guidance: evidence from normal subjects

As mentioned in section 9.3.1, several studies have shown faster times for movement initiation by the left hand (right hemisphere) compared with the right hand. Assuming that the responses are determined by cortico-cortical relay, this result would be expected from the central hypothesis of this book, that cortico-cortical delays are, overall, less in the right compared with the left hemisphere. However, this conclusion does not relate specifically to the special version of this hypothesis, referring to differences in motor co-ordination between hemispheres.

A superiority of the *left hand* for visually guided movement was shown in one study mentioned above (O'Boyle and Hoff, 1987). The greater deterioration of performance with speed in the left than the right hand (shown in several studies of Roy and co-workers) is consistent with this, because performance dependent on visual guidance would be critically dependent on the rapidity of visual feedback, whereas preplanned movement, with sensory feedback only at the end of each segment of movement would not. Likewise the lack of interference of speaking on a visually guided task (Lomas, 1980) supports the view that visual guidance uses the right rather than the left hemisphere. It should be noted however, that several studies fail to detect asymmetry in visually guided movements (Bowen *et al.*, 1972; Watson and Kimura, 1989).

Two papers report on laterality of smooth pursuit eye movements in normal subjects, both using concurrent tasks as a means of assessing the contributions of each hemisphere. Larmande *et al.* (1985) studied the effect of visual pursuit on lateralized performance in a dichotic listening task. Visual pursuit, in whatever direction, interfered with right hemisphere function in the dichotic listening task. Sava *et al.* (1988) exposed normal subjects to a variety of visual tracking tasks. Concurrently, the subjects were given a reaction time test to simple stimuli presented in either left or right visual fields. For three out of the four tracking tasks, reaction time in the left field was increased, while right field reaction time was unchanged. This indicates that the visual tracking tasks engaged the activity of the right hemisphere rather than the left. The exceptional test was the one where the movement of the tracked stimulus was most predictable. We have already seen that in manual tracking tasks right hand superiority can occur for tracking *predictable* moving targets, and is then not based on superior visual feedback but on more accurate programs to an extrapolated target position.

All the above studies involve *visu*omotor tracking, either with the hand or the eyes. They give significant support to the proposition that the right hemisphere has a special role in sensory guidance of movement, with admittedly some negative findings. In principle it is also possible to devise experiments where *auditory* signals are used to guide motor performance, in the closed-loop sense of control. This is the paradigm adopted by Sussman (1971; Sussman *et al.*, 1974; Sussman and MacNeilage, 1975). A "target tone", changing in frequency at 20 or 10 cpm, was relayed to one ear, and a "cursor tone" to the other ear. In the 1971 paper, control of the cursor tone was arranged with either the tongue or the hand executing the control movement. There was a right ear advantage at both target frequencies (i.e. performance was better when the controlled tone went to right ear). This was shown for control by the tongue, but not by the hand. Sussman *et al.* (1974) also found a right ear advantage in 24/29 subjects for mandibular tracking ($p = 0.002$) but again, not for manual tracking ($p = 0.17$). Sussman and MacNeilage (1975) repeated these observations.

The accuracy of such auditory tracking is inversely proportional to the delay between perception of the controlled tone and production of the guided motor act. The right ear advantage in the above experiments apparently points to faster relay from sensory to motor regions in the left than the right hemisphere, a conclusion in contradiction to the hypothesis advanced in Chapter 8, and the evidence for visuomotor control just reviewed. However, more detailed analysis of the Sussman paradigm suggests that the result cannot be interpreted unambiguously in terms of the hypotheses developed in Chapter 8. The reason for this is that the target tone and controlled tones go to opposite hemispheres. There are consequently two pathways whose conduction time could give rise to lateral differences in control. The comparison of target and controlled signals could be made in either the right or the left hemisphere, with the same hemisphere presumably exercising the motor control. Suppose the target tone is relayed directly to the right hemisphere. If the comparison is made in the right hemisphere, the controlled tone will be delayed with respect to the target tone due to the transcallosal conduction time from left to right. In addition, once the comparison has been made, there will be an additional delay corresponding to the relay in the right hemisphere to the motor regions which govern the guided motor act. If on the other hand the comparison is made in the left hemisphere, the controlled tone will be advanced with respect to the target tone by the callosal conduction time from right to left. In this case, once the comparison has been made, there will be an additional delay corresponding to the relay in the left hemisphere to the motor regions which govern the guided motor act.

One cannot assume that the callosal conduction time from right to left is the same as that from left to right. Indeed, there are data which suggest that left to right conduction time is longer than right to left (see Chapter 4). Therefore superior accuracy of auditory guidance by the left hemisphere vs. the right cannot be attributed unambiguously to faster conduction from sensory to motor cortical regions in that hemisphere. It could equally well be due faster callosal relay from right to left hemispheres, a conclusion which would be in agreement with the hypotheses advanced in Chapter 8 and elsewhere in this book. Thus, the evidence using the Sussman paradigm, produced so far, does not contribute to understanding laterality in auditory-motor guidance. If on the other hand, experiments were conducted in which both the target and the controlled tones were delivered to the same hemisphere, the callosal relay would be avoided, and the result would become interpretable.

One of the experiments of Wolff *et al.* (1977), mentioned earlier in this section may be one form of such an experiment. In this experiment, temporal accuracy was assessed for tapping to a metronome (for 45 sec, at 120 or 75 beats per minute). No manual asymmetry was found, in contrast to the right hand advantage in repetitive tapping of a rhythm given at the start. Both these tasks were aimed, in that the tapping had to be confined to a one-inch key. However, the first task was easier than the second. The authors own explanation of their result is that under some circumstances lateral difference between motor performance of the hands can be eliminated given sufficient practice. For instance, in the first task, children (but not adults) do show a right hand advantage (Wolff and Hurwitz, 1976). Moreover, in typing and other manual tests, lateral differences are found in untrained subjects, but not in trained typists (Provins and Glencross, 1968). Wolff *et al.* (1977) therefore suggest that the finding of a right hand advantage in the more difficult task reflects the fact that this task is less influenced by prior experience which could equalise manual performances. However, the right hand advantage in children may arise for reasons of laterality of rhythm perception, rather than motor control. The difference between the two tasks in adults can then equally well be attributed to the fact that tapping to a metronome is a form of motor performance guided moment-by-moment on the basis of sensory feedback; while tapping to a *given* rhythm requires preplanning ("modelling" in the terms of present hypotheses). If this is the case, the lack of asymmetry in the task requiring auditory guidance is congruent with the evidence of lack of asymmetry or definite right hemisphere advantage in visually guided aiming movements. Further studies of this sort are clearly required.

9.4 CLINICAL EVIDENCE FOR LATERALIZATION OF TIME-STRUCTURED MODELLING

In Chapter 8 it was predicted that, with brain damage in appropriate locations, selective loss of the modelling function of the dominant (usually left) hemisphere may occur, without loss of either the more basic components of motor programs, or the sensory discrimination of external stimuli. In such cases, it was suggested that patterns of errors and inaccuracies might betray the fact that it was a motor-sensory modelling function which was disturbed. In the following subsections the evidence relating to these predictions in brain-injured patients is reviewed. Section 9.4.1 deals with the disturbance known as apraxia, related to lateralized skills of the hands, limbs and oral regions. Section 9.4.2 deals briefly with disorders of eye movements after unilateral brain damage, while section 9.4.3 deals more extensively with disorders of spoken language. In principle, there could also be selective motor losses of a different sort after right hemisphere damage, particularly in tracking tasks where motor acts are guided at each moment by sensory stimuli (e.g. in visuomotor tracking tasks). A little evidence of this nature is available for eye tracking, but otherwise this prediction awaits definite evidence (see section 9.4.4).

9.4.1 Apraxia and preplanned motor performance after unilateral cerebral lesions

It has been known for nearly a century that hemispheric damage can lead to disturbance of complex motor functions without either hemiplegia, loss of sensory function or language comprehension. Hugo Liepmann (1900), who discovered this syndrome gave it the name "apraxia".

Four types of apraxia have been recognised (De Ajuriaguerra and Tissot, 1969; Hecaen, 1969; Brown, 1974; Heilman and Rothi, 1985): Ideomotor and ideational apraxias were in Liepmann's original description. Limb-kinetic and constructional apraxia were added by Kleist (1934).

Ideomotor apraxia is common and may affect oro-facial musculature as well as the limbs. According to Poeck (1986) the disorder has two characteristic features, as analysed by clinical examination: (i) A complex movement becomes *parapractic*, that is the individual movements which are components of the whole are incorrectly selected and are produced in incorrect sequences, despite the fact that there is no sensory impairment, locomotor ataxia or problems of understanding the task. (ii) The execution of the complex movement is impaired only when the movement is required out of its natural context. For instance, "brushing teeth" on a physician's command may be disturbed, while the same act in its normal context is performed without impediment. This form of apraxia is usually tested by asking a patient to perform a gesture to command, or to imitation (the latter procedure excluding any possibility of language comprehension impairment). Such a form of apraxia occurs in 80% of cardio-vascular accidents of the language-dominant hemisphere (Poeck, 1986), and very rarely for lesions of the opposite hemisphere.

Ideational apraxia is a motor disorder at a higher level than ideomotor apraxia. Patients typically are impaired in sequencing actions requiring the use of various objects in the correct order and sequence, in order to achieve a goal. According to Poeck (1986) this disorder is rare, but, when it occurs is chiefly in cases of lesions of the language-dominant hemisphere.

Limb kinetic apraxia involves imprecision in making a complex movement, although the components of the movement are accurately selected sequences, and correctly oriented in space (Heilman and Rothi, 1985). The disturbance is in moment-by-moment synchrony, rather than in speed, rhythm and sequence (Brown, 1974). Patients may for instance have difficulty in picking up a coin from a flat surface. Such a disturbance is unilateral after a lesion to either hemisphere (Poeck, 1986). Hecaen and Albert (1978) consider this syndrome to be intermediate between paresis and apraxia. Poeck (1986) does not regard it as true apraxia.

Constructional apraxia usually refers to impairment in tasks where physical objects of some sort are to be put together or articulated to form a single entity. A variety of tests are used to detect constructional apraxia (building with blocks to form a given design, drawing a figure to command or copying a given figure). Patients commonly show a deficit in one of these tasks but not in another. While the majority of cases occur after right hemisphere damage, a significant number can occur with dominant hemisphere lesions. A body of opinion holds that "constructional apraxia" is not a homogeneous disorder but is a somewhat arbitrary grouping of disorders with different psychological bases and different lateralization (Benson and Barton, 1970). Benton and Tranel (1993) do not classify it with apraxias, but label it "visuoconstructive disorder".

Apraxia has been defined as the inability to perform complex purposeful movements in the absence of primary motor or sensory deficits. Geschwind (1975) defines apraxia as a "disorder of learned movement." These definitions can accommodate the fact that there are disorders of learned movements produced by either left or right hemisphere damage. However, they hide the fact that there are disorders of learned movement specific to left hemisphere damage, categorically different from those resulting from right hemisphere damage, and probably confined to humans. From the above brief survey, it is clear that the central concept of apraxia which needs explaining applies to those forms of apraxia (ideomotor and ideational apraxia), which occur after left dominant lesions; and of these, the ideomotor apraxia is by far the most important. The remainder of this subsection is mainly about this form of apraxia.

Geschwind (1975) further analyses apraxia as a "disconnection syndrome", in which the important point is that the pathways which are damaged disconnect one region from another. The information processing characteristic of each damaged region, or of two regions coupled to each other is not important in this scheme. Thus, in his analysis, lateralization of apraxic syndromes derives in part from disconnection with language comprehension functions, rather than from a distinct lateralized form of motor control. In support of this concept Geschwind mentions patients who show apraxia when asked to respond to a verbal command, but not to imitation, which suggests that disconnection of language comprehension plays a part in some apraxias. However, it is more typical for impairment to be seen both to command and to imitation, and this cannot be explained simply as disconnection from language centres.

In applying the hypothesis of time-structured modelling, advanced in Chapter 8, to explain apraxia, it is necessary to go beyond this connectionist view of apraxias. In particular, for ideomotor and ideational apraxias, there are many indications that a distinct variety of motor information processing is disturbed.

For a dominant-hemisphere lesion the apraxic disorder is usually bilateral (Poeck, 1986). Geschwind (1975) points out that individuals who lose the use of their right arm for peripheral reasons (such as having the limb in plaster) quickly develop very "dextrous" use of the left hand and arm. In contrast, patients with apraxia due to a left hemisphere lesion become more clumsy than normal with the left as well as the right hand. Loring *et al.* (1989) report on a callosotomy patient and find a typical apraxia both to command and to imitation in the left hand. As already discussed the isolated right hemisphere has enough language comprehension ability to understand the imitation test, yet, in this case, was unable to perform gestures to imitation. Presumably the impairment was because the right hemisphere (and left hand) was disconnected from the left hemisphere, where manual praxis for either hand is represented. All this evidence attests to the fact that the left hemisphere (in right handers) has a special form of motor control and exerts this control over *both* upper limbs (though more characteristically over the right hand and arm). This leads to the key question: *What could be the nature of the deficit in apraxia*? Both clinical and experimental evidence from brain-damaged subjects contribute to answering this question.

Kimura and Archibald (1974) required brain-damaged patients to copy unfamiliar meaningless hand gestures, such as those used in sign language. Left hemisphere patients were impaired in this task, using either hand, but not in imitating static hand postures, or in producing single finger flexion movements, which were not part of a sequence. Visual perception of the movement to be imitated was also not impaired. The authors believed that the unique function of the left hemisphere is in motor sequencing, and this applies both to limbs and oro-facial musculature. Kimura (1982) required brain-injured subjects to perform single or triple manual or oral gestures to imitation. Right hemisphere patients were only mildly impaired on all tasks. Impairment in the left hemisphere patients was greater for the triple movements than the single ones.

The implication made in such clinical studies is that there is an impairment in sequencing of components of each movement. This was investigated by Jason (1983a,b). Brain-damaged subjects were required to reproduce, as fast as possible, a short sequence of hand gestures. No difference was found between right and left hemisphere patients, although in recall of individual gestures during the learning phase, left hemisphere patients were more impaired that right hemisphere ones. The author concludes that the left hemisphere is specialised for *memory* of the hand positions, but not for ordering of sequences of them. The latter conclusion may not however be in conflict with the suggestion from clinical examination that apraxia consists of disordered selection and sequencing, because the individual components to be sequenced in Jason's experiments were relatively long and slow ones (about 1 second for each component), whereas the complete gesture made in a clinical test of apraxia (or in fluent use of sign language) may be of much shorter duration than this. Thus, in attempting to analyse the apraxic movement disorder, it may be necessary to focus on very rapid movements.

Several studies show that in rapid reaching with the hand, left brain-damaged subjects are slower than normal controls. Wyke (1967) used a rapid aiming task, and measured time for the total movement (including "initiation time"). Left brain-damaged patients were slower than controls with both hands, while right hemisphere patients were slower than controls only with the left hand. Similar results were derived from a speeded tapping task. Fisk and Goodale (1988), testing patients with the hand ipsilateral to brain injury, showed that right hemisphere patients had normal movement times, but took longer to initiate a movement than controls. Left hemisphere patients showed the reverse patterns, that is, normal initiation times but longer movement times. Haaland and Harrington (1989b), again testing patients with the hand ipsilateral to their brain injury, found that reaction time as well as the time for the first phase of an aiming movement were longer in left brain-injured patients than in controls. The time for the "corrective" late phase of the movement was not impaired in left hemisphere patients. Right hemisphere patients had no impairment in any phase of the movement.

Despite disagreement on some points, these studies support the conclusion derived from normal subjects, that the left hemisphere has greater involvement than the right in co-ordination of rapid and accurate reaching movement. In particular, a common theme is that the *movement time* in rapid aiming movements is more impaired with the left hemisphere than right hemisphere damage. As with studies of normal subjects, this leaves open the question of whether the basis of the difference is loss of sensory guidance, or loss of pre-programmed actions by the left hemisphere. Haaland and Harrington (1994) provide evidence relating to the "loss of sensory guidance" hypothesis. Their task involved alternate tapping of two targets varying in size from 0.5–6.5 cm. For left hemisphere patients, there was no impairment in tapping speed compared to controls on the smallest targets, when tapping was slowest. For larger targets, when both controls' and left hemisphere patients' tapping speeds were faster, the patients were impaired in speed relative to controls. Apraxic but not the non-apraxic left hemisphere patients also showed more errors on the smaller targets, compared to controls. For right hemisphere patients, performance was unimpaired, and speed did not increase compared to controls as target size increased. One of the hypotheses considered by the authors is that small targets require a larger component of visual (closed-loop) guidance than large ones, for which ballistic (open-loop) programs will suffice. The increasing impairment of the left hemisphere patients on the speed measure as target size increases may therefore indicate that the intact right hemisphere is better for closed-loop control, which dominates over open-loop control with small targets. With the larger targets, advance programming (open-loop control) of the movement comes to dominate performance in controls, but this requires an intact left hemisphere. However, other evidence they cite (see section 9.3 for data from normal subjects) suggests that availability of cues for visual guidance benefits each hemisphere equally. The authors recognise that the evidence at present available does provide a decisive answer to the role of visual guidance in the motor deficit after left hemisphere injury.

Fisk and Goodale (1988) suggest that loss of visual guidance cannot account for the deficit in left hemisphere patients. They refer to a previous study from normals (Goodale *et al.* 1986) which showed that visual feedback to update the limb position can be based solely on visual information about the target, not about instantaneous limb position. This suggests that corrections are based on comparison of target position and internal (kinesthetic) representation of the limb, not on that of target position and *visualised* limb position. In addition, Cook and Diggles (1984) showed that in-flight corrections are so rapid that they cannot be based on information from the moving limb but rather on a central monitoring of the commands for movement.

Two papers provide evidence more directly relevant to loss of visual guidance after hemispheric injury. Wyke (1968) required subjects to use a pointer to make contact with a spot on a moving turntable: Right hemisphere patients were impaired with both hands, but especially with the left. Left hemisphere patients showed only right hand impairment. Haaland and Harrington (1989a) required subjects to move a pointer from a start position to a target circle. Subjects did not see their own hand, but the position of the start point, the pointer, and the target were transformed onto a video screen. Since subjects had no experience of this transformation prior to the experiments, they must have relied on visual guidance. The movements were also so slow that they could not have been ballistic (open-loop) ones. In this task, right brain-damaged patients were more impaired than left hemisphere ones. The former had longer reaction times than normals, with the left arm, and were less accurate for narrow targets. Thus, from these two studies, the right hemisphere rather than the left seems to be mainly involved in movements with sensory guidance. This conclusion is made tentatively however, because of difficulties with some of the evidence: In the previous section, dealing with normal subjects, the paradigm used in Wyke's (1968) study was interpreted as relying mainly on preplanned action, giving a left hemisphere advantage; but for brain-injured patients the results indicate a dominant role for the right hemisphere.

After tentatively excluding impairment of visual guidance as an explanation of the motor deficit after left hemisphere damage, Fisk and Goodale (1988) are left with two hypotheses both utilising proprioceptive feedback. The left hemisphere patients may be deficient in "use of proprioceptive feedback to determine the instantaneous position of the limb" (still implying that such feedback is used for closed-loop control). Alternatively the left hemisphere group may be "unable to update the program adequately as the trajectory of the movement unfolds" (implying use of proprioceptive feedback for choice of the next segment of open-loop programs).

The second of these two hypotheses is supported by other work in normals and brain-injured patients. In normals, Viviani and Terzuolo (1982) have provided evidence that a motion can be broken down into segments, such that within each segment the trajectory goes through equal angles in equal times. This "isogony principle" was shown to apply to both handwriting and drawing, and to either unconstrained movement or movement using visual or tactile guidance. In their results, the segments were of duration 100–200 msec. Transition from one segment to another occurred whenever the mismatch between an intended and actual trajectory exceeded a given threshold. The greater the curvature of the movement, the shorter the segments. Overall, "the figural aspects of a movement trajectory uniquely determine[d] the dynamics". Poizner *et al.* (1990) applied this conclusion from normal subjects to an investigation of brain-injured patients. The space-time coupling seen in normals was shown to break down in cases of apraxia. In normal subjects the velocity profile for tracking an oscillating target was often quasi-sinusoidal. With apraxia there were large departures from the sinusoidal pattern. Presumably this is a laboratory demonstration of the clinically-recognised phenomenon of parapraxia.

In a recent paper (Poizner *et al.*, 1995) three apraxic subjects with left parietal damage were studied in detail. Although the authors do not emphasise the temporal aspects of the discoordination in these patients, they illustrate marked loss of specificity of the phase relation between movement of different upper limb joints (e.g. shoulder and elbow). They point out that this temporal disorganisation is different from that seen in patients with loss of proprioceptive input. In such deafferented patients, impairment in movement occurred at particular phases of movement when angular acceleration or deceleration was greatest (and implicitly when sensory guidance was most needed). In contrast, in the apraxic subjects, discoordination was evident throughout the movement, as though a plan for a complete fluent movement had been damaged.

Study of brain-injured apraxic patients also provides an opportunity to discover which regions of the left hemisphere are involved in the motor control for which that hemisphere is specialised. Wyke (1967) found that a reduced speed of arm movements to a stationary target was associated most consistently with left parietal damage, but was found in almost as high a proportion of cases with left frontal and sometimes in those with left temporal damage. Right hemisphere damage was not associated with a higher prevalence of impairment than in normal control subjects. For movement to a moving target (Wyke, 1968), impairment was found most often found with frontal damage, but still quite commonly with parieto-temporal damage. Kolb and Milner (1981a) found that, for imitating gestures with either arm, the main deficit occurred with left parietal patients, with milder deficit for damage to either frontal lobe. In imitating facial gestures, left parietal damage caused no impairment, but marked deficits were found with damage to either frontal lobe. Kimura (1982), in a study already mentioned, required brain-injured subjects to make single or triple oral or manual gestures to imitation. Frontal and parietal lesions on the left side were mainly responsible for deficits. Both regions needed to be intact for unimpaired performance in the more demanding triple movement tasks. Central, temporal and occipital lesions gave rise to little impairment. The minimal role of the temporal lobe in the motor control tasks investigated, does not exclude a role for this region in motor-auditory control, such as might be involved in speaking. Impairment of single oral gestures was attributed almost entirely to left frontal damage, while impairment of single manual gestures was caused by left parietal damage (most markedly) but also by left frontal damage (a non-significant trend). The "single gestures" included such things as tongue protrusion or blowing (oral) or imitating one of five static hand signs (manual). Production of these may in fact have required a preplanned motor sequence, like that for the multiple movements.

In summary, it can be said that the evidence on the motor deficit in apraxia does not give such a clear indication of the special role of the left hemisphere in actions involving time-structured planning as does the data from normal subjects. Indeed there is one report (Haaland *et al.*, 1980) in which the most severe deficit associated with apraxia is in *static* maintenance of a precise hand position. Nevertheless, the balance of the clinical evidence suggests that rapid unguided ballistic movement is better with the left hemisphere. The suggestion that sensory guidance is superior with the right hemisphere receives more tentative support. Impairment in the former of these functions is correlated with the clinical syndrome of apraxia. A little recent evidence suggests that in apraxia there is a breakdown in normal segmentation of movements, such as might be expected when the hypothetical combinatorial function needed for preplanning an action is disturbed. It may be very significant that the duration of the segments of movement identified in normals is approximately the same as that of syllables in speaking. The available clinical evidence does not provide decisive evidence that the left hemisphere constructs programs as time-structured models defined by motor-kinesthetic association. Nevertheless there is no evidence against this hypothesis, and some suggestive indicators that it is correct.

Apraxic disorders of the limbs result largely from damage to the frontal and/or parietal lobe on the left side, while frontal but not parietal lesions giving rise to oral apraxia. This regional evidence is broadly in line with the hypotheses advanced in Chapter 8, because frontal regions are the origin and parietal regions are the termination of pathways mediating motor-kinesthetic association. The greater involvement of frontal rather than the parietal lesions in oral apraxia may arise because, in initial discovery of the programs for oral movement, visual monitoring by the subject of his or her own movements is not likely to be implicated as it is for programs of manual or limb movement. Visual inputs are known to be included within the associations performed by the parietal lobe, and this may be related to the greater role of this region in manual and limb co-ordination, as revealed from the study of apraxia. A further prediction on this subject is made in sections 9.4.3 and 9.4.5.

It may be asked why the impairments of apraxia are shown only to command and imitation, and not when the same movement is carried out in its natural context. Once the time-structured program has been laid down, there must be various alternative ways of activating it. An explicit instruction, given either verbally or by imitation, may contain less redundancy than one generated internally in response to the individual's continual integrated stream of motivations and intentions. The movement produced to explicit instructions may therefore be more vulnerable to failure if the network of connections giving the spatio-temporal template of consequences of it is damaged.

9.4.2 Eye movements and unilateral cerebral injury

Two papers reporting on eye movement disorders after unilateral hemispheric damage contribute to defining the special roles of the two hemispheres in motor control. Gaymard *et al.* (1993) studied three patients with lesion of the right supplementary motor area, and three with damage to the corresponding area on the left. When asked to perform single saccades to previously memorised positions neither group of patients were impaired. In a task involving sequential saccades to three positions pre-ordered in memory, many more errors were found for the left hemisphere patients than the right hemisphere ones. This result is similar to that of Kimura (1982) for sequences of manual or oral gestures. Given that saccades are an equivalent to ballistic, open-loop control of the limbs, the result supports the hypothesised role of the left hemisphere in such movements.

Complementary results were reported by Bogousslavsky and Regli (1986) who studied smooth pursuit eye movement in brain-injured patients. Impairment was found for all lesions in the parieto-occipital lobes, but was severe only for lesions in the non-dominant hemisphere. The lateralized deficit was limited to contralateral pursuit. (Ipsilateral pursuit was similarly impaired by lesions in either hemisphere). The impairment with non-dominant hemisphere lesions was not due to neglect alone, because voluntary saccades were preserved. Since smooth pursuit eye tracking is a closed-loop form of movement, this result strengthens the evidence that the right (non-dominant) hemisphere can perform such movements better than the left.

9.4.3 Speech performance and unilateral cerebral injury

Speech impairment after injury to the left hemisphere (in right handers) is known as *aphasia*. Aphasia has been a subject of scientific study for longer than has apraxia. Correspondingly the phenomenology of aphasia is known in greater detail than that of apraxia. Several distinct aphasic syndromes have been separated, occurring after damage to different regions of the dominant hemisphere. Separate apraxic syndromes, occurring after damage to different regions of the left hemisphere have not been recognised in such an explicit manner*.

In the following subsection, in order to relate the detail known about aphasia to the hypotheses developed in Chapter 8, a different approach must be taken to that adopted for apraxia, which emphasises the morphological correlates of each syndrome. Most of the evidence on aphasia reviewed below is clinical description rather than experimental studies. Much of this concerns cases of dominant hemispheric injury, which will be reviewed first, though there is a little data from callosotomy and dominant hemispherectomy cases also to be considered.

For left hemisphere injury, the first aphasic syndrome to be defined was that occurring after inferior frontal damage, the syndrome now known as *Broca's aphasia*. The following description is based on Benson (1979). The hall-mark of this disorder is the non-fluency of the speech. Patients can utter single words, especially monosyllables, but cannot program the fluent concatenation of the different speech sounds into a flowing phrase or sentence. Nevertheless, they know what they want to say, and the fact that they cannot say it may be a source of great frustration to them. Verbal output is almost entirely of substantive words (e.g. nouns, action verbs, adjectives). Connecting words and prosody are lacking. Comprehension of language, although not always normal, is much closer to normality than is language expression. Repetition of spoken language is abnormal and fails in the same respects as spontaneous speech (loss of the fluency, prosody and connecting words, but not the substantives). Speech may become more fluent for familiar or automatic speech (days of the week, months, counting, common phrases). Confrontation naming is invariably impaired, but can often be overcome by appropriate cueing, (e.g. with the initial syllable) in which case properly articulated names are produced.

In interpreting this syndrome, it is clear that the intention behind speaking (which should trigger the utterance) is preserved, but it cannot activate the time-structured chain of speech sounds needed for its realisation. The failure appears to lie in the temporal co-ordination in the executive part of the program, rather than in the checking of the auditory consequences of that program. Speech thus becomes a series of isolated sounds rather than a fluent stream. Substantives have a close relation to the intended expression, and can be produced one by one, even when the fluent stream is broken. The function words on the other hand have little role, except as part of a fluent stream. One must presume that Broca's area is centrally involved in the temporal chains of neural activation by which the executive part of a program for speech is planned. Broca's area is part of the prefrontal area, which also includes the supplementary motor area. In comparable fashion, we have already seen that left prefrontal lesions can impair production of sequences of hand movements, and that lesions of the left supplementary motor can impair sequences of eye movements.

A comment is needed on the fact that the impairment in Broca's aphasia can to some extent be overcome if sentences are familiar ones or if there is appropriate cueing: This

* It is a very interesting question to ask whether the apraxias can be subdivided into syndromes with morphological correlates similar to those of the aphasias. This might be expected if both aphasia and apraxia are disorders of the same aspect of motor control (called "time-structured modelling" in Chapter 8). No certain answer to this question is yet available, though it has been considered by several writers. Their discussions are briefly reviewed in section 9.4.5.

appears to be a similar phenomenon to that seen in apraxia: Apraxic impairments are found only when actions are required in response to command or imitation, and not when the same movement is carried out in its natural context. Both for Broca's aphasia and for apraxia, the essential part of a time-structured chain of neural activation may, by itself, transmit relatively securely in the intact brain, due to connections in the part of the brain where it is most strongly localised (e.g. Broca's area). However, these chains of activity (as for cell assemblies generally), are not likely to be exclusively localised in a single region. Thus, when Broca's area is damaged, components of the same chain of activation can sometimes be triggered, sufficient to elicit the required utterance, but this requires much more favourable circumstances, such as a very familiar (i.e. strongly represented) motor sequence, the right context, or specific cues to enhance the security of the chain.

A small body of experimental data has attempted to analyse the problem with articulation in Broca's aphasia. Ziegler and von Cramon (1985) provided evidence, from single patients with anterior aphasia, that the relative timing of different articulators involved in producing speech sounds was disturbed. Methodological aspects of this issue have been debated in a number of subsequent papers (Sussman *et al.*, 1988; Katz, 1988; Ziegler, 1989; Katz *et al.*, 1990). A paper by Baum *et al.* (1990) provides new data, which suggest that anterior aphasics are indeed impaired in the rapid co-ordination of different articulators, for instance in relative timing of the production of the voiced and unvoiced part of a syllable. More data is needed to resolve this issue.

The second most clearly-defined aphasic syndrome, occurring after damage to the posterior superior part of the temporal lobe is known as *Wernicke's aphasia* (see Benson, 1979). Speech is fluent, often more fluent than normal, produced without effort or hesitancy, with adequate grammatical structure and use of function words, and with normal phrase length and prosody. However, there is a deficiency of meaningful substantive words. As a result, the patient's speech, though fluent, becomes empty. There may be "paraphasias" of various sorts. These may include the substitution of an inappropriate word (semantic paraphasia) or syllable (literal paraphasia), or the formation of neologisms. Taken together these deficits may cause speech to degenerate into an incomprehensible "jargon aphasia". Comprehension of the language of other people is deficient, often severely so. There may also be an inability to discriminate elementary speech sounds. In sharp contrast to Broca's aphasia, patients with Wernicke's aphasia commonly have little understanding of the fact that they have a language impairment (Benson, 1979, Chapter 17). Repetition of spoken language is impaired, in so far as their comprehension of it is also deficient. Recitation of familiar speech (such as days of the week) can be normal, and without paraphasias, once the patient has established the right "set". Naming to confrontation is often totally absent, or attempts to find the right word are disrupted by severe paraphasias. The failure to find the right start to articulating a word (as found in Broca's aphasia) is not a characteristic of this syndrome.

In interpreting this syndrome it is clear that the temporal constraints in executive aspects of programming speech are intact. The loss of substantive words from speech suggests that their representation is based strongly in the tissue in the damaged auditory association area. This corresponds to Wernicke's own view, that the area named after him held the "sensory images of words". On this, Luria (1973) disagrees, regarding the loss of phonemic representation as more fundamental. This hardly seems a point worth debating, since both phonemes and words are presumably represented by temporal chains of neurones, with their essential members in Wernicke's area, and others distributed in other regions of the cortex. Paraphasias imply ineffectiveness at the planning stage in the utterance of each word or phrase, and reveal a selective impairment in the combinatorial processes involved in that preparatory modelling. The fact that paraphasias are uttered, rather than never even formulated at the planning stage also requires comment:

As explained in section 8.3.2.5, the condition of *matching* between a modelled speech sound and a memory of it designates a program as accurate. This matching is envisaged to be produced in Wernicke's and adjacent areas. The correctness of the plan is then communicated back to the later components of the motor sequence represented in the executive part of the system (i.e. in Broca's area), by pathways depicted in Figure 8.7. This allows the implementation of the later parts of the speech program. The fact that paraphasias occur in Wernicke's aphasia implies that Broca's area cannot, as usually happens, be informed when the preparatory model of speech was inaccurate. The deficit in monitoring the patient's own speech also results in deficits in comprehension of other people's speech. The typical lack of insight into their disorder, shown by patients with Wernicke's aphasia is probably based on the fact that they fail to monitor their own linguistic output, just as they fail to comprehend that of other people. The severe difficulty in naming presumably arises in part because the sensory image corresponding to the object to which each word refers (represented in these other cortical regions) no longer has its locus of convergence with the auditory image of the word. In addition, naming normally must depend on a pathway from Wernicke's to Broca's area. Normally, the perceptions (usually visual) of the thing a word signifies will activate the sensory image of a word in Wernicke's area, which in turn can activate the program for articulation of the word, via pathways habitually used in activating correctly modelled speech programs. This chain of events will be broken by a lesion in Wernicke's area. The fact that such a lesion is reflected in loss of speech output in naming is another indication that the connections between Broca's and Wernicke's areas are from the sensory to motor area as well as in the opposite direction. The fact that repetition can be performed adequately for very familiar material may reflect the fact that the time-structured chain which forms the executive part of a speech program may become so strongly represented that it can be carried out with no further checking by the modelling process.

The third syndrome to be dealt with here, *conduction aphasia*, was predicted by Wernicke, on the basis that there must be connections between Broca's and Wernicke's area, which could be damaged without damage to either of the cortical regions. Nevertheless, this variety of aphasia is not universally recognised, or its occurrence is said to be rare. Benson (1979) however, regards it as a distinct syndrome occurring in 10–15% of cases of aphasia. As a discrete entity this syndrome can be produced by relatively small lesions of the inferior parietal lobe. Lesions confined to Wernicke's area can sometimes produce the same syndrome. For the inferior parietal lesions, Benson admits that empirically it is undecided whether these lesions need only affect the white matter bundle known as the arcuate fasciculus, or also require, as an essential component, damage to the cortical tissue in this region. According to Benson (1979), in conduction aphasia, comprehension is normal or almost so. Speech is fairly fluent, and is not excessive. It may have a broken, dysprosodic character, with pauses and hesitations, apparently for word finding. Paraphasias occur, as in Wernicke's aphasia. There is however a serious problem in repeating spoken language, and the hallmark of conduction aphasia is the great difference between this and the almost intact comprehension. Patients who fail in a repetition test may convey the meaning of the sentence they are supposed to repeat by creating a paraphrase with similar meaning. In addition to repetition deficits, there is a deficit in confrontation naming.

If the hypotheses in Chapter 8 are correct, damage to connections between Broca's and Wernicke's area, without either area itself being damaged, should permit time-structured executive messages to be emitted from the executive part of the language system, their sensory consequences to be perceived when the sound is uttered, but the matching against the spatio-temporal template of the sound should be deficient. Hence the same paraphasias as in Wernicke's aphasia should occur, betraying a difficulty in preparatory modelling of words in advance of their utterance. The defect in repetition, and in naming, is likely to be due, as in Wernicke's aphasia, to deficient conduction of messages from the auditory association area, where the sentences are decoded (or the name identified) and the premotor area, where the program for articulation of the sentence or name is produced. Since comprehension is intact, a paraphrase of the sentence to be repeated can be constructed. Presumably this is achieved by far less direct routes to Broca's area than the arcuate fasciculus, for instance utilising the semantic representation which is possible in the right hemisphere. Conduction aphasia following lesions restricted to Wernicke's area may arise because in these cases comprehension is adequately served by the right auditory association area, and so remains intact, though disconnected from Broca's area. Thus, a prediction can be made for future PET studies. In conduction aphasia, spontaneous speech should show the usual activation of anterior regions such as Broca's area, but the activation of Wernicke's area, described for normal subjects in section 9.2.1, should not occur.

Experimental studies contribute to the understanding of aphasias. A report already mentioned, provides support for a role of posterior peri-Sylvian cortex in modelling of motor acts related to speech: Ojemann and Mateer (1979) performed experiments during neurosurgery, in which they charted the regions of the left hemisphere at which different linguistic functions could be disrupted by application of electrical stimuli. Subjects were asked to produce sequential oro-facial movements (e.g. a sequence of elements, such as lip protrusion, initiated by showing a series of slides depicting each action). This task was disrupted not only by stimuli applied to the inferior frontal region, but also by those applied in the superior temporal/inferior parietal area. Sequential oro-facial movements and phonetic identification were both disrupted by stimulation of both of these regions. This is a good demonstration of the interchange of functions usually attributed individually to these two regions (see Figure 8.8 and section 8.2.3.8 for theory underlying this).

Experimental studies also contribute to understanding the perceptual side of language. Tallal and Newcombe (1978) found that impairment in the perceptual task involving recognition of CVC stimuli correlated highly with the degree of language comprehension impairment.

A number of studies report on delayed auditory feedback (DAF) effects in aphasic patients. Early studies, mentioned by Boller *et al.* (1978) find abnormal DAF effects, with one study (Alajouanine *et al.*, 1964) finding complete absence of a DAF effect in fluent aphasics. Nagafuchi (1980) tested delayed auditory feedback in a population of mixed aphasic (left hemisphere) and right hemisphere-damaged patients. The disruptive effect was decreased if the delayed signal was fed back to the damaged compared with the intact hemisphere. This was however found equally for right and left hemisphere-damaged patients. No differences were attributed to Broca *vs.* Wernicke aphasia. However, the disruption by DAF was clearly a normal effect, requiring intact hemispheres. More specific findings were reported by Boller *et al.* (1978), and by Chapin *et al.* (1981). These studies found that the disruptive effect of 180 msec-delayed auditory feedback was greater than normal for Broca's (nonfluent) aphasics, less than normal for Wernicke's (fluent) aphasics, and much less than normal for conduction aphasics. Presumably for Broca's aphasia the difficulties in articulating connected speech were already so difficult that any added hindrance (such as an imposed mismatch between what is uttered and what is expected to be heard) would have a completely disruptive effect. For Wernicke's and conduction aphasia, the executive part of the speech program is presumably as fluent as in normal individuals, but is not subject to the segment-by-segment checking. Delayed feedback then does not produce a mismatch, any more than ordinary feedback produces a match, and so there is no disruption in the former case.

Apart from the classic aphasia syndromes, the suggestion has been raised (section 8.3.2.8) that the speech production areas (Broca's and surrounding areas) might play a part in speech perception, especially when it demands precise phonetic identification. Evidence for this proposition from electrical stimulation of the brain surface was mentioned in section 9.2.2. Further evidence comes from study of left hemisphere-damaged subjects. Blumstein *et al.* (1977) studied phonological discrimination in a variety of left hemisphere-damaged patients. The task involved same/different judgements in real or nonsense words differing in only a single consonant. There was significant impairment in patients with anterior as well as posterior lesions. The impairment did not co-vary with comprehension deficits. For instance, many patients who were aphasic as a result of anterior lesions had worse phoneme discrimination, but better speech comprehension than patients with posterior lesions and Wernicke's aphasia. Presumably speech comprehension is often possible without exact phonetic identification, as we have seen in considering language comprehension by the right hemisphere. The prefrontal cortex appears to play an important role in such phonetic identification. The exact location of the lesions was not reported however, so more precise analysis of these result was not possible. Efron (1963b) reported experimental evidence of relevance in this context. Subjects were presented with a sequence of two 10 msec tones of low and high pitch binaurally, with a temporal order judgement required. Non-aphasic head-injury patients (presumed to have a right hemisphere injury) achieved 100% correct responses when the temporal separation was little more than 100 msec. Aphasic patients (with presumed left hemisphere injury) required a much longer separation than this, under some conditions two- to three times as long. The expressive aphasic patients were more impaired on this test than the receptive aphasic ones, though both subgroups were clearly impaired compared to the non-aphasic patients. This may also be an indication of the importance of left anterior cortical tissue in phonological discriminations.

Section 8.3.2.8 also raised the possibility that phonetic identification by the frontal lobe could be used for addressing semantic content, giving such regions a role in semantic representation usually attributed to more posterior regions. PET evidence in normal subjects compatible with this (Petersen *et al.*, 1988) was mentioned in section 9.2.2. Further evidence is available from the study of aphasic patients. Milberg *et al.* (1987) conducted semantic priming experiments (see sections 7.3 and 7.4) in left hemisphere-damaged patients. Wernicke's aphasia patients took longer to make word/non-word decisions than did controls, but semantic priming (speeding of the decision by semantically related "primes") was nevertheless intact. Patients with Broca's aphasia, on the other hand, had lost the capacity for semantic priming. The precise interpretation of these results is uncertain. However, the authors suggest that the role of Broca's and surrounding areas in these experiments may not be in semantic representation *per se*, but rather in the use of phonetic representation to access the appropriate word meaning.

9.4.4 The right hemisphere and motor control by sensory guidance: clinical evidence

Fisk and Goodale (1988) showed that patients with right hemisphere damage had normal movement times, but took longer to initiate a movement than did controls. This supports the conclusion (section 9.3.4) that the right hemisphere determines speed of reaction, and is superior in this compared with the left. However, this tells us little about the specialisation of the right hemisphere in motor *co-ordination*.

From evidence mentioned above for brain-injured subjects (Wyke, 1968; Haaland and Harrington, 1989a), the suggestion that sensory guidance of limb movement is superior with the right hemisphere receives some support. Admittedly the paradigm of Wyke (1968), involving aiming at a target on a moving turntable is rather problematical because it appears to depend on open-loop control in normal subjects, but visual guidance in the brain-injured subjects. This discrepancy may however result from the fact that the task in normals requires extrapolation of target position, as well as open-loop control to the target. Brain injury may impair the process of extrapolation, and thus change the task from an open-loop to a visually-guided control task.

In a smooth pursuit eye movement task Bogousslavsky and Regli (1986) found impairment for all lesions in the parieto-occipital lobes, but this was severe only for lesions in non-dominant hemisphere. The lateralized deficit was limited to contralateral pursuit. (Ipsilateral pursuit was similarly impaired by lesions in either hemisphere). The impairment with non-dominant hemisphere lesions was not due to neglect alone, because voluntary saccades were preserved. Since smooth pursuit eye tracking is a closed-loop form of movement, this result strengthens the evidence that the right (non-dominant) hemisphere can perform such movements better than the left.

These reports, together with the data from normal subjects summarised in section 9.3.4, give substantial support to the view that visuomotor control involving sensory guidance is carried out better with the right than the left hemisphere. This conclusion applies to visuo-oculomotor control (smooth pursuit) as well as limb and hand control by visual guidance. The data on auditory-motor control from normal subjects was far less satisfactory, though it did not decisively refute the idea that the right hemisphere has a superiority in sensory guidance of motor responses. It is interesting in this context to mention evidence of speech abnormalities in cases of right hemisphere damage. The distinguishing feature of the speech of such right hemisphere patients is the abnormality of prosody, rather than word and sentence production (e.g. Tucker *et al.*, 1977b; Borod *et al.*, 1985). Ross (1981) has defined various classes of aprosodia (motor, sensory etc.) after right hemisphere lesions, which he believes correspond to, have the same localisations as, and are as common as the different types of aphasia seen after left hemisphere damage. Amongst these cases of aprosodia Ross reports one case, with right temporal lobe lesions, in which he suspects that inappropriate and excessive prosody were present (corresponding, Ross believes, to fluent empty speech of Wernicke's aphasia). Such inappropriate and abnormal prosody after right hemisphere damage is also described by Shapiro and Danley (1985) and by Shields (1991).

The interpretation of the abnormal prosodies after right hemisphere damage must remain uncertain until further experimental analysis of this type of deficit has been carried out. One of the possible interpretations is that the right hemisphere may be more important than the left in representation of emotions. This topic is considered in more detail in Chapter 10. It is relevant to point out here, however, that the abnormalities of prosody after right hemisphere lesions apply to prosody used for semantic purposes (e.g. emphasis) in non-emotional utterances, as well as for expressing emotion (Ross and Mesulam, 1979; Weintraub *et al.*, 1981; Shapiro and Danley, 1985).

Apart from this, it is worth mentioning that the temporal structure of prosody is on a much longer time scale than that of the verbal production. In both the right hemisphere/prosodic syndrome and in Wernicke aphasia, there appears to be a failure to monitor the actual speech output in comparison with a central model of it. However, in the case of the left hemisphere lesion, the monitoring fails at the phonemic and syllabic level; whereas for right hemispheric lesions, it fails with respect to aspects of speech output on a much longer time scale. It is plausible to suggest that prosodic elaboration of a sentence forms part of the model constructed in and directed by the left hemisphere. However, in actual execution of the prosody, it is possible that auditory-motor guidance is involved. Detail from one of the cases reported by Ross and Mesulam (1979), with right fronto-parietal damage supports this view. The patient, a school teacher had, along with absence of prosody, a difficulty in projecting her voice in the classroom. Although words and sentences could be articulated clearly, she could not match the intensity and acoustic qualities of her voice to the acoustics of a large room. This seems to imply a loss of immediate auditory feedback, which did not affect word production (as it does in Wernicke aphasia) which is modelled segment-by-segment, but does affect melodic properties of speech controlled on an instant-by-instant basis. Abnormalities of prosody after right hemisphere damage may thus reflect a failure in auditory-motor guidance. If this could be proved, it would provide further support for the generalisation that the right hemisphere is specialised for sensory guidance.

9.4.5 Summary and discussion of clinical evidence

For those forms of apraxia giving rise to bilateral symptoms after left hemisphere damage, the balance of evidence suggests a disruption of preplanning of those rapid movements which are normally performed without moment-by-moment sensory guidance. Kimura (1977) suggests that the aspect of motor control which is left-lateralized is the "rapid positioning of a limb or its parts in terms of kinesthetic control towards internalised representations of target loci". The same was envisaged to apply in apraxia of oro-facial musculature. These conclusions are close to the hypothesis put forward in Chapter 8, that in the left hemisphere the motor program consists of time-structured models in which a spatio-temporal template of the expected consequences of an action can be matched against a memory trace of kinesthetic feedback from the actual movement. Parapractic movements in apraxia indicate some loss of the correct selection and sequencing of components of movement, which is compatible with failure in preplanning and testing of combinations of elementary movements before the action is carried out.

Attempts have been made to find an animal model of apraxia. Deuel (1977) found that monkeys with lesions of anterior cortex (premotor cortex and arcuate sulcus) are poor at carrying out sequences of actions which they can correctly perform individually. Moll and Kuypers (1977) found that removal of the premotor cortex in monkeys renders them incapable of a variety of "detour problems", that is reaching round a transparent object to retrieve a food pellet. The tasks in both of these studies involved modelling an action in advance of performance. However, it was not required that the tasks be performed at speed, and so it is unlikely that *time-structured* modelling was involved. Thus it is doubtful that the impairments shown are really models of the classic syndrome of apraxia in humans.

Descriptions of more complex phenomenology are available concerning aphasias. Broca's aphasia appears to be dominated by an inability to produce the executive part of a speech program with normal fluency, as might be expected when the chaining of motor elements into a flowing stream is disrupted. Wernicke's aphasia seems to be dominated by inadequate registering of the mismatch between the auditory implications of a planned motor program and the intended speech sounds. Paraphasias (found in conduction as well as Wernicke's aphasia) may indicate failure to check the correctness of the combinations in the plan of an utterance against the actual utterance. In this context it can also be mentioned that *spoonerism* in normal adults is known to obey certain rules (MacNeilage and Davis, 1990) which indicate the constraints on the combinatorial processes which are systematically used in creating programs for multisyllabic words.

Apart from the interpretations of aphasia given above, there is a possibility that some of the phenomenology reflects the capacity of the intact *right* hemisphere for speech. Kinsbourne (1971) has provided evidence using the Wada test that aphasic speech is sometimes generated by the right hemisphere. There is indeed a similarity between the speech of the isolated right hemisphere (in commissurotomy or hemispherectomy cases) and Broca's aphasia, both of which have short phrase length and lack of fluency (see reports of two cases of dominant hemispherectomy reported by Gott (1973) and Smith (1966)). However, this idea cannot be developed in more detail without further evidence similar to that provided by Kinsbourne.

Overall the clinical evidence provides support for the idea that the left hemisphere's superiority in motor control, both for manual praxis and for speech, is in time-structured modelling. One study of a sequential saccadic task in patients with unilateral lesions of the supplementary motor area shows that this concept may also apply to eye movements. The enhanced disruptive effect of delayed auditory feedback in Broca's aphasia, and its reduced effect in Wernicke's and conduction aphasia, provide strong support for the interpretation suggested for the different aphasic syndromes, based on the hypothesis of time-structured modelling during the planning stage for speech.

These conclusions suggest that apraxia and aphasia are similar deficits of time-structured modelling. From this one might predict that separate apraxic syndromes similar to those of aphasia may exist. Admittedly such a parallel cannot be complete: The modelling whose disturbance is envisaged to lead to apraxia is based on strictly proprioceptive sensory information, which cannot be part of a social communication system. Thus there can be no counterpart to a linguistic comprehension deficit when considering apraxia.

Hence a distinction amongst apraxias corresponding to that between Wernicke's and conduction aphasia is unlikely to be found. However, apraxias corresponding to Broca *vs.* Wernicke aphasia might be distinguishable with (respectively) frontal or parietal lesion localisations.

Kimura and Archibald (1974) believe that the unique function of the left hemisphere is in motor sequencing, and this applies both to speech and non-speech. Accordingly, the left hemisphere has, in their view, no special function in symbolic or language function. Thus, the region of the body which shows apraxia is also the one where there may be sequencing problems, whether in spoken language or gestural language. A few studies report on the co-variation of apraxic and communicative motor deficits, for oral and limb effectors.

De Renzi *et al.* (1966) have reported that performance in oral apraxia tests (whistling, yawning, teeth chattering etc.) are independent of limb apraxia, but correlate with Broca's aphasia. Tognola and Vignolo (1980) also found that oral apraxia (failure to imitate facial gestures) was almost always associated with aphasia (in 27 out of 28 cases of oral apraxia). Some cases of aphasia (usually fluent aphasia) occurred without oral apraxia (4 out of 12 cases of left hemisphere lesion without oral apraxia). However, these had smaller lesions, without involvement of frontal lobe. Thus, oral apraxia, though correlated with aphasia, appeared to be indicative of a more profound disturbance than aphasia, seen only with quite large lesions, and usually accompanied by a global aphasia.

In comparison, for limb co-ordination, Kimura and Archibald (1974) have found a strong relation, in left hemisphere patients, between the ability to make meaningful verbally instructed motor gestures (the traditional test of apraxia), and the ability to copy sequences of hand movements (such as might be used in sign language). These correlated deficits are found using either hand. They are not explained by deficits in visual perception of movement to be imitated. Imitating static hand postures, or single finger flexion did not correlate with the apraxia test.

Thus, these studies find correlations between different tests of skill with the same effectors, but not between analogous tests with different effectors. However, although deficits to different effectors may be dissociated by lesions in different premotor regions, these deficits may still have a close formal analogy. Luria (1973, pp. 180–185) makes an explicit comparison between Broca's aphasia (due to damage to the part of the left premotor cortex representing the vocal organs) and the motor deficits of limb co-ordination (after damage to the more medial parts of the premotor cortex). The premotor cortex as a whole is seen as responsible for the "conversion of individual motor impulses into consecutive kinetic melodies". Hence damage to each part of the premotor cortex leads to a disturbance of skilled movements in the corresponding body part, which are no longer performed smoothly, so that "each component now requires its own isolated impulse." With left sided damage this is shown in both upper limbs, and in speech. In each case, the intention to perform the movement or vocal gesture remains intact, as well as the general plan, but the actual execution of the plan is disrupted by "inertia of motor stereotypes" and "motor perseveration" (cyclic repetitive movements). These two features are seen in both Broca's aphasia and the limb co-ordination difficulties which arise from lesions of the premotor cortex.

Luria (1973, pp. 170–175) also identifies a special syndrome arising from the disturbance of the afferent organisation of movement ("afferent apraxia"). In such cases, movements are carried out, but in a coarse manner lacking precision. This is found with lesions of the parietal lobe, and has an obvious correspondence to Wernicke's aphasia. Luria does describe in these terms the disorder of speech resulting from damage to the *left* inferior parietal lobe, but does not describe any left-lateralized form of "afferent apraxia" for the limbs. There are however suggestions that apraxia of the limbs is sometimes comparable to disturbances of motor co-ordination in Wernicke's aphasia. For instance paraphasias may be a linguistic parallel to parapraxias, seen in apraxic disorders. The similarity of these two betrays disruption of a common underlying programming mechanism, hypothetically the previously-elaborated time-structured models. It would be interesting to seek further support for Luria's elegant and precise clinical descriptions with a more quantitative experimental approach to these issues.

A further way in which the analogy between aphasias and limb apraxias can be explored is by study of aphasia in deaf persons who communicate with sign language. The few cases of sign language aphasia which have been reported are sufficient to tentatively define equivalents of Wernicke's and Broca's aphasias. Chiarello *et al.* (1982) reports one case and Bellugi *et al.* (1983) report two more with lesions mainly in the parietal region where sign language is retained with normal fluency and grammatical structure, but with paraphasias (identified as mainly literal in one case and semantic in another). Comprehension of signed commands was impaired in one of Bellugi *et al.*'s (1983) cases. These are presumably the sign-language analogy of Wernicke's aphasia.

In contrast, a third patient of Bellugi *et al.* (1983), with a massive prefrontal lesion showed a severe deficit in fluency of signing. This particularly affected phrase length and spontaneous sign articulation, but not repetition of signed communication. Comprehension was normal and there were no paraphasias. This case may be a sign language analogue of Broca's aphasia.

The correlation between sign language aphasia and apraxia appears to be variable. Kimura *et al.* (1976) studied hand movements in an aphasic deaf signer. They found no impairment in single isolated hand movements, but a deficiency in imitation of complex movements. The patient of Chiarello *et al.* (1982) showed abnormality in copying finger gestures, but not in typical tests of apraxia using the whole arm. Corina *et al.* (1992b) found preservation of non-linguistic gestures, despite pronounced sign language aphasia. Further evidence that sign language and manual praxis are closely related comes from a study of Vaid *et al.* (1989) in normal signers: In speeded signing, right handers did better with right hand, but in left handers signing superiority also shifted to the left hand. The variations in the relation between sign language aphasia and apraxia probably depends on the exact lesion location.

If the mechanisms of normal manual praxis and speech production are analogous, an important prediction can be made for cerebral activation studies (e.g. with PET) in normals. When a time-structured motor program of the limb (such as a rapid reaching movement) is being prepared, it should activate not only the frontal regions, but also parietal regions where modelled kinesthetic consequences are envisaged to be represented. This would be a finding analogous to the activation of Wernicke's area during imagined speech (see section 9.2.1). One regional cerebral blood flow study in the literature does use the task of imagining a hand movement (Roland *et al.* 1980). In this study only the supplementary motor area was activated, and not the parietal association cortex. However, in an EEG study (Beisteiner *et al.*, 1995) both imagined as well as real hand movements were found to be associated with increased negativity in the parieto-temporal (as well as frontal) regions of the left hemisphere. It would be interesting to make further investigations with such imagined movements using PET.

Drawing such parallels leaves one with some uncertainties and queries. For instance, why is Wernicke's aphasia excessively fluent? There appears to be no comparable hyperkinesis in apraxias after parietal lesions. A possible reason for the excessive fluency in some cases of Wernicke's aphasia, but not in parietal apraxia could be that for production of paraphasic speech there are few mechanical constraints, once the constraints of matching auditory feedback to perceptual images is lost. On the other hand, excessive parapractic movement may run into difficulties because the limbs move best only when subject to optimal motor control. An additional reason for excessive fluency sometimes found in Wernicke's aphasia is that speech production may in itself have a rewarding component, so that when the normal constraints imposed by perceptual feedback are lost, an increase in volume of paraphasic speech may occur.

An additional theoretical question is why the apraxic deficit from a dominant hemisphere lesion applies to both hands. Assume, according to the hypotheses of Chapter 8, that the superiority of the left hemisphere arises from a richer repertoire of delay lines from premotor to parietal cortex in the left than the right hemisphere. Assume also that the premotor region of the left hemisphere can exert some influence over the left (ipsilateral) hand as well as over the right. For a left-handed movement initiated in the right premotor cortex, kinesthetic feedback will arrive in the right parietal region too late to be capable of being matched with any motor-sensory template transmitted there by cortico-cortical pathways, since these pathways are rapidly conducting ones (according to hypothesis). However, for a left-handed movement, initiated from the ipsilateral (left) premotor cortex, kinesthetic feedback reaching the right parietal area can be transmitted via the corpus callosum to the left parietal region, and, once there, is able to form a match with the corresponding motor-sensory template, just as for a right handed movement. These relationships are depicted in Figure 9.1. The essence of this figure is that, for the purposes of time-structured modelling it matters little whether the peripheral loop is ipsilateral or contralateral, provided that, on the left side, there are motor-sensory delay lines slow enough to coincide with the returning kinesthetic signal.

9.5 OVERALL SUMMARY OF MOTOR LATERALIZATION

The predictions of Chapter 8 receive strong support from the evidence reviewed in this chapter. Clear evidence exists from the study of language acquisition suggesting the slow formation of time-structured models. This includes data on acquisition of sign language as well as speech. A suggestion is made that similar phenomena may be found during the acquisition of dextrous motor control by normal subjects. Also in accord with the predictions of Chapter 8, there is good evidence from functional brain imaging and other methods for a reciprocal interchange of functions between premotor and sensory association areas involved in production and perception of speech. From the survey of normal subjects, there is very strong support for the view that the right hand has an advantage in rapid aiming movements based on more accurate open-loop programs, rather than the more accurate visual guidance. Evidence from brain-damaged subjects, though not so strong, still points mainly to the same conclusion. Lateralization of saccadic eye control

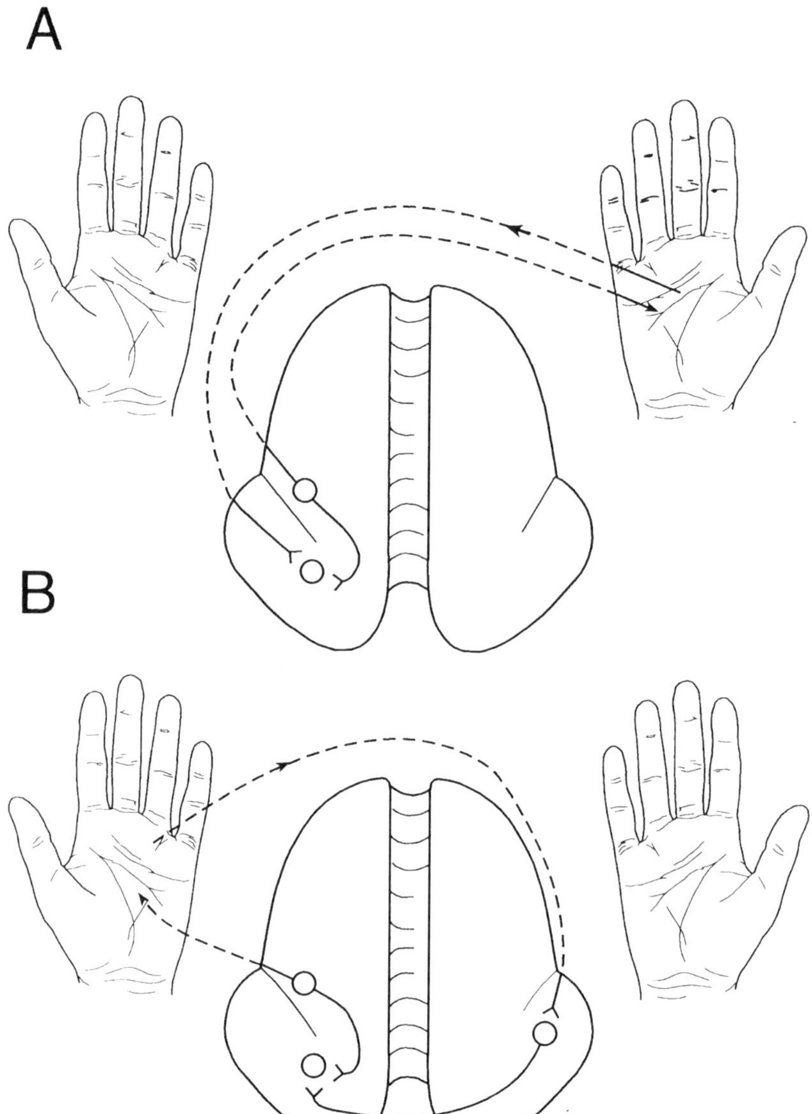

Figure 9.1 Control of Left (as well as right) Hand by Left Hemisphere. In **A** the left hemisphere controls the right (contralateral) hand. The left hemisphere/right hand superiority arises from the fact that the motor-sensory connections from hand motor cortex to hand parietal cortex in the left hemisphere are slow enough to be able to coincide with the kinesthetic feedback signal from the controlled hand. This is not possible for right hemisphere/left hand combinations, where cortico-cortical delays are less. Thus, a time-structured model of the hand movement can be constructed only in the left hemisphere (see Chapter 8). In **B** the left hemisphere controls the left (ipsilateral) hand. Coincidence between the motor-sensory signal and the kinesthetic feedback signal is still possible, given that the latter is relayed from right to left parietal region along relatively fast conducting axons, and the same set of motor-sensory delay lines are used as in **A**.

in normals does not adequately evaluate the hypothesis of superior open-loop programs in the left hemisphere, though clinical evidence supports this. Evidence using the delayed auditory feedback paradigm gives some support to the prediction that matching of an internal and a feedback representation of auditory consequences of an action is important for preplanned performance, especially of speech, though the effects shown are capricious. Study of aphasias, by both clinical and experimental methods can be interpreted in terms of motor-sensory cortico-cortical delay lines being the substrate for preplanning of speech programs. Some of the evidence using delayed auditory feedback in aphasic subjects is particularly striking in this respect.

There is also substantial evidence supporting the prediction that the right hemisphere is superior in motor acts co-ordinated by sensory guidance, using the visual sense for eye or hand co-ordination, though there are also some negative findings. Clinical evidence tends to support this, including intriguing details of the clinical evidence of abnormal prosody after right hemisphere lesions, suggesting that the abnormalities involve loss of auditory-motor guidance.

Future work in this area may be able to clarify the hemispheric involvement in auditory-motor control, and strengthen the small body of evidence for lateralization of eye movement control. Clinical and PET cerebral activation research may be able to reveal the extent to which manual praxis in normal subjects and the deficit of this in the apraxias has parallel phenomenology and underlying mechanisms to normal speech and the abnormalities seen in the aphasias. It is also becoming possible to investigate some of these issues with EEG methodology. Asymmetries of the EEG during processing of visually presented word *vs.* non-word stimuli have recently been reported (Lutzenberger *et al.*, 1994). Similar methods might be applicable during speech production.

Finally, it seems likely that the full range of motor specialisations of the right hemisphere has not yet been discovered. Control of the human motor apparatus has been considered by Bernstein (e.g. 1967) to consist of complex synergies between effectors acting at different joints, and different limbs. This would seem to be an aspect of motor control for which the right hemisphere should be superior, just as it is for representing the more complex perceptual Gestalts. This has never been studied in a systematic way. However, it has been noted that for most facial gestures the left face of right handers is more expressive than the right face (Chaurasia and Goswami, 1975; Borod and Caron, 1980), a possible sign that the right hemisphere can recruit more complex synergies than the left. It has also been noted that right brain-damaged subjects may lose the ability to use gesture and body language as part of their communication style (Ross and Mesulam, 1979). This could be an indication of loss of the more complex motor synergies.

10 Laterality Effects for Higher Cognitive Processes: Short Term Memory, Attention and Alertness, and Emotion

10.1 INTRODUCTION

Chapters 5 to 9 dealt with lateralization of basic processes of perception and motor co-ordination, such as could be studied with stimuli or motor tasks lasting a small fraction of a second. Considerable support was found for the idea that perception and motor control achieved by the left hemisphere incorporates temporal structure, contained in the stimulus or defined by the task requirements, in a superior manner to that achieved by the right. The right hemisphere on the other hand is capable of synthesising Gestalts, which are spatially complex, though temporally unstructured, better than the left hemisphere and is also faster in reaction time tasks, and in tasks involving moment-by-moment sensory guidance. These generalisations can be derived from the central hypothesis put forward in section 3.6, according to the specific predictions derived in Chapters 5 and 8.

Processing of incoming signals does not stop after the first few hundred milliseconds required for perception. Higher level integrations are also possible, achieved over longer intervals of time than the rapid temporal integrations of perceptual psychology. Likewise, preparatory modelling of motor output is not limited to the few hundred milliseconds ahead of the time when the model is constructed: It may extend further forward in time by much longer durations. Proceeding from the conclusions of earlier chapters, one might therefore expect that the differences in basic processes of perception and motor co-ordination between the hemispheres might have implications for higher level functions of each hemisphere, involving integration and modelling of information over longer spans of time. The specific topics to be dealt with in the present chapter are *short term (working) memory, attention and alertness, and emotional representation* in its various forms. All of these involve information processing on a longer time scale than immediate perception or immediate motor preplanning.

These longer intervals are generally classed as "short-term memory" (STM) intervals. As a psychological function STM is believed to have a limited capacity. The exact time span of STM has different definitions according to different authors. For the purposes of this book (following Blumenthal, 1977) STM intervals are taken as ranging from about 1 second up to a few tens of seconds. STM is closely related conceptually to "working memory". The implication of the latter term is that comparisons, and other integrations can be carried out between components of the input arriving at different times within the duration of STM. Such operations make up part of the "work" of working memory. In addition, forward planning of motor performance, for forthcoming times which lie within the duration of STM, can also be included as examples of the work of working memory. Lateralization of the process of short-term memory and the work of working memory are discussed in section 10.2.

Of course, every component stimulus which can be integrated within short-term memory will have its own characteristics which may confer lateralization even before STM comes into operation. This makes it difficult to identify laterality effects in STM itself. Therefore, empirical investigations of laterality within the ambit of STM might address two questions.

The *process* of STM might differ between the hemispheres. If this were to be the case, there should be examples where the hemispheric advantage to be expected from the nature of the stimuli themselves is reversed when an added STM component is introduced into the task. The latter pattern would indicate a laterality of STM *per se*, either in its capacity or its duration. Alternatively, there may be no laterality effect for the process of STM as such, in which case short term memory hemispheric advantages will be governed exclusively by the hemispheric advantage which applies to the component stimuli when used without an STM component.

Apart from the process, one may compare *what is achieved* by the integrations of short term memory in each hemisphere, in other words by the *work* of working memory. Given that STM can achieve an integration between components of input up to many seconds apart in time, the products achieved by STM may be expected to be strongly influenced by whether the components to be integrated have a temporal structure to start with.

The subjects of short-term and working memory are closely related to the concept of *attention*, though that is an all-embracing term covering several identifiable functions, whose distinction is often obscured by use of the single the term "attention". Attention is commonly defined as achieving a *selection* between competing channels of information, enhancing the capacity of the brain to deal with the class of information so selected. It is however related, at least in popular word usage, to concepts such as *arousal* or *alertness*, which also refer to an enhancement of information processing capacity, but do not have the implication of a selective enhancement. This complex of high level processes is dealt with in section 10.3. It is found that there are important differences between the hemispheres in mechanisms of general arousal, which can be related directly to the central hypothesis of this book. Apart from this issue, the possible lateralization of the *selective* aspects of attention is discussed. A viewpoint is mentioned that hemispheric differences are not rooted in permanent structural differences between the hemispheres, but reflect lateralization of attentional focus to each hemisphere. This view, which contrast with that advocated in the present work, has been defended by M. Kinsbourne (and several others) whose work is discussed below.

Finally in this chapter (section 10.4) the contentious questions of lateralization of perceived emotion and of normal variations in mood states are discussed, together with the influence of mood and emotional changes on lateral differences in perception. Since emotional perception and expression wax and wane on a much longer time scale than immediate sensory perception or motor co-ordination it is appropriate to discuss these issues at this stage. However, it should be pointed out that the questions of emotional representation are at present only partly resolved, though they may eventually find their correct interpretation in terms of the lateralization of the other high level cognitive functions discussed in this chapter.

10.2 SHORT TERM MEMORY IN THE TWO HEMISPHERES

10.2.1 Theory of short term memory

The brain mechanisms upon which the psychological function of short term memory is based are not fully understood. An electrophysiological observation which might be an indication of its underlying biology, (already mentioned in section 8.3.1.5) is that many neurones in the frontal regions of cortex can maintain elevated levels of activity in tasks where there is a delay of several seconds between events. In humans, various anticipatory potentials (such as the "contingent negative variation" (CNV) or the "readiness potential (Bereitschaftspotential)) are recordable with gross scalp electrodes. These probably represent the summed potentials in many of the frontal delay neurones such as are recorded in single unit studies in experimental animals. These anticipatory responses can be maintained throughout the delay between events, even though no overt stimulus is driving them during that period.

The factors which lead to such maintained firing are not established yet. However, the suggestion has been made (Miller and Wickens, 1991) that such firing depends on reverberatory activity in cortico-striatal loops. The mechanism proposed could register the sequence of events within short-term memory intervals, but not their exact timing. Thus, a good example of the sort of association this mechanism would be suitable for representing is the expectancy of thunder, once a lightning flash has occurred. Such registration of sequence is seen as a more general aspect of the same process which registers relations between a behavioural act and its consequences, as was discussed in Chapter 8 (see also: Miller and Wickens, 1991).

According to Miller and Wickens (1991) the linking together of the sequence of events within short term memory is envisaged to depend on biophysical processes in the striatal output neurones. To recapitulate material from Chapter 8, activation of such neurones is supposed to leave a trace within the synapses generating the activation, which lasts for a period of the order of 1 second. Significant events occurring regularly, in the period before that trace has been dissipated, are envisaged to lead to strengthening of the synapses rendered "eligible" when the trace is set up. Such strengthened cortico-striatal synapses can potentially lead to reverberatory activity in cortico-striatal loops of connections. In more detail, these loops would consist of selections of neurones in the cortex projecting in turn to others in the striatum, pallidum, thalamus and back to the cortex. These dynamic configurations of neurones were referred to as "cortico-striatal cell assemblies" by Miller and Wickens (1991). Such self-activating loops are set up by virtue of the regular sequential occurrence of two events. The first establishes the trace in striatal synapses, the second, a more "salient" event, activates a reinforcement mechanism which strengthens the connections which led to the establishment of the trace. Strengthening of these synapses means that, in future, the first event will set into activity a self-maintaining circuit of activity during a delay. The pattern of reverberatory activity represents an expectancy that the second event will occur following the first. What is more, once the pattern of reverberatory activity is set up it can be maintained much longer than the duration of the trace which was critical in first registering the sequence of events. It should nevertheless be emphasised that this mechanism sets up representations of sequence, not of precise timing. Figure 8.1 depicted a version of this self-maintaining circuit.

The central hypothesis of the present book is however not just about sequences, but about their exact timing relationships, initially on a smaller time scale than short term memory, but by extension to longer time periods. Lateralization of function, postulated on the basis of differing axonal calibre spectra in the two hemispheres, has no implications for this proposed mechanism of short term memory. If any difference between sides does exist in the *process* of short term memory, it must therefore require an additional and quite separate postulate, and the theoretical side of cerebral lateralization would correspondingly lack parsimony. Nevertheless, there may be important differences between the hemispheres in the *work achieved* in short term memory, even if the underlying process may be the same on each side. This is discussed next.

Short term memory is supposed to register sequence but not accurate timing. However, in the left hemisphere accurate temporal structure can be represented over intervals which, at their longest, encroach upon short term memory intervals. These are the intervals between salient events whose sequence is envisaged to be registered by the quite different mechanism involving cortico-striatal interplay. Therefore the work of short-term memory in the left hemisphere could potentially include the building-up of long accurately-timed linear chains of information, lasting much longer than the time-structured integrations and models dealt with so far, that is for the many seconds over which STM intervals endure. The same should hold true whether the basic material is linguistic or not, and whether in visual or auditory modality.

In the right hemisphere however, there appears to be a much weaker capacity for representing fine temporal structure, and accurately timed sequences cannot be constructed which are long enough to reach the duration of short-term memory intervals. What *can* be represented in right hemisphere perception are Gestalts where the various parts occur simultaneously. There may also be a richer capacity in the right hemisphere than the left for association between such Gestalts in a manner not determined solely by matters of sequence, though this remains to be proven. In any event, it may be envisaged that each such right-hemisphere Gestalt be held in a short term memory form, which is physically realised by reverberatory cortico-striatal loops which are static, rather than time-structured. Within short term memory, associations can be formed between different Gestalts so represented. Thus, it is possible to link Gestalts activated at times separated by several seconds. However, as the different Gestalts become linked, there is no mechanism for registering the precise timing of their sequence of activation. Thus, when short-term memory processes are combined with the basic perceptual mechanisms of the right hemisphere a quite different sort of integrative activity will be achieved from that in the left hemisphere. One would predict that short term memory in the right hemisphere should specialise in large scale meanings, synthesised from Gestalts and other meanings, each activated as much as several seconds apart. Such large scale meanings would not however contain precise *temporal* structure such as is represented in the left hemisphere.

In an earlier chapter a crucial temporal relationship was specified, upon which many aspects of hemispheric lateralization depended, especially in perceptual tasks. This was the relationship between the neuronal integration time controlling Hebbian synaptic modification (in the range 10 to 20 msec), and the effective delay times commonly occurring in cortico-cortical axons. If the latter was much greater than the former, chains of neuronal activity could be set up which would have precise temporal structure within them. If the two intervals were more nearly comparable, the chains of neuronal activity which could be set up would contain spatial information, but not temporal structure.

In the previous paragraphs we have just mentioned a second temporal comparison which has an equally decisive influence on lateralization, especially of higher integrative processes. If temporal chains can be established which are of long enough duration to span the interval between salient events as registered in the short term memory mechanism, one has a basis for representing temporal chains of events many seconds in duration. Several "chunks" of neuronal activation which each have fine temporal structure can then be joined together by the short term memory mechanism, to form a continuous temporal chain much longer than is possible without the STM mechanism. If on the other hand, the capacity for temporal chaining is much weaker, or is restricted to intervals much shorter than the duration usually recognised by STM mechanisms, STM integration will contain quite different sorts of information, where there is little temporal structure, on either the large or small scale. Rather the integration is achieved from summing up the constant features in a succession of complex "snapshots" of the world, whose exact timing is irrelevant.

10.2.2 The process of short term memory: is there a difference between hemispheres?

Overall, there is no substantial evidence for lateralization of the process of STM. Most tests of STM using linguistic stimuli (or other stimuli where left hemisphere dominance would be expected) in normal subjects show a right ear or right visual field advantage (Hines *et al.*, 1969, 1973: visual digits; Hines *et al.*, 1976: visual letters; De Renzi and Nichelli, 1975: various verbal and digit tasks; Hannay and Malone, 1976: nonsense words), or no laterality effect (Williams, 1981, Belmore, 1981: word lists, monaural condition). Studies with such stimuli in unilaterally brain-damaged subjects (Samuels *et al.*, 1980: auditory trigrams, five note tonal sequences) show greater impairment for left than right hemisphere damage.

For STM tests using non-linguistic stimuli there is either an advantage to the right hemisphere (left sensory surface) (Dee and Fontenot, 1973: complex nonsense figures; Kumar, 1977: tactile spatial configurations; Suberi and McKeever, 1977: emotionally expressive faces; Polich, 1982: complex random visual forms), or no laterality effect (Lewinsohn *et al.*, 1972; Hannay and Smith, 1979: tactile nonsense shapes; Wilkinson and Carr, 1987: Braille test). Unilaterally brain-damaged subjects are impaired with right but not left hemisphere damage in this class of tests (Samuels *et al.*, 1980: facial recognition; Cremonini *et al.*, 1980: learning picture sequences to criterion). One apparent exception to the above generalisation for right hemisphere superiority is the result of Cremonini (1980) where recognition of picture sequences was more impaired for left hemisphere injury patients. However, performance by these subjects was shown to be correlated with performance on a linguistic test, and so was probably mediated by verbal associations of the pictures. Likewise, Malone and Hannay (1978) found that a right visual field advantage in colour memory in normal subjects disappeared when verbal mediation was prevented.

Admittedly some papers show that the laterality effect which applies for an immediate recognition test disappears when there is an added STM component (Belmore, 1981: word lists, dichotic condition). Some of the above studies reveal an increase in hemispheric advantage for recognition after a delay (Hines *et al.*, 1969, 1973: letters; Dee and Fontenot, 1973: complex figures; Yeni-Komshian and Gordon, 1974: three-word sequences; Hannay and Malone, 1976: nonsense words; Samuels *et al.*, 1980: five-note tonal patterns). However, since none of these results *reverses* the laterality to be expected from the stimulus elements themselves, it is likely that this effect, as well as all other results, is derived from interaction of lateralized representation of each stimulus with a non-lateralized short term memory. Thus *from the papers just cited, there is no evidence for lateral differences in duration of short term memory, nor in its capacity.* The work achieved by working memory in each hemisphere is however a different question, considered next.

10.2.3 Evidence of right hemisphere specialisations in the work of working memory

10.2.3.1 Perception of musical "contour"

Music is a complex form of sensory input, combining many perceptual, motor, cognitive and emotional features, in quite different ways from their combination in spoken language. It should not be a surprise that different aspects of music are lateralized in different ways. We have already found that perception of sounds which are spectrally complex is better using the right than the left hemisphere, although lateralization of the perception of pitch of pure sounds has not always been found to be lateralized (e.g. Gordon, 1978b; Sidtis, 1980 in normal subjects; cf. Robin *et al.*, 1990, in brain-damaged patients). Melody is, amongst other things, a temporal sequence of pitches, and so might exhibit the same lateralization as speech or non-speech patterns of similar duration. Indications of left hemisphere dominance for such strictly temporal aspects of short passages of melody are considered below (section 10.2.4.1). However, there are aspects of a melody other than simple temporal sequence which can be encoded by the brain, especially when the mechanisms of short-term memory are recruited. The principles determining the lateralization of these other aspects need not be congruent with those for shorter temporal sequences. In fact, the perception of the overall "contour" of a melody appears to be achieved better with the right hemisphere.

Several early studies (e.g. Shankweiler, 1966) had shown an overall right hemisphere advantage for melody perception as a whole. Bever and Chiarello (1974) performed an important experiment on this subject, which analysed melody perception into separate functions. The task in their experiment involved tonal melodies of 12–18 notes, each note lasting 300 msec. Two seconds after the end of the melody, a two-note phrase was sounded, which was an excerpt from the previous melody in three quarters of cases. Subjects had to say whether the two note phrase was an excerpt. They also had to say whether the melody they heard was familiar (i.e. heard in an earlier part of the experiment). Non-musicians showed, as expected, a left ear advantage for the second of these two tasks, and they could not perform the first. Musicians however showed a right ear advantage for the second task and could perform the first, equally with either ear. The authors attributed the right ear advantage of the experienced musicians to the use of a different strategy from that used by non-musicians, in recognising what they called melodic "contour". They suggested that this strategy was a more "analytical" one, and was also reflected in the recognition of two-note intervals as excerpts. This subject is pursued further in section 10.2.4.1. The definition of the word "contour" (as that of the word "analytic") was not made clear. Bever and Chiarello assumed that musical "contour" was dealt with by the right hemisphere. Subsequent work has given the word "contour" a variety of different operational meanings.

Peretz (1990) defined "contour" as the sequence of successive rises and falls of pitch along the melodic line, (i.e. regardless of the exact size of pitch changes). The task in her experiment involved same/different judgements for two melodies, the difference between which could be one of transposition of pitch of the overall melody, contour change, and pitch change with contour preservation. In normal and left hemisphere-lesioned subjects the judgement was easier if changes were made to contour than to the other features. For right hemisphere lesioned subjects, impairments were overall greater, than in other groups. However, in contrast to other groups, there was not an additional impairment when the task was made more difficult by preserving the contour of the melodies. Given Peretz' definition of contour, this result supports the contention of Bever and Chiarello (1974), that the right hemisphere is more important for representation of musical contour.

Peretz and Babai (1992) used normal subjects and separated left and right hemisphere functions in melody perception in another way: Melodic phrases were constructed so that they could be subdivided into parts by "contour boundaries". In other words, some parts of the melody were separated by larger pitch intervals than those found *within* each such part. In their experiments, subjects (experienced musicians) were required to recognise probes that lay entirely within contour boundaries, or alternatively crossed such contour boundaries. The former task was accomplished better by the left ear, the latter by the right ear. Thus, a left hemisphere advantage accrued for recognition of distinctive two-note sequences which could not be assimilated as part of a "contour", but a right hemisphere one appeared for recognition of sequences within a contour.

One study in normal subjects provides a more precise analysis of the right hemisphere's contribution. Charbonneau and Risset (1975) presented two melodies in sequence, each 1.5 sec long. Such pairs could be "tonal" melodies, each consisting of 6 tones, of different frequencies. Alternatively they could be "spectral" melodies consisting of three tones related to each other by octave intervals. Subjects made "same/different" judgements in each pair. For the tonal melodies, a right ear advantage was found, while a left ear advantage was found for spectral melodies. The former task placed greater demands on coding of temporal sequences, while the latter could be solved by representation of the sonority implied if the tones were combined. Such a combined representation is not possible by the perceptual processes considered in Chapter 6, because it is extended over STM intervals, but does become possible when the integrations of STM are considered. The result indicates that while the left hemisphere has a superiority in representing the sequential pitch relationships of notes, the right is superior in representing the sonority which would be produced if their pitches were to be combined within STM, thus forming a new kind of Gestalt.

A study in subjects with unilateral head injury appears to provide corroboratory evidence. Samson and Zatorre (1988) used a test involving recognition of an altered note in a three-note sequence. Overall, the patients with right temporal and right fronto-temporal lesions were impaired, compared to controls or left lesioned patients. Looking at the results in more detail, a more specific pattern can be found which is compatible with the data reviewed above. For three-note melodies which would be consonant if all notes were sounded together, there were indeed greater error rates in the subjects with right temporal or fronto-temporal lesions compared with other groups. For three note melodies containing dissonances, there was no significant difference between normal subjects, those with left temporal, right temporal or right fronto-temporal lesions. Thus the right hemisphere superiority was probably not determined by the temporal structure of the melodies but by the consonance of the constituent intervals. The result is similar to that of Charbonneau and Risset (1975), and is probably another aspect of the right hemisphere superiority in non-temporal integration within STM.

In summary, it is clear that laterality effects in melody perception are of more than one type. There is a left ear advantage for perception of overall melodic contour. Although "contour" is given different detailed definitions in different studies, it should be pointed out that the integration required to recognise "contour" is, in all cases, on a longer time scale than most of the cases where a right ear advantage is seen for auditory temporal patterns. All definitions of "contour" thus involve integration over intervals in the STM range. However, they do not imply representation of precise temporal structure of a melody. There is also a left ear advantage for melodic fragments whose notes form a consonant combination*. This laterality effect is superficially similar to the right hemisphere advantage for recognition of spectrally complex tones or chords. However, it is not possible to derive the former effect from the perceptual lateralization implicated in the latter. It can however be explained in terms of the lateralization of the integration within STM, discussed above.

10.2.3.2 Perception of prosody

Prosody may be defined as fluctuations of tempo, stress or frequency in spoken language other than those specified by each word taken by itself. Such fluctuations enrich the meaning which can be conveyed by the spoken word, especially at the level of phrases, sentences and longer. They may subserve functions of emotional expression, as well as indicating whether a sentence is an interrogation, a command or a statement. Prosody can also have a variety of more subtle semantic functions rarely analysed experimentally. The domi-

* The perception of contrapuntal music is of some interest here. This is music for several parts (or voices) where the interest is on the melodic line for each instrument (or voice) rather than on the chordal combinations made by the simultaneously sounded notes. However, in typical classical counterpoint, the melodic lines of each voice also blend together as harmonic consonances. Thus, such music may be capable of recruiting both the left hemisphere for perceiving the melodic lines, and the right hemisphere for perceiving the harmonic patterns. I report my own subjective impression from repeated listening to passages of such music: Initially I am aware of the type of chord occurring at each measure of the music; but with increasing familiarity with the piece of music, my perception changes and I can hear the individual parts as separate melodic lines. One may speculate that this change represents a shift from right to left hemisphere involvement. Experimental studies may be able to define such perceptual shifts objectively.

nant role of the right hemisphere in prosodic expression was considered in section 9.4.4. For prosodic perception, many studies, mostly in cases of unilateral brain damage, show that, on the whole, comprehension of prosody is also accomplished mainly by the right hemisphere (e.g. Tucker *et al.*, 1977b; also below). These include one study using physiological indicators of laterality: Papanicolou *et al.* (1983) recorded the evoked potentials to click stimuli, used as probes, whilst subjects performed tasks involving identification of prosody or phonetic features of speech. The evoked potential was reduced in amplitude more on the right side during the prosodic task, and more on the left during the phonetic task.

However, some studies of unilateral brain damage report equal impairment in comprehending prosody after left or right hemisphere damage (Schlanger *et al.*, 1976; Tompkins and Flowers, 1987; Cancelliere and Kertesz, 1990). Moreover, since emotional expression is perhaps the commonest use of prosody, the question of lateralization of prosody is confounded by that of lateralization of emotion *per se*, a subject considered later in this chapter. Furthermore, in some studies reporting a right hemisphere dominance for non-verbal sounds communicated from one person to another (e.g. crying, shrieking, laughing in the study of Carmon and Nachshon, 1973), the spectral complexity of individual sounds (dealt with in Chapter 6) is confounded with other temporally-extended aspects of prosody. The paragraphs below focus on analysing these potentially confounding factors.

A number of studies have attempted to distinguish between emotional and phonological attributes of prosody by using sentences with prosody used for purposes other than emotional communication, or conveying emotional messages by spoken language devoid of prosody. Weintraub *et al.* (1981) required subjects to interpret the meaning of emotionally neutral sentences from intonation and stress (e.g. implying the interrogative in "Margo plays the piano?"). Right brain-damaged subjects were impaired with respect to normal controls. Weniger (1984) presented sentences with alternative meanings, implied by prosody alone and subjects were required to match each sentence with a corresponding picture. Left hemisphere patients were similar to normal controls, but right brain damage led to significant impairment. Heilman *et al.* (1983, 1984) found that right brain-damaged patients were deficient in comprehending both emotional and non-emotional prosody, while left brain-damaged patients were deficient only in non-emotional prosody. This result could indicate a role for the right hemisphere in emotion *per se* rather than prosody. However, the same group has later found that the deficit in comprehending non-emotional prosody (question *vs.* statement) is statistically significant only in the right hemisphere group, not in the left hemisphere one (Blonder *et al.*, 1991). Tompkins and Mateer (1985) used passages of several sentences where the last sentence was either consistent with the first few, or was not, and so would appropriately be read with different (e.g. "sarcastic") prosody. The last sentences were spliced off the tape recordings of each utterance. Using these fragments alone, subjects were required to judge whether words and prosody were or were not consistent. For the passages where words and prosody were consistent, patients with right temporal lobe epilepsy made significantly more errors than left temporal lobe epilepsy patients or controls. However, in answering factual or inferential questions about written narratives, the right temporal patients were unimpaired. The impairment in the first task was therefore likely to be due to failure in interpretation of prosody. Overall these results provide little support for the idea of right hemisphere specialisation for representation of emotion. The idea is discussed in more detail later.

Several papers provide more detail on the acoustic characteristics of prosodic perception whose representation is lateralized in the right hemisphere. Weintraub *et al.* (1981) required brain-damaged patients to grasp the meaning of a word using only word stress (e.g. the distinction between "green house" *vs.* "greenhouse"). Right brain-damaged patients were impaired with respect to normal controls. A test equivalent to this was carried out by Behrens (1985) in normal subjects. When subjects distinguished word pairs such "hot dog" *vs.* "hotdog", a right ear advantage was shown. When both semantic and phonetic cues were filtered out, leaving only word stress, there was a left ear advantage. When word pairs were phonetically normal but semantically nonsense ("botgog" *vs* "bot gog") ears were equal. The implication of both results is that perception of word stress is right hemisphere dominant.

Tompkins and Flowers (1985) presented subjects with phrases which had been low-pass filtered, to obscure the higher frequency features usually important in phonetic discrimination. In a "same/different" judgement between two phrases, right hemisphere patients were impaired compared to left hemisphere ones or normal controls. Lalande *et al.* (1992) asked presenters to express the emotion in a sentence by humming it with a closed mouth, and experimental subjects were required to identify the implicit emotion. Right hemisphere patients were again significantly impaired with respect to normal controls or left hemisphere patients. The authors conclude that the right hemisphere is pro bably dominant for vowel trains, and inferences made by integrating them. Van Lancker and Sidtis (1992) performed an acoustic analysis of the features permitting discrimination of the emotional tone of sentences emotionally charged in a number of ways (happy, sad, angry, surprised). In brain-damaged subjects it was found that those with intact right hemispheres could still use information about frequency fluctuation, but if the right hemisphere was damaged, the intact left hemisphere relied mainly on information on modulation of syllable duration.

In summary, right lateralization of prosody is not simply a consequence of emotional lateralization. Moreover, perception of prosody is not completely lateralized to the right hemisphere: Aspects of prosody involving recognition of precise temporal structure appear to be left lateralized. Aspects involving slower patterns of frequency change, and meaning inferred from extended patterns of such change are however right lateralized. These conclusions are what might be expected from the predictions made above. In particular, some of the features of prosody lateralized to the right hemisphere are similar to the feature of musical contour, which is also right lateralized.

10.2.3.3 Right hemisphere advantages at the short-term memory level in the visual sense

In section 6.6.2.3, the result was mentioned of Carmon (1978), who found a left hemisphere advantage in perception of short (two item) sequences of visual stimuli, which shifted to a right hemisphere advantage as the sequence became longer. The authors explained the latter result as a function of increasing spatial complexity. However, according to the theoretical ideas presented here, conversion of a temporal sequence to a spatial Gestalt is not a trivial process, but requires the integrations possible in short term memory. Thus, although the interval during which the sequential components were presented in this experiment were less than 1 second, the right hemisphere advantage seen for the longer sequence may be regarded as somewhat equivalent to the STM integrations seen in the auditory mode. Such a right hemisphere advantage may be likened to that for recognition of sequences of tones if they happen to form a consonance, or for recognition of longer sequences of Morse code (see Papcun *et al.*, 1974 and section 6.6.1.3), or certain aspects of prosody.

Another example of right hemisphere specialisation for visual tasks at the STM level comes from Zaidel and Sperry (1973). They tested each hemisphere of commissurotomy patients with a version of Raven's Progressive Matrices task. Each part of this test consists of a series figures, arranged in a systematic order according to some unstated criterion, but with one figure in the series missing. Below this arrangement are a number of other figures, one of which has to be selected as completing the figure series. This task, a visual STM task without detailed temporal structure in any of its components, was performed better by the right than the left hemisphere. The left hemisphere could perform the task however, though with lesser efficiency, and the authors suspected it was using a different, less effective strategy compared with the right.

10.2.3.4 Comprehension of extended messages: humour, inference and indirect statements

Apart from short term memory operations where slow perceptual features of the physical stimulus trains are integrated, there are a number of other instances where short term memory accomplishes an integration of meaning over a period of many seconds. The individual components of meaning which are integrated in this way may have a special representation in the right hemisphere, if they are concrete or visually imageable, and right hemisphere representation of prosody may add to the compilation of meaning. However, the right hemisphere superiority applies to more complex aspects of linguistic comprehension. This goes beyond what can be inferred from the right hemisphere specialisation for concrete words or prosody. It is also in sharp contrast with the usual left hemisphere advantage for simpler aspects of language. Examples are given below.

Zaidel (1977) provided general evidence for an overall role of the right hemisphere in complex aspects of language. Right hemispherectomy patients were considerably impaired in the Token test, in which subjects followed instructions such as "Put the green box on top of the yellow triangle".

Such lateralization has been also shown in unilaterally brain-damaged patients in a variety of tasks. Understanding *humour* appears to be mainly right lateralized. According to Gardner *et al.* (1975) there is some selective preservation of linguistic humour in right brain-damaged subjects and of non-linguistic humour in left brain-damaged subjects. Beyond this however, the left brain-damaged patients showed much better integration of cognitive and affective aspects of a joke; whereas the right brain-damaged subjects laughed indiscriminately at all cartoons, or at none of them, and regardless of whether they actually understood the joke. In all these respects the right brain-damaged patients lacked the ability to acquire the overall form or Gestalt of a connected piece of language. Wapner *et al.* (1981) found that assessment of the plausibility of elements within a story, in choice of the most suitable "punch-line" for a joke and in integrating elements of a story into a coherent narrative was deficient in right brain-damaged subjects. Brownell *et al.* (1983) also used a task in which subjects had to select the most humorous punchline for a joke. Right brain-damaged persons selected options which included the element of surprise which is part of a joke, but which nevertheless were "non sequiturs", that is, failed to make the joke a coherent whole.

There have also been suggestions that a similar right hemisphere advantage applies to the achievement of an *inference* (Brownell *et al.*, 1986), or the realisation of the implications of an indirect request (see Weylman *et al.*, 1989). Such lateralization also applies to the comprehension of metaphor: Winer and Gardner (1977) tested brain-damaged patients in comprehension of metaphor, by asking them to match a metaphorical statement with pictures which might represent the literal or the metaphorical meaning of the picture. Left brain-damaged subjects chose the metaphorical depiction but could not explain the basis of their choice verbally. Right brain-damaged subjects chose literal pictures, and explained their choice in literal terms, or objected that the metaphor made no sense.

Curiously, the right hemisphere dominance for inferences drawn from a connected utterance also appears to apply to *mental arithmetic*. Many aspects of calculation are left hemisphere dominant, for instance acalculia related to alexia or agraphia for numbers. On the other hand acalculia related to disordered spatial ability is right hemisphere dominant (Levin and Spiers, 1985). Such lateralizations do not go beyond the perceptual lateralization discussed in Chapter 6. However, it appears that in *mental arithmetic* there is also a right hemisphere dominance not predictable from principles of perceptual lateralization: Dimond and Beaumont (1972) presented three double-digit numbers visually to normal subjects (640 msec per number), who had to say whether the third number was the sum (or the difference) of the first two, with reaction time as the response measure. There was a significant left field advantage, larger for the subtraction than the addition task. Qureshi and Dimond (1979) used a task similar to this in brain-damaged subjects, who were required to respond by nodding or shaking their head, with digit-pairs flashed to either left or right visual fields. Right brain-damaged subjects were much more impaired than left brain-damaged ones, regardless of the hemisphere to which stimuli were directed.

Not all varieties of inference give right hemisphere dominance however. Phelps and Gazzaniga (1992), in a study of two commissurotomy patients, tested memory of a picture of a complex scene, using "yes/no" questions. The right hemisphere responded according to the physical aspects of the picture, while the left hemisphere was vulnerable to making false positive errors which were plausible, given the picture as a whole. The left hemisphere seemed to have a more inferential representation of the scene than the right, which was more literal.

10.2.4 Evidence of left hemisphere specialisations in the work of working memory

10.2.4.1 Temporal structure in music

The study of Bever and Chiarello (1974) was described above, showing that, for tasks involving identification of specific temporal sequences within a melody, experienced musicians succeeded where non-musicians failed. The musicians also used a left hemisphere strategy for recognising melodies. Although the right ear advantage in this experiment was found only in experienced musicians, later work has shown that under some circumstances it can also occur in non-musicians. It has also demonstrated more clearly what is the strategy used when a left-hemisphere advantage in melody perception is found. Deutsch and Roll (1976) used pairs of sequences of five dichotically presented tones (800 or 400 Hz), with each tone having a duration of 250 msec, and the interval between them being 450 msec. Sequences delivered to the right ear were reported better than those to the left. Gates and Bradshaw (1977) found that modifications of the rhythm of single notes in a melody was detected more accurately by the right ear than the left, despite the left ear advantage in reaction time. Gordon (1978b) confirmed this, both for musicians and non-musicians. Peretz and Morais (1980) asked non-musician subjects to identify which of two brief melodies, played dichotically, corresponded to that which had been played binaurally a moment earlier. Overall there was no lateral difference. However, some subjects consistently gave a right ear advantage, and they also described that the strategy they used was to focus on critical constituents of the melody rather than its overall "shape". Later studies showed that a right ear advantage in tests of this sort could be encouraged by asking subjects to make a same/different response as fast as possible (Peretz and Morais, 1987) or by advising subjects to focus on local critical notes as opposed to overall contour (Peretz *et al.* (1987). Clearly the strategy used by the left hemisphere in such tests involves focusing on fragments of melody so small that accurate analysis of temporal sequence is possible. In the most recent study of this (Messerli *et al.*, 1995) recognition of popular melodies presented dichotically gave no ear asymmetry in either musicians or non-musicians when the melodies were played at normal speed. When the same melody-pairs were presented at four times normal speed (pitch being adjusted to be the same as in the slower presentations), overall performance deteriorated, but less so in the musicians for right ear presentations. Thus a right ear advantage appeared for recognition of the very fast melodies, in the musicians. Presumably this was based on the superiority of the left hemisphere in analysis of extended temporal patterns.

One of the most subtle demonstrations of the role of the left hemisphere in registering music in short term memory was performed by Dowling (1978). He used a task in which subjects had to determine whether dichotic presentation of two-part music was a "canon" (i.e. the part which started second was an exact replica of that which started first). The delay between the start of the parts was either 2, 4 or 8 sec. With the right ear leading there was a much closer dependence of subject's performance on this lag than with the left ear leading. As a result, the right ear was better for a 2 sec lag, and was marginally worse for an 8 sec lag. The authors suggest that the right ear (left hemisphere) can store chunks of auditory information for a limited period of time, and can use this stored information for comparison with the left ear input. Implicitly these chunks contain accurate representation of temporal structure. With the left ear leading a different (unspecified) strategy must be used, which does not depend on lag. The experimental design clearly involves time intervals longer than those catered for by the hypothesis of axonal conduction delay, ones which must be classed as short-term memory intervals.

Two studies appear to show a *right* hemisphere advantage in registration of temporal sequences, which is unexpected from theory and from other experimental data. Papcun *et al.* (1974) showed a left ear advantage for longer sequences of Morse code in naive subjects, presumably involving short-term memory. Kester *et al.* (1991) found that patients with right temporal lobectomy showed a differential decline in tasks involving perception of metre and tempo, but *not* melody, harmony or tonal processes. Left temporal lobectomy patients showed no selective decline. The metre and tempo task involved intervals over one or two "bars" of music, long enough to recruit short term memory mechanisms. The explanation of these discrepancies is not clear, but it is possible that an aspect of rhythmic sequences other than their exact temporal structure permits their recognition, and may confer a right hemisphere advantage.

10.2.4.2 Precise temporal structure of spoken sentences and other non-musical auditory stimuli

Recall of sentences is a complex task calling on both the capacities of the left hemisphere for primary representation of spoken language, and those of the right hemisphere for representation of more complex linguistic and semantic features. Most studies of recall of sentences or digit sequences obtain either a right ear advantage (e.g. Jarvella and Herman, 1973; Frankfurter and Honeck, 1973; Bever, 1971; Geffen, 1978; Taylor and Heilman 1982), or no significant ear advantage (Bakker, 1968). These results do not go beyond what would be expected from the type of stimulus used. One paper that gives a hint that the left hemisphere may be specialised for the accurate representation of long verbal trains is that of Newcombe and Marshall (1967). Their test involved verbatim repetition of sentences. For sentences constrained by the need to "make sense" there was no difference between left or right brain-damaged subjects. However, if the sentences defied the usual syntactic or semantic constraints (that is if the repetition depended simply on STM for phonological sequence, with no other clues as a guide) men with left hemisphere damage were significantly impaired compared to right hemisphere patients or normal controls.

Another indication of a special role of the left hemisphere in representation of precise temporal structure of sentences comes from Clark (cited by Zurif, 1974). He used pairs of sentences such as:

"In her hope of marrying, Anna was surely impractical"

and

"Your hope of marrying Anna was surely impractical"

The very different meaning of these two similar sentences depends critically on whether the sentence is segmented before or after the word "Anna". Clark presented two such sentences, one of each pair to each ear of normal subjects. He also presented clicks (in the style of the experiments of Bever described in section 6.1.1), which were timed to occur in the middle of the word "Anna". There was a perceptual distortion in the timing of the click, such that it was heard more often separating the segments of the sentence given to the right ear, rather than that to the left ear. This suggests that the left hemisphere is involved in segmenting sentences into its constituent phrases. The result is comparable to the right ear advantage seen by Peretz and Babai (1992) in recognising fragments of melody which crossed contour boundaries.

10.2.4.3 Serial order effects and manipulation of mental images in the visual sense

Hines and co-workers have conducted a series of experiments (Hines *et al.*, 1969; Hines & Satz, 1971; Hines *et al.*, 1973; Hines *et al.*, 1976) which hint at a left hemisphere advantage in representation of sequential information presented visually. Typically they used sequences of 3–5 digit pairs, one of each pair shown to each visual field, with the pairs following one another in rapid succession within 0.5–1.0 sec. Subsequently, subjects were required to report all they saw from both fields, with controls in some studies to eliminate "order of report" effects. The last pair to be presented was most accurately reported, but without a lateral differences. For the earlier pairs, there was a significant right field advantage. The authors suggest that this occurred because the response was linguistic, and so would be better for stimuli presented to the left hemisphere. However, this fails to explain the lack of field differences for the last pair to be presented. An alternative explanation is that the left hemisphere can store sequential visual information with less interference between items compared to the right. Though this suggestion is inferred rather than directly demonstrated, it is a specialisation comparable to that shown more directly in the auditory modality. The inference needs to be checked with a non-linguistic response, and with the dependent measure being direct reports of visual sequence rather than merely serial recall primacy effects.

Apart from this there are a number of papers from the "imagery" literature which demonstrate a left hemisphere superiority for active construction and manipulation of visual images. Farah (1984) asked brain-injured subjects a series of questions involving imagery (for instance: "What is the colour of the stars on the American flag?") Patients with left posterior lesions were most impaired on such a task, although with simpler questions (e.g. "What is the colour of the sky?") they were not impaired. In another study (Farah *et al.*, 1985), using a commissurotomy subject, tasks were devised using letters which involved imagery (e.g. envisaging the features of a lower case letter, given its upper case equivalent), with simple perceptual discrimination based on the same features serving as a control. The left hemisphere was clearly superior in the tasks involving imagery, while the right hemisphere performed at chance levels. A more thorough study, involving a large number of imagery tasks, with two split-brain subjects (Kosslyn *et al.*, 1985) concluded that the left hemisphere but not the right one is capable of constructing and inspecting an image made up of several parts, while the right hemisphere can construct and inspect only simple images. A recent addition to this line of research is from Cook *et al.* (1994), employing the task of assessing whether two geometrical figures are the same, but at different orientations, or are different. This task involves mental construction and mental rotation of an image. In normal subjects it is carried out better with the right fields (left hemisphere). In addition when two geometrical figures were presented for comparison, one to each field, and one of these had a fixed position, while the other was "gravitationally unstable", the comparison could be made more quickly if the mobile figure was directed to the left hemisphere. It was again concluded that the left hemisphere can manipulate images, whereas the right hemisphere is better for forming a static "reference" image.

10.2.5 Evidence of lateralization of motor planning within short-term memory

The previous two sections show that the work achieved by short-term memory differs between the hemispheres. In principle, one of the inputs which can be integrated within the span of STM consists of the models for motor performance discussed in Chapters 8 and 9. Therefore, one might expect that there be differences between the hemispheres in motor planning on a scale longer than that discussed in Chapter 9, namely the scale of

STM operations. A little evidence for this exists, referring entirely to the specialisation of the left hemisphere. As for right hemisphere specialisation in motor planning functions at the short-term memory level, no information is available.

For two cases of dominant hemispherectomy (one in a 12 year-old girl (Gott, 1973), the other in a 47 year old man (Smith, 1966)) it was reported that the remaining right hemisphere is particularly deficient in stringing words together into sentences. Speaking was thus limited to single words or short phrases, or to expletives (Smith, 1966) which are nevertheless well articulated. In the months after the operation, the length of phrases which could be repeated to command gradually increased, and could be aided by cueing or other auditory assistance. Occasionally, in Smith's patient, complete sentences were uttered, well articulated and appropriate to the occasion, but this probably relied on strong situational cues. Despite these occasional performances, in both cases, expressive speech remained severely impaired. In contrast to these reports, dominant hemispherectomy carried out at an early age has less severe deficits. Two cases are reported by Dennis and Whitaker (1976), children of age 8 and 9, who were operated on before acquisition of speech. In these children there was no deficit in either sensory discrimination or articulation of phonemes, and there was no selective loss of expressive *vs.* receptive aspects of language. The deficit that was revealed, in both cases, was a relatively subtle one, which does not correspond to any of the classic aphasic syndromes. There was a selective inability to either comprehend, or to produce, sentences with normal syntactic structure. The children were thus deficient in detecting and correcting errors in syntax, or in repeating sentences with the more unusual constructions (e.g. use of the passive form of verbs). They were also impaired in comprehending or repeating a sentence whose meaning depended on word order.

The evidence from the older hemispherectomy patients gives some support to the idea that the left hemisphere superiority is in modelling of *extended* speech, rather than in phoneme articulation. However, in these cases, the deficit after dominant hemispherectomy, though profound, was not absolute. In children who have had to acquire language *de novo* with the right hemisphere, the deficit in these matters is negligible. However, the more subtle syntactic deficit that does remain in such children is again in the construction and interpretation of sequential structure in whole sentences. Thus the right hemisphere by itself may develop some ability to produce the time-structured programs for individual speech elements, as mentioned earlier, but, even after years of independent functioning, it is impaired in the time-structured modelling of *extended* speech. It is likely that this impairment is also one of the contributory factors to the impairment in grammatical construction seen in Broca's aphasia. This has been analysed by Zurif *et al.* (1972), Zurif and Blumstein (1978) and by Kolk (1978). Zurif and Blumstein (1978) suggest that the agrammatism found in some cases of Broca's aphasia arises because subjects have access to only limited routines which rely more heavily than in the normal case on an ability to access word meanings. In contrast, correct grammatical construction of extended speech relies heavily on correct sequencing of small words and word endings, and the impairment of this function after left hemisphere lesions presumably reflects loss of the ability to construct long and precise sequences within short-term memory.

A recent electrographic study (Pulvermuller *et al.*, 1995) gives some support to this interpretation. "Function words" and "content words" were used in a lexical decision task, during EEG recording, with the time of stimulus presentation being used for averaging the EEG signal. In the time window 150–180 msec after the stimulus, content words produced almost symmetrical negativity in the left and right hemispheres. On the other hand, function words (those whose primary role is in organizing extended strings of language), produced greater negativity in the left hemisphere than the right.

A dramatic example of the left hemisphere specialisation in motor planning at the short-term memory level integrations comes from a recently reported single case study (S. Clarke *et al.*, 1993). A highly motivated student suffered a brain tumour in the right hemisphere, which after surgical removal led to a large lesion in the posterior part of this hemisphere. Although some right hemisphere function would have remained intact, there was no visual sense in the right hemisphere. This subject continued studying, in architecture. However, after the operation, instead of representing the world in two dimensional space, her route maps for finding her way around the town were transformed into strictly one-dimensional linear maps. Likewise, a planned timetable for the week's activities was not structured in two dimensions (day-by-day *vs.* hour-by-hour) but became strictly linear. The compensatory strategy adopted by this subject reveals the role of the left hemisphere for planning within short-term memory, and corresponds to the role of the left hemisphere in more immediate planning shown by Balfour *et al.* (1991; section 9.3.1)

10.2.6 Discussion of short term memory and laterality

In summary, no difference between hemispheres has been documented in the basic process of short term memory, but the work achieved within STM appears to differ considerably between hemisphere. Temporal succession is better represented in the left hemisphere, as shown in certain aspects of melody representation, sentence memory and parsing, and possibly also in perception of sequential visual information. The left hemisphere can use visual images which undergo changes (such as mental rotation, or construction of a whole from several parts). The left hemisphere also appears to be superior to the right in long-range motor planning, including the planning of correctly grammatical sentences.

The right hemisphere, on the other hand, is specialised for integrations over time which do not emphasise exact temporal structure. The right hemisphere may record low-resolution temporal information, as in encoding the overall temporal contour of a melody, or the sequence of slow pitch and stress changes in a sentence. It may also hold a static visual image, while the left hemisphere "manipulates" an image to achieve a comparison with this reference. Apart from this the right hemisphere is specialised for semantic integration at the level of phrases, sentences and longer. The "meaning" obtained from such extended messages is in no way temporal in its structure.

These differences between the hemispheres are broadly what would be expected from the theoretical analysis given in section 10.2.1. It is difficult to say that the specialisation of the left hemisphere with regard to visual imagery involves a precise temporal measure. Nevertheless, if the left hemisphere can form extended time-structured chains of neural activity, mental images which change over time, or can be manipulated in the mind, would find better representation in that hemisphere than in the right.

A generalisation can be made about several areas of research reviewed above: A number of cognitive psychological functions such as "spoken language", "prosody", "melody", "imagery", "semantic meaning", "calculation" appear on first examination to have a clear lateralization in one hemisphere or the other. However, on closer scrutiny each of these turns out to be a complex function, some of whose components are left-lateralized, others right-lateralized. This subdivision of psychological functions is carried out in terms of the spatio-temporal structure of the stimuli used. A more comprehensive account of cerebral lateralization thus becomes possible, based ultimately on concepts which are more akin to those of sensory and motor physiology than cognitive psychology.

10.3 LATERALITY IN RELATION TO ALERTNESS AND SELECTIVE ATTENTION

Closely related to short term memory are the concepts of attention and alertness. Everyday usage of these terms is loose and misleading.

As scientific concepts, the term *alertness* (roughly synonymous with *arousal*) implies a general preparedness to analyse or respond to a wide range of information types, regardless of the special category or subset to which the information belongs. The term *attention* is usually used to refer to the selection of one category or subset of information, to which the organism thereby becomes specially prepared to analyse or respond.

On the subject of selective attention, the present author has previously made two quite different suggestions about the physical basis of this selective process. These two are in no way incompatible, but refer to selected categories of information defined at different levels.

At the *lower level*, the selective aspect of selective attention is seen to be a product of the cortico-striatal interaction referred to in the previous section. Within the striatum, the majority of neurones are mutually inhibitory, and in principle should compete fiercely. Motivationally significant concepts or programs of motor or behavioural output are envisaged to be represented by circuits of cortico-striatal connections (referred to as "cortico-striatal cell assemblies" in Miller and Wickens, 1991). If several such circuits might be activated at once, it is conjectured that only one will emerge as the "winner" because of the competitive interaction of the striatal neurones within each such circuit. This is another aspect of the model referred to in Chapter 8 and in the section of this chapter dealing with short term memory. Clearly this aspect of selective attention is very closely related to short term memory (or working memory).

At a *higher level*, selective attention may involve the selection of different "contexts" (frameworks for interpreting large sets of data, rather than the data sets themselves). Such information processing is seen as involving interplay between neocortex and hippocampus (Miller, 1989; 1991; 1994). In the following sections, this aspect of attention will not be discussed further. Admittedly, there may well be interesting aspects of laterality in which contexts and neocortico-hippocampal interplay are involved, but there is insufficient data to develop the subject to any depth.

However, there *is* considerable evidence relating laterality to both the generalised arousal (or alertness) and the lower level aspect of selective attention (hypothetically involving cortico-striatal interplay). These two topics are discussed below.

10.3.1 Laterality and alertness (vigilance)

10.3.1.1 Evidence

It is well known from the study of patients with unilateral cerebral injury that patients commonly suffer from "neglect" syndromes. Such "neglect" syndromes consist of a systematic failure to notice stimuli. A common test of neglect is "simultaneous bilateral" stimulation, when the patient fails to notice the stimulus contralateral to the injured (usually right) hemisphere (extinction). Often the neglect is multimodal, though more restricted forms of neglect have also been described in recent years (Halligan and Marshall, 1991; Tegner and Levander, 1991; Mennemeier *et al.*, 1992; Liu *et al.*, 1992; Binder *et al.*, 1992; Guariglia and Antonucci, 1992; Guariglia *et al.*, 1993; see review by Heilman and Valenstein, 1979). Most cases of neglect are hemispatial neglect, and this occurs with almost equal frequency for right and left hemisphere damage, usually from inferior parietal damage (Valler and Perani, 1987). However, the basis of such neglect is asymmetrical because neglect resulting from right hemisphere damage is far more severe and persistent than that resulting from left sided damage (e.g. Albert, 1973).

A number of experimental studies in brain-injured subjects indicate that the right hemisphere has a more important role than the left in maintaining general arousal. Howes and Boller (1975) found that simple reaction time (with hand ipsilateral to the lesion) was slowed very significantly in right hemisphere patients (from about 200 msec to about 700 msec), whereas only a small increase in reaction time occurred with left hemisphere patients (from about 200 msec to about 300 msec), this occurring whether or not they were aphasic. Posner *et al.* (1987), Yokoyama *et al.* (1987) and Ladavas *et al.* (1989) all confirmed this result. Ladavas *et al.* (1989) found that the slowing of reaction time for right hemisphere damage occurred both for parietal and temporal lesions. Posner *et al.* (1987) found that provision of a cue stimulus indicating that the reaction time stimulus would soon be given, and its position, abolished the deficit of the right hemisphere patients. All studies attributed their findings to a disruption of mechanisms, localised in the right hemisphere, for maintaining alertness. Heilman *et al.* (1978) showed that uncomfortable forearm stimulation given ipsilateral to the side of hemispheric injury elicited a much smaller arousal response (as judged by the Galvanic skin response from fingers on the same side) for right hemisphere patients with neglect than for left hemisphere patients with aphasia. Yokoyama *et al.* (1987) showed that the increase in heart rate, which occurred in normal subjects, if a reaction time task was anticipated by a warning signal 6 or 10 sec earlier, was almost abolished in patients with right brain damage. On the other hand in patients with left brain damage, the anticipatory heart rate increase was greater than normal. Wilkins *et al.* (1987) asked their subjects to count auditory or tactile stimuli given at 1 Hz or 7 Hz. The subjects were normal controls, or those who had received left or right frontal and/or temporal lobectomies for intractable epilepsy. In two separate studies patients with removal of the right frontal lobe were impaired at counting 1 Hz stimuli, compared with other groups. For the 7 Hz stimuli the number of stimuli was underestimated by all subjects, and the impairment was greater for right brain-damaged subjects than other groups, in one of the studies, but not the other. There were no impairments in the task for temporal lobe removals on either side. Coslett *et al.* (1987) tested maintained arousal using two concurrent attention-demanding tasks: Subjects were asked to perform a reaction time task to an irregular repeated tone, or they were asked to sort coins of two types into piles, or they were asked to do both tasks concurrently. Patients with right brain damage were impaired, compared to other groups, in the reaction time task (as found in other studies). However, more importantly, they also suffered more severe worsening of performance than left hemisphere patients or normals when both tasks had to be performed together.

It is less well known that a somewhat similar syndrome can be seen in callosotomy patients, for stimuli activating the left hemisphere. Dimond (1979) found, in callosotomy patients, that for all tests of sustained attention, the right hemisphere was substantially better than the left. When testing the left hemisphere there were often gaps of several seconds when responses to stimuli were omitted. This failure to notice stimuli delivered to the left hemisphere did not change after a switch to a different type of stimulus, though after a switch to the other hemisphere the neglect was no longer observed. In a similar vein, Loring *et al.* (1989) found that in a callosotomy patient tested with double bilateral stimuli (in visual and tactile modalities) there was often left hemisphere extinction. Similarly the "continuous performance test" (a test of vigilance) was carried out better with the left hand (right hemisphere). Ladavas *et al.* (1994) tested five callosotomy patients in a lateralized visual reaction time task, and found right hemisphere dominance for maintained alertness.

The usual interpretation of such evidence is that there is a mechanism for maintaining alertness located in the right hemisphere. However, a more precise interpretation is that the left hemisphere, when isolated by callosotomy, or when connected to an injured right hemisphere, has a greater vulnerability to neglect that the isolated right hemisphere, or the intact right hemisphere when connected to an injured left hemisphere.

Neglect after cerebral lesions can also be seen in experimental primates. However, in monkeys, neglect is shorter-lived than in humans (recovery occurs within a month), is less severe, and apparently is not lateralized (Welch and Stuteville, 1958; Valenstein *et al.*, 1982; Heilman *et al.*, 1970; Crowne *et al.*, 1981; Watson *et al.*, 1973, 1984). Another significant finding is that, in such lesioned monkeys, callosotomy accentuates neglect or reinstates it after recovery (Crowne *et al.*, 1981; Watson *et al.*, 1984).

These data in brain-injured or callosotomy subjects can be related to a small body of data in normal human subjects. Heilman and van den Abell (1979) presented a tone as an alerting signal, followed by a delay of 12, 18 or 22 sec, and then a warning signal (red light), itself to be followed 1 or 2 sec later by a green light to which the subject gave a simple reaction time. When the warning signal was given to the left field, reaction time was reduced more than when it was given to the right field. In other words, the right hemisphere could benefit more from the cerebral activation provided by the warning signal. A subsequent EEG study (Heilman and van den Abell, 1980) supported this conclusion: It was found that desynchronisation of the alpha rhythm in the right hemisphere could be elicited strongly with stimuli to ipsilateral as well as contralateral visual fields, while that in the left hemisphere was strongly influenced only from the contralateral field.

There is also asymmetry in the capacity of the two hemispheres to maintain vigilance for long periods. For instance, Dimond and Beaumont (1973) conducted vigilance tasks involving detection of brief brightening of a signal light, throughout an 80 minute period. At the start of this period the left hemisphere detected a higher proportion of signals than the right, probably because the signals lasted 200 msec, supraoptimal for temporal integration in the left hemisphere, as discussed in Chapter 6. They also found that signal detection by the left hemisphere showed a gradual decline over time, not shown by the right hemisphere. Such tests of vigilance have been used by Pardo *et al.* (1991) in a study of cerebral activation using PET. Two vigilance tasks were used. A somatosensory task involved detecting 1–3 sec pauses in a sequence of touches to the right or left hallux. A visual vigilance task involved detection of luminance changes in a dim light. The increase in cerebral blood flow associated with these vigilance tasks was found in the right frontal and right superior parietal lobes, for vigilance test to either toe or either visual field. The left hemisphere was not activated in this way. This provides direct evidence that the right hemisphere is activated more in tests of sustained alertness than is the left.

10.3.1.2 Discussion of evidence

There is clear evidence from brain-injured patients, and from tests of general alertness in normal subjects that the right hemisphere can be more easily activated, and can sustain its activation better than the left hemisphere. This raises the question of the basis of this asymmetry. In particular, can such asymmetry be derived from the central hypothesis of this book, or does it require a totally different explanatory premise?

One such alternative mechanism was proposed by Posner and Petersen (1990). These authors note a variety of evidence that the diffusely-distributed central neurotransmitter noradrenaline might play a role in alerting reactions. In addition they cite a paper by Robinson (1985), which shows that in rats, lesions of the right hemisphere (but not the left) lead to cortical noradrenaline depletion (by mechanisms which are unknown). They also cite evidence that drugs acting on noradrenergic systems can produce attentional shifts in humans. Their proposal therefore is that asymmetry in the distribution of noradrenaline, or its response to brain injury can be the basis of the right hemisphere dominance for alertness in humans.

The idea that diffusely distributed neurotransmitters, such as noradrenaline, may have a role in cerebral laterality and alerting mechanisms is entirely plausible. However, the detailed suggestion they make is not well supported. Asymmetrical changes in amine levels after brain lesions in rats may have nothing to do with the situation which obtains in humans. No data is available of a more dense noradrenergic innervation of the right cerebral cortex than the left. There is one report of asymmetrical noradrenaline distribution in the human brain, referring to the thalamus (Oke *et al.*, 1978). However, in this report, the amine had a higher level in the *left* pulvinar, and the *right* somatosensory thalamus. The pulvinar projects to the parietal association areas, commonly implicated in human neglect syndromes, so the asymmetry of thalamic noradrenaline in this case is the opposite of what would be expected to support Posner and Peterson's idea. Thus, while the concept that asymmetrical distribution of neurotransmitters may play a role in human lateralization is plausible, their suggestion must be regarded as an unsupported conjecture at this stage.

The central hypothesis of this book provides a simpler and more parsimonious explanation of the lateralization of alerting mechanisms. In section 3.4 it was explained that a consequence of temporal dispersion of signals in a neural network is a corresponding loss of signal amplification, at least for brief signals. Detailed predictions of this for hemispheric asymmetry were considered in Chapter 5. Given the idea that the left hemisphere has a repertoire of cortico-cortical delay lines providing greater temporal dispersion than in the right, one would expect that signal amplification in the left hemisphere would be lower than in the right, for equivalent brief stimuli. Empirical evidence for these predictions was reviewed in sections of Chapter 6. It follows that the right hemisphere would generally require less intense input stimuli than the left to activate it. This should be demonstrable not only in tests of perception, but also in tests of alertness and vigilance. In other words, the left hemisphere, by itself, is more vulnerable to neglect than the right; and it should have greater difficulty in maintaining vigilance under conditions where sustained alertness is required. From the studies in monkeys, it is clear that callosal traffic contributes to minimising neglect due to a cerebral lesion. It is therefore likely that impulse traffic across the corpus callosum in humans could minimise the differences in amplification factor in the normal brain. However, after a right hemisphere lesion, or after callosotomy, the vulnerability of the left hemisphere due to its low amplification factor would be unmasked, and neglect syndromes would appear.

From the reasoning in Chapter 5, the superiority of the right hemisphere in detection of stimuli should apply only when they are quite brief. In theory therefore, one would expect that the left hemisphere should *not* be at a disadvantage in vigilance tests if the to-be-attended stimulus is a temporally-extended one, or is immediately preceded by a warning stimulus. Some of the evidence reviewed above fits this pattern: Posner *et al.* (1987) found that the right hemisphere advantage in reaction time tests of alertness disappeared if warning stimuli preceded the tests stimuli. This was found if the warning stimulus was 100 msec ahead of the RT stimulus, but not if it was 1000 msec ahead. In addition, when high-alertness conditions were instated by keeping the inter-trial interval very short (500 msec) a clear left hemisphere advantage was found in all conditions. One may conclude that a very recent cue can transiently bring the vigilance levels of the left hemisphere up to those of the right hemisphere; and that the brain also has access to mechanisms for *maintaining* high levels of alertness in the left hemisphere, but only under special circumstances. Normally the left hemisphere is vulnerable to lapses of attention. In the study of Wilkins *et al.* (1987) the greater impairment of right brain-damaged subjects in a counting task was a more robust finding for counting of slow (1 Hz) stimuli than for rapid (7 Hz) stimuli. This again indicates that, given very recent priming, the left hemisphere is as good as the right in vigilance tasks. Likewise Eling (1983) has noted that the right ear advantage for detection of features of words in dichotic lists is much more consistent for rapid word lists (1.1 sec per pair) compared with slower lists (2 sec per pair). Alertness-holding manoeuvres appear to be needed for the left hemisphere to show its full capacity. However, rather specific conditions seem to be necessary to show these things: In Heilman and van den Abell's (1979) paper, priming of a reaction task with a warning stimulus 1 or 2 seconds previously still gave a right hemisphere advantage. Presumably in this study the warning stimulus was too far ahead of the target stimulus for arousal levels to remain high in the left hemisphere at the time of the target stimulus.

10.3.2 Laterality and selective attention

10.3.2.1 Evidence

The relation between laterality and selective attention has been raised a number of times especially by Kinsbourne and colleagues (e.g. Kinsbourne, 1970, 1973; Ledlow *et al.*, 1978; Kinsbourne and Byrd, 1985). Kinsbourne argues against a "structural" view of cerebral laterality, which he claims implies an unalterable and "robust" differentiation of functions between the hemispheres. He has also produced a good deal of evidence showing that lateralization of different functions in the two hemispheres is often not at all a robust finding, but influenced in subtle ways by many of the factors which influence attention. It is not necessary to present here a full coverage of this evidence. However, some examples should be given to allow discussion of the viewpoints taken by Kinsbourne (and others). The examples below are chosen because they show interactive effects between *specific lateralized categories* of stimulus, rather than effects of activation by non-specific stimuli.

One of the ways in which laterality can be changed from the perceptual styles characteristic of each hemisphere (discussed in Chapter 6), is by requiring the subject to engage in another ("concurrent") task. In one of Kinsbourne's early papers on this subject (Kinsbourne, 1970) the basic task was detection of a gap in a square, presented tachistoscopically for 6–16 msec (individually adjusted to give 30% error rate). For the concurrent task, subjects were also required to rehearse a series of words "subvocally". In the control condition (i.e. without rehearsal), there was no significant laterality effect. With verbal rehearsal as the concurrent task, overall performance did not deteriorate, but a right field advantage appeared ($p < 0.01$). In a later paper (Kinsbourne, 1973) this result was repeated, and it was made clear that subvocalisation of words not only improved right field performance but also worsened left field performance. Kinsbourne also discusses more tentative evidence suggesting that subvocal rehearsal of music shifts the advantage to the right hemisphere.

Later work with concurrent tasks has often used an explicit memory task, where memory for the series of words is assessed, as well as performance in the primary tachistoscopic viewing task. These experiments have documented not only improvement of performance in the primary task ("priming"), as discussed above, but also deterioration of it ("interference"), as a result of the concurrent task. In cases where different degrees of memory load are used in the same experiment, it is generally found that small memory loads prime a response, while larger ones interfere with it (e.g. Hellige and Cox, 1976; Kinsbourne and Byrd, 1985). As an example of interference effects, Hellige *et al.* (1979) used as the main task a word recognition task, with accuracy of report and reaction time as the response measures. A right field advantage for a verbal task with no load could be shifted to a left field advantage with a concurrent verbal memory load. This effect was due to both improvement of the left field RT as well as worsening of the right field RT. Hass and Whipple (1985) used as one of their basic tasks the categorisation of words. Verbal memory load improved the left field RT and worsened the right field RT for words, so that the right field advantage for words, seen in control conditions, was abolished.

Given the inverted U-function relating performance to concurrent memory load, one might expect that the effects of a concurrent memory task engaging the left hemisphere would apply to left hemisphere-type primary tasks, and those of a concurrent task engaging the right hemisphere would influence a right hemisphere primary task. Generally speaking this is so (though exceptions have also been reported). In the experiments of Hellige *et al.* (1979) (just mentioned), use of a nonverbal memory load (memory for visual patterns) had no effect on the word recognition task, but impaired performance in the form recognition task. Hass and Whipple (1985) used as basic tasks the categorisation of not only words but also pictures. Verbal memory load abolished the right field advantage for words, but increased the left field advantage for pictures. Picture memory load as the concurrent task reduced the left field advantage for picture categorisation, but enhanced right field advantage for words. In other words, in these studies with concurrent memory load, interference was demonstrated for stimuli best dealt with in the same hemisphere as the load. Most of the shifts of laterality as a result of the concurrent task were produced in the context of a decline in performance in both hemispheres, though in one case there was improvement in one, deterioration in the other hemisphere.

As another example, Kinsbourne and Byrd (1985) used shape recognition as the primary task, and found, as expected, that there was a left field advantage without any load. With a mild verbal memory load there was an significant improvement in right field performance, which led to a switch to an apparent right field advantage in some groups. (Whether this reversed advantage was statistically significant was not made clear.) Under heavy verbal load both hemispheres declined in performance. This result is interesting because it shows that mild memory load improves performance of the corresponding hemisphere not only for its preferred stimulus category, but also for that preferred by the opposite hemisphere.

Another way in which attentional influences on laterality have been demonstrated is by assessing the effect of predictability of the task. For instance, Ledlow *et al.* (1978) used a task similar to those discussed in section 6.1.2 in which subjects were presented with letter pairs (for 50 msec) in either visual field, and were required to make a speeded judgement about their physical identity or name identity. The right field advantage expected for name matches (see section 6.1.2) was found only when location was randomised, and was reduced in size when the type of match was predictable. Predictability of the task removed the perceptual laterality otherwise found. Geffen (1988) conducted an experiment in the auditory modality which also involved unpredicted changes of stimuli, but obtained more complex results than Ledlow *et al.* (1978). Subjects were presented with dichotic word lists, and subjects had to respond to target words if they were spoken in the voice of a particular gender (male or female), in the "focused attention" condition, or to either gender in the "divided attention" condition. At random points during the list, the side of male or female voices was reversed. This transiently disrupted performance in the focused attention condition, more for the left ear than the right, but transiently improved it in the divided attention condition, more for the right ear than the left. Geffen summed up her results with the dictum "if in doubt attend to the right ear". In other words, disruption of attention was taken to favour the left hemisphere, and by implication enhancement of attention favours the right.

In another approach, a warning signal can be given to allow preparation of a response. This approach developed from earlier work (e.g. Lansing *et al.* 1959) which showed that reaction time could be reduced by a warning signal, provided it occurred long enough before the stimulus. (This effect increased as the "foreperiods" were increased from 100 to 200 msec, and was still strong when the foreperiod was extended to 1000 msec) Several laterality studies have been reported by Mondor and Bryden using this paradigm. For instance in their 1991 paper, consonant-vowel syllables (lasting 350 msec) were presented dichotically, with, on some trials, a preceding warning tone to indicate which ear to report. When the warning tone was advanced ahead of the target syllable by only 150 msec, the expected right ear advantage was found, but as the foreperiod was lengthened, left ear but not right ear performance improved, so that the right ear advantage was significant only for the shortest foreperiod. This result was confirmed by Mondor and Bryden (1992), who also showed that the left hemisphere disadvantage for emotional quality of speech cannot be overcome under similar conditions. These experiments show that the right hemisphere can benefit more than the left from a warning signal in this task. As in Geffen's experiment, supposed enhancement of attention favours the right hemisphere.

Not all studies of the factors modifying attention show a modification of laterality effects. Nicholls (1994b) studied the detection of a temporal gap in a continuous light.

The basic finding, already referred to in section 6.6.2.2, was that the left hemisphere could detect shorter gaps than the right. Additional signals were introduced preceding the gap-detection signals, either attracting or distracting attention from one field or the other, and their effectiveness in modifying attention could be shown by modifications of reaction time. Nevertheless they had no influence on this laterality effect for visual temporal gap detection. This laterality effect thus seems to be a more robust finding than Kinsbourne would have us believe.

10.3.2.2 Discussion of evidence

Kinsbourne and colleagues are clearly correct in their assertion that laterality effects can be influenced by so many subtle variations of method that a structural explanation of perceptual laterality cannot be the whole story. Kinsbourne (1973) writes:

"It is clear that concurrent lateralized hemispheric activity can bias attention so as to yield perceptual asymmetries comparable to those that have frequently been found empirically. If concurrent lateralized activity can have this effect, it becomes conceivable that the stimuli and task themselves are capable of eliciting lateralized cognitive activity sufficient to generate the range of laterality effects."

This statement implies that attentional biases could be the whole explanation of perceptual lateralization, with *no* requirement for inherent structural differences between hemispheres. In support of this view Lupien *et al.* (1993) showed a very good correlation amongst individuals between the degree of perceptual bias and a measure assumed to indicate attentional bias, in a visual task. Subjects were presented with two pairs of squares, one pair in each field, for 25 msec. On some trials, the members of the pair on one side were unequal in size. Subjects were required to identify, as fast as possible, the pair whose members were unequal in size. Reaction time for this task was generally long, about 1000 msec, with no laterality effect observed overall. Nevertheless, some subjects habitually showed a perceptual advantage in reaction time for detecting inequality in one field or the other, some subjects showing a consistent right field advantage, others a left field advantage. In addition, some subjects reported inequality even when there was none. This occurred for stimuli in the field to which they also had a perceptual advantage. Thus what appears to be an attentional bias correlated, over individuals, with their perceptual bias.

Thus, attentional bias might sometimes be the sole determinant of perceptual bias, as Kinsbourne's arguments suggest. However, it is clear that the circumstances in which attentional effects are important are usually those where perception is not studied in a pure form. For instance, attentional effects are shown in complex experiments using concurrent tasks. They are also shown when there is plenty of time to establish preparedness for a particular task or category of stimulus. This may be achieved by making the task predictable or by sounding a warning signal, preceding the stimulus by a long enough interval for attentional mechanisms to be activated. The experiment of Lupien *et al.* (1993), was a complex reaction time task, with relatively long response time, presumably allowing more complex processes to occur in addition to the initial perception. On the other hand, in the perceptual experiments discussed in Chapter 6, conditions of the experiment were generally simpler.

Thus it is not correct to argue against a structural explanation of laterality on the grounds that laterality effects are not robust. Perceptual asymmetries *are* fairly reliable, provided the right conditions are used where perceptual rather than attentional effects dominate (see Chapter 6). The fact that, under other conditions, many other influences can modify such perceptual laterality effects, does not refute the idea that they have a structural basis. Indeed, experiments of Ley and Bryden (1982), and Bulman-Fleming and Bryden (1994) show simultaneous lateralization to both hemispheres for different aspects of the same stimuli. In the first of these papers, emotionally-toned sentences were paired dichotically with monotone sentences. A left ear advantage was found for recognising emotional tone, with a simultaneous right ear advantage for verbal content. Bulman-Fleming and Bryden (1994) used pairs of words presented dichotically, each word being read in one of four emotional tones. Subjects had to identify words which matched a target with regard to both word and emotional tone. A "false positive", when matches were made on the basis of only one of these features, was taken as a measure of lateralization. Under these conditions it was shown that there could be simultaneously a left hemisphere advantage for words and a right hemisphere advantage for emotions. Since opposite lateralization was shown simultaneously for different features of the same stimuli, it is unlikely that a lateralized switch of the mechanisms of selective attention, specific to one or other task can explain these lateralizations.

It is also clear that, to support his attentional version of perceptual laterality, Kinsbourne assumes that there is a pre-existing laterality at other levels, namely a left hemisphere dominance for language, and a right hemisphere dominance for spatial information processing. Only by making such prior assumptions can he account for the differential hemispheric activation in perceptual tasks where lateralization is found. Thus, in his view, cognitive activity of either type shifts attention to the corresponding hemisphere. However, he makes no attempt either to explain, or to analyse these fundamental hemispheric specialisations. We have already many reasons for questioning the simple view that the left hemisphere is specialised for language, the right for spatial operations. Dissecting "language" according to its perceptual characteristics shows that several aspects of language are actually mediated better by the right hemisphere. Spatial operations show a right hemisphere advantage only when there is some degree of spatial *complexity*. The differences between the hemispheres can be described in more fundamental terms than are captured by Kinsbourne's assumptions. Moreover, the central hypothesis of this book provides a simple global structural premise from which can be derived the assumptions about laterality which Kinsbourne makes.

Apart from the issue about whether attentional factors are the real mediators of asymmetry, the most obvious comment to make from the above brief sampling of the literature on attentional influences on laterality is the great diversity of findings. Concurrent tasks can either prime or interfere with performance (according to the size of memory load). Attentional influences produce opposite changes in performance of the two hemispheres in a few studies, implying competition between hemispheres; but the more general finding is changes in the same direction (i.e. decline of performance in both). Occasionally a mild memory load can abolish a performance asymmetry by improving performance of the hemisphere opposite to that which the load deals with best*. All these effects are given an explanation in terms of "attention".

One theme on which there is some agreement in the above studies (Geffen, 1988; Mondor and Bryden, 1991, 1992) is that enhancement of attention improves performance

* A dramatic example of this was mentioned in section 6.5: Smith *et al.* (1977) demonstrated a left hand advantage for Braille reading, but when music was played to the left ear (but not the right) this asymmetry was reversed. The right hand then was not only better than the left, but was as good as the left hand in the "no-music" condition.

by the right hemisphere more than the left, and so may abolish left hemisphere superiority. This is a finding opposite to that made for general alertness: For that function, the right hemisphere is more vigilant under resting conditions, with the left hemisphere improving towards right hemisphere levels only when general alertness is enhanced (Posner *et al.*, 1987; Wilkins *et al.*, 1987). However, Geffen, and Mondor and Bryden used consonant or word stimuli, while Posner *et al.*, and Wilkins *et al.* use much simpler stimuli with no relation to language. Thus it is possible that, under baseline conditions, each hemisphere has near-optimal conditions for dealing with its own preferred type of stimulus; and only under conditions of enhanced alertness does the opposite hemisphere come to approach the performance of the hemisphere normally dominant for that type of material. The data of Kinsbourne and Byrd (1985) also fit this rule: In a visuo-spatial task, a right hemisphere advantage was found under control conditions, but this was eliminated and possibly reversed under mild activation.

If this rule held true, the instances where the right hemisphere was brought up to the performance levels of the left would fit a generalisation about alertness, rather than one about selective attention. Control of alertness would then be symmetrical provided one tested each hemisphere with material appropriate to that hemisphere. However, there are also plenty of exceptions which refute this rule: Eling (1983) used words, and found that the "normal" right ear advantage depended on conditions of alertness (high presentation rate). Mondor and Bryden (1992) found that the left hemisphere disadvantage for recognising emotions of words under baseline conditions could not be eliminated with warning cues to increase "attention"*. Hellige and Cox (1976) found effects which speak of general activation, alertness or arousal by verbal and non-verbal material alike, of both cerebral hemispheres, rather than activation of each hemisphere by its preferred type of material. They found that mild concurrent verbal memory load can activate the left hemisphere even for non-verbal tasks, and mild concurrent visuo-spatial memory load can activate the right hemisphere even for verbal tasks.

It is also worth commenting that attentional accounts of laterality seem to use reasoning which is rather difficult to pin down. Sometimes an attentional bias in circumstances without the experimental influence is inferred because of the *disappearance* of asymmetry when the influence is introduced; while in other circumstances the asymmetrical attention is inferred in the experimental situation because of the *appearance* of asymmetrical performance when the modification is introduced. This unsatisfactory situation, and the diversity of the findings for which an attentional explanation is given, arise because attention, as a psychological concept, can mean almost anything one wants it to mean. Attentional explanations can be invented *post hoc*, without there being any measure of attention other than the result one is trying to explain. This situation is likely to prevail as long as attention is defined in the shifting sands of purely psychological concepts. If however, a model of attention can be found (and becomes accepted) which is based on biological mechanisms identified at a lower level of organisation, the circularity of arguments could be avoided, and truly independent measures could be developed for correlation with the performance measures.

* When the emotion task was made more difficult, by reduction of sound intensity, there *was* a switch to right ear dominance, and the effect of a warning tone became similar to that seen with consonant identification. This probably occurred because the easy task on emotion identification depended on vowel identification, but this switched to consonant identification when the sound intensity was reduced.

The theory of cerebral function developed with the aid of our central hypothesis may be able to provide such a model. Certainly, it can accommodate very easily a combination of structural and functional influences on laterality. The structural assumption — that the left hemisphere has a richer repertoire of slow delay lines than the right — does not operate free from the principles of neuronal physiology. From those principles we know that the firing of a single neurone is influenced not only by the number of co-active afferents (related to structural matters of convergence ratios, and temporal dispersion of signals, as discussed earlier). The membrane potential of each neurone also plays a crucial part in determining the number of co-active afferents required to fire a neurone, as well as the latency of firing. A variable like membrane potential in a single neurone may have a macroscopic counterpart, the readiness of a large population of neurones to be activated. This is likely to be the physical basis of both general alertness, and the selective preparedness we refer to as selective attention, and thus can correspond to the "functional" or "attentional" determinants of laterality. In fact, in the model of selective attention of Miller and Wickens (1991) it is exactly this parameter which allows the selection of one cell assembly rather than another. Admittedly one has a long way to go before the complex of results related to "attention" can be explained in these terms. However, the attempt to analyse attentional effects on laterality in terms of anatomy, and sensory and neuronal physiology, may allow more incisive theorising about such effects than is possible using "psychological" concepts, which are not properly defined in terms of the basic variables used in the rest of science.

Given these general comments about structural versus functional (attentional) explanations for laterality, there is still the important question to ask about whether there is any lateralization of *selective* attention. Selective attention applies not only to stimulus categories which are lateralized but to many other stimulus categories as well. One of the most dramatic demonstrations of selective attention is the Stroop test. In this test, subjects are presented with a word designating a particular colour, printed in ink which may be the same colour (congruent condition) or a different colour (non-congruent condition). In such a test it is well known that congruence between the named colour and colour of the ink accelerates reaction time compared to words with neutral meaning, and noncongruence slows down the reaction time (e.g. Dalrymple-Alford, 1972). These effects are similar to those seen when a concurrent task activates a hemisphere and primes that hemisphere's performance for tasks it prefers (e.g. for a reaction time measure), and impairs the opposite hemisphere in the same task (not preferred by that hemisphere). In so far as priming and interference can be seen equally in either hemisphere when the concurrent material is that preferred by the corresponding hemisphere, *there is no real evidence of lateralization of selective attention.* Thus, no extra premise is needed to account for laterality effects related to selective attention in the human brain. The theory of selective attention in terms of cortico-striatal interplay, like that of short term memory, need not be amended, and so the hypotheses used here to explain laterality effects in the human cerebral cortices remain parsimonious.

10.3.3 Interhemispheric interaction

In section 10.3.1.2, dealing with general arousal, a premise was needed, that the two hemispheres can co-operate, so that callosal traffic from the right hemisphere could minimise the relative impairment in vigilance of the left hemisphere. In section 10.3.2.2, when dealing with selective attention, it was mentioned that Kinsbourne used a contrasting premise of interhemispheric competition to explain some of his psychological data. This clash of underlying premises requires more detailed discussion.

N.D. Cook (1986; also Cook and Beech, 1990) has put forward a variety of evidence in favour of interhemispheric inhibition (and therefore favouring the competitive variety of interaction). Some of this evidence is physiological: For instance, when the cerebral cortex is electrically stimulated on one side, the response of the homotypical region of the other side, although a mixture of excitation and inhibition, is predominantly inhibitory. Other aspects of his case come from the experimental psychology literature: Laterality effects (such as the perceptual ones discussed in Chapter 6 of this book) are shown best with bilateral stimulation (dichotic auditory stimuli, stimulation of both visual fields together, or dichaptic tactile stimuli). It is thus envisaged that when competition is possible between the two hemispheres, minor differences between hemispheres become exaggerated, to produce detectable laterality effects. Moreover, in some of Kinsbourne's work (e.g. 1970 and 1973 papers), while right field performance improved with moderate priming conditions, left field performance declined. The opposite direction of the changes in the two hemispheres indicates some form of interhemispheric competition. Further evidence adduced by Cook is clinical: In subjects with congenital agenesis of the corpus callosum laterality indices for a dichotic phoneme recognition tasks are less than in normal controls (see Cook *et al.*, 1990). In addition, it is well known that the isolated right hemisphere has a much greater linguistic capacity than the right hemisphere of subjects rendered aphasic by partial damage to the left hemisphere (see Regard *et al.*, 1994). Apparently, the remaining parts of the left hemisphere inhibit the right hemisphere in the latter situation, and prevent it exhibiting its full linguistic repertoire.

Cook and others have also produced a variety of more technical evidence in favour of interhemispheric inhibition. In one study (Wey *et al.*, 1993) subjects were required to make lexical decisions for words presented to either visual field. When such a word was presented to the left field, and a nonsense word was presented simultaneously to the right field, right hemisphere performance was significantly impaired. A mask of the right field nonsense word immediately after its presentation prevented this effect. The authors interpret this as implying inhibition from left to right hemispheres when the additional left hemisphere activation is introduced. In another study (Regard *et al.*, 1994) subjects with unilateral temporal lobe epileptic foci were studied. When left-sided seizures were in progress (as detected by EEG recording) word recognition by the left hemisphere was impaired, but at the same time word recognition by the right hemisphere was *improved*. Evidence from another group supports the notion of interhemispheric inhibition: J.M. Clarke *et al.* (1993) asked subjects to identify each of two dichotically presented CVC syllables, and correlated the laterality indices across individuals with morphological evidence of the cross-section area of the corpus callosum (obtained from MRI scans). The performance in the dichotic test by the left ear was significantly negatively correlated with several measures of callosal area, and even when the correlations were not significant, all such measures still showed negative absolute correlation values. It is implied that the greater the number of callosal axons, the greater can be the inhibition of the right hemisphere function by the left hemisphere. Similar conclusions were derived by Meyer *et al.* (1995) from experiments in which the motor cortex on one side was activated by transcranial magnetic stimulation. In normal subjects such stimulation suppressed tonic voluntary electromyographic activity in the ipsilateral hand, implying transcallosal inhibition of the cortex controlling that hand. In subjects with abnormalities in the anterior callosum (agenesis, hypoplasia, lesion etc.), this inhibitory effect was reduced or absent.

Cook and colleagues use such various data to argue that cerebral laterality is mainly not to be attributed to stable structural differences between hemispheres. Admittedly they do allow that there be initial minor differences between the hemispheres. For instance left hemisphere speech dominance may arise because of unequal access of the two hemispheres to the mid-line motor apparatus for speech. Moreover it is allowed that the left hemisphere has a speciality for the literal and denotative aspects of language; whereas the right has a predilection for contextual aspects (by which is meant implications, connotations, metaphorical uses, internal consistency, and inference, such as discussed in section 10.2.3.4). The possible mechanisms behind the latter "minor" difference are not analysed. Nevertheless, it is strongly argued that these differences between the hemispheres become exaggerated by interhemispheric inhibition to produce the wide range of other laterality effects which have been described.

Available evidence consistently points to the fact that the actual callosal axons which connect the two hemisphere are excitatory in function, not inhibitory: Neurochemical evidence suggests that these axons utilise excitatory amino acids as their transmitter substance (Sah and Nicoll, 1991; Dinopoulos *et al.*, 1989). Cytological evidence of callosal synaptic terminals reveals that over 95% of synapses made by transcallosal axons have ultrastructural profiles typical of excitatory junctions (Hughes and Peters, 1992). The predominance of inhibitory influences produced in the opposite hemisphere after electrical stimulation on one side does not reflect direct inhibitory callosal influence, since the smaller excitatory component is of shorter latency than the inhibitory one. Do doubt the inhibition is indirect, produced locally as a result of the direct trans-callosal excitation. Such inhibition is the typical result of any sort of synchronous excitation of the cerebral cortex.

Moreover, some of the evidence cited by Cook in favour of interhemispheric inhibition is capable of other, better interpretations. The paradoxical greater linguistic capability of the isolated right hemisphere (compared with the right hemisphere in the intact brain) can be explained by the argument of section 8.2: Differential reinforcement in the right hemisphere of an instrumental task such as speech is only possible when the right hemisphere is isolated. Thus, only in this case can the right hemisphere display its full potential for speech.

Despite these amendments to Cook's arguments, most of the psychological evidence for interhemispheric competition has to be accepted, even if it is not used to support the existence of direct interhemispheric inhibitory pathways. (J.M. Clarke *et al.* (1993) use the more accommodating phrase "functional inhibition".) Therefore, an alternative interpretation of these data needs to be provided. Such an alternative can be found if one explores some of the implications of the "cell assembly" concept for function of the cerebral cortex.

As discussed in section 3.1 cell assemblies are envisaged to be groups of distributed neurones which are more strongly connected with each other than with surrounding neurones. Thus, activation of a portion of the assembly may activate the assembly as a whole. Such assemblies are envisaged to be the conformations used for representing meaning in the cortex. In the human cortex, all the lateralized categories of information considered so far should be represented by cell assemblies. The emphasis so far has been on the *differences* between the hemispheres. However, for other purposes it is possible to think of the two hemispheres as part of a single "super-cortex", with callosal connections being but one subset of all cortico-cortical connections. From this viewpoint, cell assemblies may have component neurones in either hemisphere. This need not undermine concepts of hemispheric specialisation: A cell assembly representing a lateralized category of information is likely to have the majority of its constituent neurones in one hemisphere, with a smaller fraction in the opposite hemisphere connected via callosal rather than ipsilateral axons. Such a perspective can be used to provide an explanation of interhemispheric competition, without recourse to the unlikely postulate of direct interhemispheric inhibition. Let us consider an example from the evidence mentioned above in this context: Laterality effects in perceptual experiments are shown best when there is stimulation of both hemispheres by different stimuli.

Envisage that each stimulus is represented by a cell assembly. Each assembly is located predominantly in the hemisphere to which the corresponding stimulus is directed, with smaller parts in the opposite hemisphere. Moreover, when either cell assembly is activated, it can be assumed that this causes an inhibition of most other neurones in the same locality, which are not part of the assembly. It follows that if the two stimuli are of a type

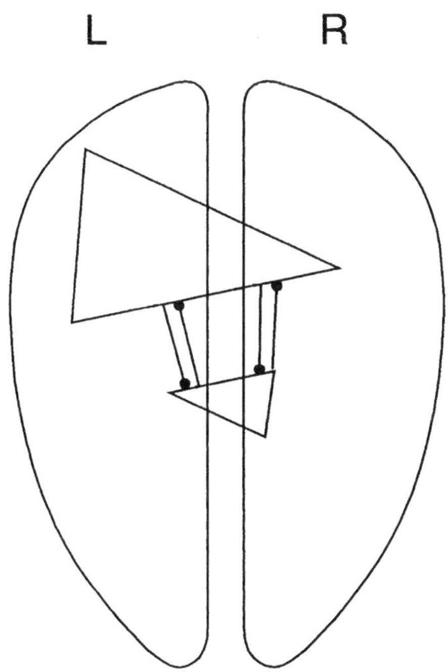

Figure 10.1 Local Inhibition Between Bilateral but Asymmetrical Cell Assemblies can Produce "Functional Competition" Between Hemispheres. This diagram depicts in schematic form the situation of stimulation with two stimuli (one to each side) whose perception is lateralized. Each stimulus activates a cell assembly (triangle) which is mainly in the hemisphere to which the stimulus is directed, but partly also in the opposite hemisphere. As a result of the lateralization, the cell assembly on one side is larger than that on the other (triangles of unequal size). Within each hemisphere there can be mutual inhibition between cell assemblies (connections between portions of triangles within a single hemisphere). The minor fraction of the left hemisphere assembly within the right hemisphere is larger than the major fraction of the right hemisphere assembly in that hemisphere. Hence the right hemisphere assembly will overall be inhibited by the left hemisphere one, even though no assumption is made about direct transcallosal inhibition.

which is represented better in one hemisphere than the other, the local inhibition by the dominant assembly will tend to suppress the few members of the other cell assembly which are located in the same hemisphere. This local inhibition reduces activity in that assembly as a whole, including its majority of neurones in the opposite hemisphere. The effect produced appears to be competition between hemispheres, and, as Cooks argues, this exaggerates any primary differences between the hemispheres. Similar reasoning can be used for other results mentioned above. However, the apparent interhemispheric competition is achieved by *local* inhibitory processes together with a reduction of transcallosal activation, rather than by direct transcallosal inhibition. This is depicted schematically in Figure 10.1.

Apart from local competitive effects between parts of bilateral cell assemblies, there are other possibilities to account for the psychological evidence of interhemispheric competition. There is for instance some evidence of interference in attentional mechanisms between sides at a subcortical level (see Lambert, 1993). This could include competition at the striatal level. This might seem implausible, since, according to the model of Miller and Wickens (1991) competition can only occur locally within a striatal domain (see also Wickens, 1991; Wickens *et al.* 1991), and the competing signals would be relayed to striata on opposite sides. However, there are some crossed cortico-striatal fibres. Moreover, a cell assembly activated predominantly in the right hemisphere might have some of its neurones in the opposite hemisphere, involving callosal connections, so that uncrossed cortico-striatal connections of two different cell assemblies could compete within the striatum of one side. In summary competitive effects between hemispheres can be incorporated into the basic biological model used here, without using them as an almost completely different premise for lateralization.

10.4 LATERALITY AND THE REPRESENTATION OF THE EMOTION

10.4.1 Introduction

Cerebral representation of emotion is a complex subject. A major division of the subject, along basic sensorimotor lines, is between *perception* and *expression* of emotion. Most of the next section is about different aspects of emotional perception.

Perception of emotion may be influenced by two quite different sorts of stimuli: *Neutral stimuli* may secondarily acquire emotional meaning because of what they are used to signify. Examples of this include the auditory stimuli used to represent emotion (e.g. by means of prosody in speech, or of melodic line or harmony in music), or the perception of visual stimuli used to convey emotion (e.g. body gesture). Similarly, emotion may be conveyed in the visual or auditory sense by explicit word meanings; or it may be merely inferred without ever using a word of direct emotional meaning. Quite apart from neutral stimuli which acquire secondary emotional meaning, emotional experience may be affected by *primary emotional stimuli*, in other words stimuli used as rewards or punishments. The primary and secondary emotional attributes of stimuli are not sharply separate in practice, and sometimes, it is not clear whether the emotional attributes of a stimulus are primary or secondary. For instance with facial expressions conveying emotion, it is difficult to say whether the stimulus is a primary emotional one, or one which has secondarily acquired emotional value.

Registration of primary or secondary emotional stimuli may be expected to occur not only at the level of sensory perception, but also because of the emotional state the stimulus elicits. For the secondary emotional stimuli, it is likely that the sensory perceptual process is dominant. Thus, an experimental subject may be able to identify the emotion in a face or tone of voice on the basis of its sensory features, without the stimulus actually evoking the corresponding emotional state. However, one cannot exclude the possibility that the emotional states elicited by such stimuli may help in their identification.

For the primary emotional stimuli, the elicited emotional state may assume much greater importance. Indeed, these states may be explicitly manipulated experimentally by use of positive and negative emotional stimuli in various reinforcement contingencies. However, these states may also be controlled in less explicit fashion by techniques of hypnosis or self-suggestion. Usually the emotional state so created lasts many minutes or longer, in which case they could be referred to as "moods". However, with specially devised methods using very sudden reinforcement contingencies, quite transient emotional states can be controlled experimentally.

The present subsection deals with lateralization of both the sensory perceptual aspects of emotion, and the emotional states produced by sensory input. Motor expression of emotion is left to a later subsection. In addition, no mention will be made of lateralization of mood as inferred from study of endogenous mood disorders. While this topic is undoubtedly relevant, it raises so many new issues in a subject already full of complexity, that the thrust of the overall argument would be lost.

Even when one confines oneself to lateralization of emotion within neutral stimuli, or of emotional states induced by various means, the evidence is quite complex. In the past, experiments have been guided by two main hypotheses:

(i) The right hemisphere plays a dominant role in perception of all emotional attributes of stimuli, regardless of the type of emotion.

(ii) Emotions may be categorised according to their "valence" (that is whether they are subjectively pleasant i.e. "positive", or unpleasant i.e. "negative"). According to this "valence hypothesis" the right hemisphere is often regarded as dominant for the negative emotional aspects of stimuli, the left for the positive emotional attributes. Whether this interaction between hemisphere and emotional valence applies to emotion attributed secondarily to neutral sensory stimuli, or to the emotional states generated by primary emotional stimuli is an important point for discussion in the sections below. If the valence hypothesis does apply to *sensory perception* of secondary emotional stimuli it is be expected that identical stimuli presented to either hemisphere would be recognised more readily by the right hemisphere if they are also of negative emotional valence, and by the left hemisphere if they are also of positive emotional valence. If the valence hypothesis applies to the *emotional states*, resulting from primary emotional stimuli, one would expect that self-report of affective state induced by stimuli directed to one or other hemisphere would betray the lateralization of valence; and slow autonomic or EEG indicators of affective state might give further evidence of this.

There is another possible implication of the valence hypothesis which should be mentioned, because it has been experimentally tested: Induction of an emotional state of positive or negative valence might have asymmetrical effects on the two hemisphere's ability to process emotionally neutral stimuli. Evidence for such an effect, and its implications, are discussed below.

No firm predictions are offered from this discussion with regard to the above two hypotheses. However, preliminary and tentative suggestions can be made in the light of the central hypothesis, and other consequences of it discussed in previous chapters. These will at least form a basis for organising the evidence about lateralization of perception of emotions and emotional states. Perception of sensory stimuli (such as prosody, musical contours of characteristic form, facial expressions, and the sound of certain spoken words or the sight of certain written words) may be lateralized to one side or the other, according to perceptual principles already analysed in previous chapters. All such stimuli might have identifiable emotional attributes as well as sensory perceptual features. There is no reason to expect any change in perceptual lateralization just because the stimuli become the vehicle for emotional signals. However, this may not apply if we also consider associated emotional states and how they are arise:

From Chapters 8 and 9 we have found that certain types of instrumental acts are dealt with better by one or other hemisphere. The left hemisphere is better for performing instrumental tasks involved temporally-structured planning, for instance in rapid reaching movements or in speech; while the right hemisphere is apparently superior for instrumental tasks involving sensory guidance of limbs, and possibly in auditory guidance of prosody.

Instrumental actions are goal-directed, and the success of each in fulfilling its intended goal is a central part of the process by which the program for each action is acquired. When an instrumental task is carried out successfully, it is rewarded, and a corresponding positive emotional state is generated. When performance is unsuccessful, a negative emotional state may be generated. Thus, activity in the left hemisphere plays a dominant role in successful speech performance, or dextrous limb movements, and this will be accompanied by positive emotional states. When the right hemisphere attempts to program the same actions, it is less likely to be successful, and therefore more likely to be accompanied by negative emotional states. In principle, one may also envisage that motor acts involving sensory guidance would be more likely to be successful when controlled from the right hemisphere, which would then be accompanied by positive emotional states; whereas similar acts controlled from the left hemisphere would be less successful, and so would be accompanied by negative emotional states.

If one assumes that the most common and important human instrumental acts are left hemisphere ones (speech, manual and limb praxis), one can generate a prediction that the left hemisphere would be preferentially associated with positive emotional states, the right with negative ones. However, the associated prediction is that under special circumstances (e.g. for instrumental tasks related to sensory guidance), the reverse lateralization of emotional states might emerge. It is also worth pointing out that the theory of instrumental action developed in the first part of Chapter 8 implicates cortico-striatal interplay, and the anterior parts of the cortex are much more important than the posterior parts in this interplay. Therefore, one might expect that such lateralization of emotional representation according to valence would apply mainly to the anterior regions of cortex.

Given these arguments, one should revisit the topic of lateralization of neutral stimuli which acquire secondary emotional value. Often, such stimuli will have acquired that emotional value because when they appeared, it signified success in an instrumental task. For instance, in conversation, one may know that one has expressed one's intention successfully, if the other person smiles or laughs. In principle the same might be true if the person with whom one converses responds by employing certain styles of prosody, or tone of voice. Therefore, in principle, neutral stimuli which have acquired emotional significance (such as facial expression or tone of voice in another person), might have done so in association with the motor programs of the first person, which elicited them. If those motor programs are elaborated preferentially in one hemisphere, the sensory stimuli in the other person which are their consequences would have a rewarding effect only with respect to that hemisphere. Perception of the rewarding properties of these stimuli would then become preferentially lateralized in the same sense as the motor programs which, in previous experience, have elicited them. It should be stressed that these suggestions are tentative: Whether (and to which sort of stimulus) they might apply cannot be predicted precisely. The empirical evidence must decide this.

The rest of this section reviews this evidence in relation to the above hypotheses. Section 10.4.2 deals with emotion in relation to sensory perception. Many studies show that in a general sense the right hemisphere is specialised for perception of emotion. As an example, Bear (1983) used a non-verbal test of emotional discrimination, where emotion was conveyed by facial expression, body language or non-verbal vocal expression. Right brain-injured subjects were impaired in discriminating emotions conveyed in multiple ways. However, the results of all tests used in this study of emotional discrimination can readily be explained in perceptual terms, without recourse to the hypothesis that there is something special about the right hemisphere for emotion *per se*. Thus, in section 10.4.2 we start from the supposition that there is no lateralization of emotion contained within sensory stimuli, except in so far as it is determined by perceptual principles already discussed, and seek to falsify this idea. Some of the evidence shows that there is lateralization of emotion independent of perceptual lateralization. Evidence for the valence hypothesis from such perceptual experiments is weaker and not so clear cut. Section 10.4.3 discusses the evidence on the lateralization of emotional states. Here there is stronger support for the valence hypothesis. In the process of evaluating hypotheses of progressively increasing specificity, evidence will not be discussed comprehensively, but will focus on the more discriminating evidence, which indicates principles of lateralization going beyond those already discussed.

10.4.2 Lateralization of emotion in sensory perception

10.4.2.1 Perception of emotion in auditory stimuli

Two experiments in the auditory modality show that when adequate controls are used, no lateralization specific to the emotional attributes of auditory stimuli exist. Safer and Leventhal (1977) presented subjects with taped passages, whose emotionality could be varied by altering two parameters: The degree of emotional intonation could be set at three levels, and the strength of emotional word content could also be set at three levels. Subjects rated the passages as positive or negative emotionally. With left ear presentations 29 out of 36 subjects based their judgements on voice tone. With right ear presentations, 21 out of 36 based them on word content. This is a highly significant difference

($p < 0.01$). In other words, emotion *per se* is not lateralized. Rather, its lateralization depends on the vehicle by which it is presented. Blonder *et al.* (1991) provided further evidence for this conclusion, using unilaterally brain-damaged patients. Right brain-damaged patients were impaired on recognition of prosody *per se*, and on several tasks where the emotional associations of prosody were tested. However, no evidence was obtained for a right hemisphere advantage for emotional representation in speech separate from prosody. Moreover, in a task where the emotion of a sentence had to be inferred with no cues from prosody, or from words with direct emotional meaning, the right brain-damaged patients were not impaired compared with left brain-damaged or healthy control subjects. These results again indicate that the right hemisphere is not specialised for emotional representation as such, but rather for the links between prosody and emotions. Since prosodic cues commonly occur in association with other cues to emotion (gesture, word content, etc.), it is not surprising that representation of emotional significance of speech is right hemisphere dominant, because it is "grafted" on to prosodic representation, which is already represented in the right hemisphere.

For musical stimuli, Bryden *et al.* (1982) found a robust left ear advantage for identification of emotional tone in seven-note tonal sequences in either major or minor mode. However, this example of lateralization is probably again not based on perception of emotion, but on perception of contour or non-temporal patterns of intervals, in a manner similar to that found by Charbonneau and Risset (1975) and Samson and Zatorre (1988). In all these cases one would predict a right hemisphere dominance, which has nothing to do with lateralization of emotion *per se*.

Regardless of this issue, there is also no evidence for the valence hypothesis from auditory perceptual experiments. Blonder *et al.* (1991) find no differential lateralization of positive versus negative emotions as detected by prosody. This is confirmed by Bulman-Fleming and Bryden (1994). In a dichotic test, a right hemisphere advantage for emotional vocal tone was obtained for positive as well as negative emotions.

10.4.2.2 Perception of emotion in visual stimuli

Several papers report on laterality effects for recognition of emotion in faces. Generally these find similar right hemisphere superiority for recognition of both facial identity and emotion. However, there are several indications that independent right hemisphere processes are used for the two sorts of task. In some cases the right hemisphere advantage for recognition of emotion in faces is of greater degree than that for recognising facial identity. These include the study of Suberi and McKeever (1977) who showed, in a short-term memory test, a significantly greater left field advantage for emotional than for non-emotional faces. Similarly De Kosky *et al.* (1980) found that right brain-damaged subjects had a greater impairment, relative to left hemisphere patients or controls in recognising emotion in faces than in recognising facial identity. Buchtel *et al.* (1978) found that discrimination of emotional faces gave a right hemisphere advantage, whilst discrimination of neutral faces was carried out better with the left hemisphere. Ley and Bryden (1979) using tachistoscopic presentation to normal subjects of faces differing in character and emotion, found a significant left field advantage for both face identity and emotion. However, statistically, these were separate effects. Bowers *et al.* (1985) compared right hemisphere-damaged patients to normals and left hemisphere-damaged patients in identification of faces and their emotions. The right hemisphere patients were impaired compared to the other groups on both aspects of the task, but were also impaired on recognition of emotions when groups were matched for identification of faces. In other studies, the difference between judgements of facial identity and facial emotion is suggested by the fact that reaction time is faster for the former than the latter (Strauss and Moscovitch, 1981; Etcoff, 1984). In another study (Ladavas *et al.*, 1980) it is noted that visuo-spatial judgements usually give a bigger right hemisphere advantage in male subjects, but for judgements of facial emotion, effects are bigger in females. In all these studies of facial emotions both positive and negative valence were tested and the left field advantage for emotion applied to both.

It has been suggested (Ley and Bryden, 1979) that evidence for right hemisphere specialisation for stimuli with any variety of emotional attribute might arise because emotional stimuli have a higher associability than neutral stimuli, and so require a higher degree of multimodal integration than do neutral ones. This could well arise because primary emotional stimuli are a powerful focus for formation of classical (Pavlovian type) associations. Since the right hemisphere may have a richer semantic network than the left, a wider variety of stimuli could become associated with primary emotional ones in the right than the left hemisphere. This conjecture is testable using word stimuli. If it were true, it is predicted that emotional words should *not* show a right lateralization compared with non-emotional words matched for associability.

Most of the papers on lateralization of perception of facial emotion give no support to the valence hypothesis. For instance, Heller and Levy (1981) used facial composites, with one half-face smiling, the other half not smiling, presented for 150 msec. Right handers perceived the face as happier if the left side (left visual field) was smiling than if right side was. Since this result is in the direction opposite to the valence hypothesis, it gives that hypothesis no support.

However, apart from papers showing right lateralization for facial emotions whatever their valence, a few other reports on perception of emotion in faces appear to support the "valence" hypothesis. Reuter-Lorenz and Davidson (1981) and Reuter-Lorenz *et al.* (1983) tachistoscopically presented faces. Right handers gave faster reaction times for right-field happy faces and for left-field sad faces. Natale *et al.* (1983) using faces presented for 200 msec, found that left-field presentations were judged emotionally more negative than right-field ones. With shorter durations (20–80 msec) individually set to allow merely the detection of a face, left-field presentations allowed more accurate discrimination of negative from positive affect (more extreme ratings for both judgements). They conclude that there are two independent effects: There is a right hemisphere advantage for perception of both valencies of facial emotions, for short duration stimuli, as well as a left hemisphere bias towards positive aspects of emotional stimuli with longer exposure. Borod *et al.* (1986) found that right brain-damaged subjects were impaired with respect to left brain-damaged ones for identification of negative but not positive facial emotions.

A physiological study gives support to this evidence for an interaction between perceived facial emotion and valence: Gur *et al.* (1994) measured regional cerebral blood flow while subjects discriminated faces on the basis of either their age or their affective expression. During age discrimination the right temporal lobe was activated, in correlation with performance in the discrimination. During discrimination of facial affect, the left frontal lobe (and, marginally, the right parietal lobe) were activated for positive emotions, while the right parietal lobe alone was activated during discrimination of negative emotions.

One report casts some doubt on the inference from the latter experiments that the left hemisphere really is specialised for perception of positive emotional valence in faces. Bryson *et al.* (1991) replicated earlier tachistoscopic work showing a left field advantage for perception of emotion of happy faces, and a right field one for that of sad faces. However, these workers exploit the fact that faces, including emotional ones are not symmetrical. For instance, Sackheim and Gur (1978) have found that chimeric facial stimuli (made using the left side of a posed facial expression for both sides of the chimera) were judged by other subjects to be more intense emotionally than double right-sided chimeras. Heller and Levy (1981) showed the same with chimeras in which only one side (left or right) of a smiling face was used. Borod and Caron (1980) found that the left face was more expressive than the right, but, in this study, there was no difference between positive and negative emotions for laterality of expressed facial emotion. However, there are also suggestions that usually the greater expressiveness of the left face applies to all negative emotions (disgust, fear, anger, sadness) but not for happiness (especially for "posed" emotion) (reviewed by Thompson, 1985), for which it is said that expression is symmetrical. For instance, in a later study of Borod *et al.* (1981) male right handers showed greater expressiveness of the right face for happy emotions, whereas expression of negative emotions tended to correlate negatively with handedness. Bryson *et al.* (1991) draw specifically on such evidence. In their own experiments they found that left hemisphere dominance for perception of positive emotion did not appear with inverted faces, and was reversed with mirror image face stimuli. Thus the apparent support for the valence hypothesis may be a function of the fact that the left side of the face (projecting to the left visual fields) is a stronger emotional stimulus, rather than arising from any intrinsic left hemispheric bias to perception of positive emotions. This would account for the results of Reuter-Lorenz and colleagues, Natale *et al.*, and Gur *et al.*, all of which used either normal faces, or, if using chimeric faces, failed to use mirror images amongst the stimuli. However, it does not account for the results of Borod *et al.* (1986) who used emotional pictures other than faces, and included mirror-reversed stimuli.

Apart from visual experiments using faces, one report deals with the perception of emotion in visually presented words. Wexler *et al.* (1986) report that as word meanings shifted from positive to negative value, the hemispheric advantage, derived from error scores, shifted more towards a right hemisphere advantage. This could imply that the left hemisphere represents positively valued words better than the right. However, it is not made clear that there is a true left hemisphere advantage, because only asymmetry scores are reported, not raw data for each hemisphere.

In conclusion there is some agreement that the right hemisphere advantage for emotional faces may be larger than can be explained by the right hemisphere advantage for faces as such, or is independent of it, statistically or in other ways. This result is not predicted by the arguments of section 10.1. However, it is possible that the right hemisphere advantage in perception of Gestalts can be subdivided into several processes, differing, for instance, in the level or "direction" of processing needed to abstract the relevant perceptual features, or the breadth of associations which are activated. Recognition of facial emotion may require a higher degree of processing than that of facial identity, as indicated by the longer reaction time for the latter task. Moreover it may use the schemata for recognising action-related stimuli of the parietal lobe.

Recognition of facial identity on the other hand may use the schemata for visual object recognition, localised in the temporal lobe (as indicated by the result of Gur *et al.,* 1994).

Given this, there is also some evidence that positive emotion in faces is identified better by the left hemisphere, and negative emotion better by the right. The result of Bryson *et al.* (1991) casts some doubt on this view, and suggests that such valence lateralization is actually an effect of asymmetry of the face stimuli themselves, rather than hemispheric asymmetry with respect to emotions. However, this does not explain all the evidence of interaction of valence effects with laterality for emotional visual stimuli.

10.4.3 Lateralization of positive and negative emotional states

10.4.3.1 Subjective emotional experience

A small number of reports investigate lateral differences in perception of emotional stimuli of different valence. Dimond *et al.* (1976) and Dimond and Farrington (1977) measured intensity of emotion in film sequences relayed to either visual field. The earlier paper found that left field presentations of films, especially surgical operations, were rated more "unpleasant" than right field ones. Overall ratings of "pleasantness" or "humour" showed no difference between visual fields. Dimond and Farrington (1977) repeated this and showed ratings of unpleasantness were greater for left fields with both surgical films, and humorous cartoons.

There has also been a number of reports on lateralization of emotional states during unilateral forebrain anaesthesia (the sodium amytal test of Wada and Rasmussen, 1960, sometimes carried out before neurosurgery in patients with epilepsy). These reports are not fully consistent, and some of them report on emotional expression rather than on subjective emotional experience. The most recent report of this sort (Ahern *et al.*, 1994) finds evidence of a small but definite predisposition towards negative emotional state when the amytal test is applied on the left side, with no change reported when the test is conducted on the right side. Older papers (Terzian, 1964; Lee *et al.*, 1990) support this finding in terms of observed change of emotional expression, and Sackheim *et al.* (1982) report that right hemispherectomy is associated with euphoria. However, it should be noted that Kolb and Milner (1981b) failed to find any asymmetry of emotion.

10.4.3.2 Physiological markers of emotional states

Induced emotional states have been used as the independent variable in experiments where various physiological indicators of activation are used. Three papers have used lateralized eye movements as indices, it being assumed that leftwards eye movements indicate right hemisphere activation, and *vice versa.* Schwartz *et al.* (1975) found that right handed subjects tended to look to the left when answering affective questions. All the questions they mentioned involve negative emotion, and so this result does not distinguish between the "right hemisphere for emotion" hypothesis and the "valence" hypothesis. Tucker *et al.* (1977a) studied eye movements during stressful testing (subjects were told that the answers would be used to assess intelligence or personal stability), or non-stressful testing (subjects were told that the apparatus was being calibrated). The former questions produced more leftward eye movements and fewer rightward ones than the latter. Ahern and Schwartz (1979) found that questions evoking positive emotions elicited eye movements suggestive of left hemisphere activation, while those evoking negative emotions elicited right hemisphere activation. Thus from these papers there is some support for the idea of differential hemispheric activation during induced emotional states of opposite valence.

Dimond and Farrington (1977) found that left field presentation of surgical films produced greater heart rate increases than did right field presentations. The reverse pattern of asymmetry was found for humorous cartoon film sequences. Compared to a "neutral" travel film, left field presentations of the cartoon produced lower heart rates. Right field presentations caused about equal heart rates for both the travel film and the cartoon.

A number of papers has also assessed cerebral activation with EEG measures. Usually the suppression of alpha activity has been employed as the measure of activation. Davidson and Schwartz (1976) found, in females, that right hemisphere alpha suppression occurred during self-induced emotions, both anger and relaxation. Only parietal EEG traces were analysed. Harmon and Ray (1977) asked subjects to recall events from their own past to induce positive and negative emotional states. The EEG average power in the 3–30 Hz range was used as the dependent measure. Left hemisphere activity built up gradually during the positive states, and progressively declined during the negative states. In the right hemisphere a similar but much weaker effect occurred. Their data is also compatible with the notion that EEG changes in the right hemisphere in either direction reached their stable plateau or trough level more quickly than those in the left (which is relevant to the lateral differences in alertness or vigilance tests, described in section 10.3.1). Karlin *et al.*, (1979) used a global measure of EEG activity (i.e. power averaged over all frequencies), and found that negative affect was accompanied by increase of activity in the right hemisphere. Davidson *et al.* (1979) asked subjects to give a continuous rating of the positive or negative affect of emotional film sequences. During positive affect there was relative left frontal activation (alpha suppression) while during negative affect there was a right frontal activation. The same effects were found during self-generated emotion. Parietal asymmetry did not distinguish positive and negative affect. Tucker *et al.* (1981) made EEG measurements during hypnotic induction of euphoric or depressed mood. In comparing euphoria to depression, the right hemisphere decreased activation (increased alpha power) and the left hemisphere increased activation (decreased alpha power) during euphoria. These changes were noted only in the frontal lobes. Ahern and Schwartz (1985) also found that alpha suppression in the frontal lobes was induced by positive emotion on the left side, and by negative emotion on the right. This was not seen in the parietal lobe, although differentially lateralized activation by verbal *vs.* visuo-spatial stimulation was seen there. Davidson *et al.* (1990b) induced happiness with films of animals playing, and this was accompanied by left temporal EEG activation (alpha suppression). Disgust, induced by surgical films, induced activation in right frontal and temporal regions. No differences in asymmetry which could be attributed to different emotions was seen in central and parietal regions. Sobotka *et al.* (1992a) recorded the EEG during a reaction time task, cued with warning signals. Reward was available in some trials, punishment in others (money was given or taken away), these contingencies being controlled according to reaction time in the preceding trial. There was more right frontal activation (alpha suppression) in punishment than reward, and more left frontal activation in reward than punishment. There was no sign of differences between contingencies in any other regions. The "contingent negative variation" (CNV) was also systematically larger on the left side, but this did not interact with reward/punishment contingencies. Sobotka *et al.* (1992b) used a task where accuracy of aiming with a photoelectric gun determined contingencies of reward or punishment. Event related potentials from the occipital cortex were found to give greater response in the right than the left hemisphere, but this was equal for both reward and punishment.

The most recent investigation of this topic has used measures of regional cerebral blood flow to measure lateralized activation. George *et al.* (1995) induced "transient happiness" by asking subject to recollect past incidents which for them had strong positive affect. This led to a fall in regional flow in the right frontal region.

As a whole the consensus from the most precise EEG data show that emotional stimuli activate mainly the anterior parts of the hemispheres, and possibly the temporal regions, but not central or parietal regions. Given this, there is also substantial evidence for a valence/laterality interaction with respect to alpha suppression, this occurring more in the left hemisphere with positive emotions, and the right with negative ones. Again this is mainly a frontal effect, and is not found in central or parietal EEG traces. The valence hypothesis receives further support from data using lateralized eye movements, heart rate changes to unilateral visual stimuli, or regional changes in cerebral blood flow. It is also compatible with the consensus from studies of self-report of subjective affect. Overall, the evidence is compatible with a version of the "valence" hypothesis: Emotional stimuli impinging on the right hemisphere elicit more negative experiences than they would on the left; emotional stimuli of positive or negative valence also activate preferentially the frontal lobes of the left and right hemispheres (respectively).

10.4.3.3 Effect of induced mood states on laterality of perception

There is also substantial evidence that emotional states of opposite valence influence perceptual processes, such that the performance in tasks for which a particular hemisphere is specialised are impaired by the emotional valence which activates that hemisphere.

Tucker *et al.* (1981) induced euphoric or depressed mood states in students by hypnotic means. While in this induced mood state subjects were presented dichotically with tones of various intensities, and were required to state in which ear the tone was louder. During euphoria there was little bias. During depression a right ear (left hemisphere) advantage emerged ($p < 0.025$). In addition visual imagery ratings during depression were less. This observation suggests a relative right hemisphere impairment during naturally induced negative emotional states. (However, performance of mental arithmetic, another right hemisphere function, did not change during depression.) McKeever and Dixon (1981) asked their subjects to develop in their minds negative or neutral imagery, aided by suggestion. They were then presented with a reaction time task, involving faces with neutral expression. Under the negative imagery condition, female subjects showed a left hemisphere advantage not seen under the neutral imagery conditions. Males showed no asymmetry under either condition. Ladavas *et al.* (1984) also assessed simple reaction time measured to lateralized visual stimuli in normal subjects, during self-induced negative affect (instructions were given: "You must feel sad") or while subjects imagined themselves in non-affective daily life scenes. Both types of mental activity slowed reaction times (by 70 or 40 msec in each test). However, in the sadness test there was significantly greater RT increase to left-field stimuli, again suggesting a right hemisphere impairment.

Derryberry (1989) provided subjects with reports of "success" or "failure" in the previous trial of a reaction time test, and these reports were used as independent variables influencing the next trial. Following report of "failure", reaction time to both left and right field improved, but the improvement was greater in the right field, so that the overall left hemisphere advantage increased. Following reports of "success" there was slight slowing for both fields, but more for the right field, so that overall left hemisphere advantage decreased. In a later paper (Derryberry, 1990) positive, negative or neutral feedback was provided by one of three signals indicating (deceptionally) whether speed on the previous trial had been fast, medium or slow. Increasingly negative feedback slowed reaction times, provided left field/left hand combination, and non-verbal rather than verbal feedback were used. In both studies a relative right hemisphere impairment was produced during naturally induced negative emotional states.

Two reports provide evidence in conflict with the above studies. Strien and Morpurgo (1992) presented tachistoscopic stimuli, consisting of three consonants, to either visual field, requiring subjects to report as many as possible of these letters. Normally a right field advantage was produced. If the stimuli were preceded by an emotionally threatening word, left field performance improved, to give an overall left field advantage. If preceded by an emotionally relaxing word, right field performance improved to give an enhanced right field advantage. Asbjornsen *et al.* (1992) used a dichotic listening task with CV syllables, which normally gives a right ear advantage. Three levels of arousal were used, and for each positive or negative incentive was induced. The only effect which changed the right ear advantage was high arousal combined with negative incentive (threat of electric shock). This slightly improved left ear performance, and slightly worsened right ear performance, to give overall no ear asymmetry.

Although the majority of the above studies report right hemisphere impairment as a consequence of induced negative mood, some papers report opposite effects. The reasons for these discrepancies are far from clear.

10.4.4 Lateralization of perception of emotion: discussion

There is no evidence of lateralization of emotion represented in auditory sensory stimuli. For visual stimuli, facial identity and facial emotion (both positive and negative) are represented preferentially in the right hemisphere, but there seem to be separate processes involved for identity and emotion. The latter probably requires a deeper level of processing than the former, but both may be aspects of the right hemisphere superiority in representing complex simultaneous patterns (Gestalts). Some evidence suggests opposite lateralization of positive emotional aspects of stimuli (to the left hemisphere) compared with negative ones (to the right), though it has been suggested that this result may be due to factors other than hemispheric asymmetry.

Laterality of representation of emotional states, and the consequences of such states is more complex, and has several aspects. Emotion is experienced more negatively if it is induced by identical stimuli directed in the right hemisphere than in the left. Induced negative emotion activates the right hemisphere more than the left in electrographic terms. This differentially lateralized activation applies mainly to frontal regions of the hemispheres, not to the more posterior parts. Induced negative emotion also produces a shift toward *left* hemisphere dominance in perceptual tests. Induced positive emotion activates the left hemisphere more than the right. Inactivation of the left hemisphere by amobarbital leads to a transient negative shift of mood.

These findings provoke the search for a parsimonious and comprehensive explanation, but it is difficult to provide one with any certainty. If one makes the assumption (as in section 10.4.1) that in humans, the instrumental acts which are most commonly and universally rewarded are left hemisphere ones, one can explain a left hemisphere predilection for positive emotional valence. When those same common and universal instrumental acts are attempted on the basis of right hemisphere activity, they are more likely to fail. Hence, with the same assumption, one can account for the right hemisphere predilection for emotions of negative valence.

This argument provides a simple and convincing explanation for the negative emotional change when the left hemisphere is inactivated by amobarbital, accompanied as it is by aphasia and motor impairment. It also fits well with the evidence that explicit reward activates the left hemisphere and explicit punishment activates the right. This account is particularly convincing, in view of the fact that these effects are mainly specific to the frontal lobes. These, it will be remembered, are the chief recipients of activity from the basal ganglia, where, it has been suggested (Chapter 8) reward and punishment exert their immediate effect on cortical information storage systems. Clinical evidence also gives this interpretation some support. In cases of unilateral brain damage, pathological laughing is mainly associated with right brain damage, while pathological crying is associated with left hemisphere injury (Sackheim *et al.*, 1982). However, it is significant that left hemisphere damage causing fluent aphasia (Wernicke's) may also be associated with pathological laughing.

However, the same differential lateralization of cerebral activation by emotional stimuli of opposite valences is seen when those stimuli are *not* part of explicit reward or punishment contingencies. It is seen in regional cerebral blood flow studies whilst subjects are discriminating facial affect (Gur *et al.*, 1994; section 10.4.2.2). Moreover focal epilepsy with positive emotions during the ictus (ictal laughing) is found mainly with left sided epileptic foci (Sackheim *et al.*, 1982). Such findings *might* yet be an aspect of the same process: The representation of an emotional stimulus of a particular valence, induced by experimental means other than explicit reward or punishment, or by ictal activation may somehow generalise to the representation of that same valence, as previously acquired during natural instrumental behaviour. The fact that, in the study of Gur *et al.* (1994) positive facial affect activated the left *frontal* lobe is consistent with this suggestion. The same argument, based on generalisation, might furthermore explain why emotional stimuli tend to be experienced as more negative, if directed to the right than the left hemisphere. However, these suggestions are speculation, and beg questions about the process of the generalisations, which cannot at present be answered. Nevertheless, this account makes some sense.

In view of this seemingly coherent account, it is deeply puzzling why negative emotion induced in seemingly similar ways causes a shift, in most studies, in the direction of *left* hemisphere perceptual superiority. Admittedly such perceptual processing can be anticipated to involve regions of the cortex different from the frontal regions implicated by instrumental tasks, or their associations. Nevertheless, the hypotheses advanced so far still do not provide an account of this perceptual shift.

10.4.5 Lateralization of expression of emotion

In overall terms, the right hemisphere is superior to the left in several aspects of emotional expression. For instance, as mentioned above, the left side of the face is more expressive than the right (Sackheim and Gur, 1978; Thompson, 1985). Damage to the right hemisphere is accompanied by loss of the gestures of face, limbs and body which normally go along with speech (Ross and Mesulam, 1979). Furthermore, right hemisphere damage leads to loss or abnormality of prosody, with restriction of expressed emotionality (Tucker *et al.*, 1977b) as discussed in section 9.4.4.

It is unlikely that such aspects of right hemisphere function are specific to *emotional* expression. Borod *et al.* (1985) noted that loss of emotional expressiveness in right hemisphere patients did not correlate with the degree of emotional expression conveyed by spoken words. Ross and Mesulam (1979) report that their right hemisphere patients had full insight into their inability to convey emotion by prosody and gesture, and were distressed and frustrated as a result. Shapiro and Danley (1985) found that the loss of prosody in right hemisphere patients applied as much for non-emotional as for emotional prosodic expression. The relation of the right hemisphere to facial expression and body language is more likely to reflect the right hemisphere's superiority in co-ordinating complex motor synergies, suggested in section 9.4.5 (though there is as yet no clear evidence for this type of lateralization). Loss or abnormality of prosodic expression is likely to reflect loss of the slower aspects of programs of organised speech, and/or loss of auditory/motor guidance (see section 9.4.4)

10.4.6 Summary and discussion

The evidence reviewed in the above subsections is complex, and in some respects inconclusive. While the right hemisphere is in a loose sense specialised for many aspects of the representation of emotional perception, this is mainly due to emotional attributes of prosody or facial expression which are right-lateralized, for basic perceptual reasons. More specific evidence in favour of the idea that the right hemisphere is specialised for emotions *per se* (of all types) comes from studies of facial affect rather than auditory stimuli. This is probably still an aspect of perception rather than of emotion itself, but one that requires a deeper level of processing than does facial identity.

A varied collection of evidence supports the valence hypothesis, giving the left hemisphere a special role for positive affect, the right hemisphere one for negative affect. Some of this evidence comes from studies of facial affect, but it appears likely that this is due in part at least to asymmetry of the perceived facial stimuli, rather than hemispheric asymmetry related to emotion. The best evidence for the valence hypothesis comes from experiments using hemifield presentation of emotional film sequences, and from studies of cerebral activation measured by lateralized eye movements or EEG alpha suppression.

Experiments studying the influence of induced mood changes on perceptual tasks provide further evidence of the effect of emotional valence, but the usual result observed — impairment of right hemisphere perceptual function by induced negative affect — is difficult to integrate with the physiological evidence of right hemisphere activation by negative affect.

10.5 LATERALITY EFFECTS FOR HIGHER COGNITIVE PROCESSES: SUMMARY

There is no evidence for asymmetry in the basic process of short term memory, either in its duration or its capacity. Nevertheless, because the information upon which short term memory works has different structure in the two hemisphere, the products of working memory differ significantly between the hemispheres. The right hemisphere is superior for integrating overall melodic contour, and for most aspects of prosody. There may be equivalent right hemisphere specialisations in the visual mode. The right hemisphere is also specialised for large scale semantic integrations, at the level of sentences and beyond. The right hemisphere superiority for visual imagery, prosody, and higher semantic integrations, combined with its significant capacity for acoustic (if not phonemic) representation of words gives that hemisphere a very considerable capacity for comprehension of language, which is qualitatively quite different from the left hemisphere's "dominant" role in language. Short term memory function in the left hemisphere is strongly influenced by that hemisphere's perceptual superiority in representing temporal structure. Since the capacity for temporal chaining in the left hemisphere is great enough to bridge between those salient events whose sequence is recorded by short term memory, the left hemisphere can represent the temporal structure in long temporal chains (complete melodies and sentences). This may also explain the left hemisphere's role in both production and interpretation of normal grammatical construction in sentences. Once again, the specialisation of each hemisphere with regard to the products of short term memory are those which would be expected from the central hypothesis of this book.

There is a clear asymmetry for generalised alertness: The right hemisphere is more easily activated than the left, and can maintain that activation in vigilance tasks more easily. This asymmetry is also predictable from the central hypothesis of this book.

On the other hand, selective attention shows no asymmetry between hemispheres, though asymmetries can arise because material for which one or the other hemisphere has a superiority, biases attentional processes in the corresponding hemisphere. The latter effects occur mainly in tasks involving higher-level integration than the basic perceptual tasks discussed in Chapter 6.

With regard to lateralization of emotion in sensory perception, it is far from proven that the asymmetry applies to emotion *per se*, as opposed to perception requiring high-level processing. Likewise evidence from perceptual experiments for differential lateralization of emotional cues of different valences does not withstand close scrutiny. However, there is good evidence that enduring emotional states of opposite valences are differentially localised, positive ones to the left hemisphere, and negative ones to the right, with the frontal lobes being principally involved in this distinction. This effect can be derived from the known lateralization of certain types of instrumental behavioural programs, and is thus indirectly derived from the central hypothesis. Lateralized effects of induced emotional states on perception have been described, and are difficult to relate, either directly or indirectly to the central hypothesis.

11 Correlations Between Different Aspects of Lateralization, and with Gender

11.1 INTRODUCTION

We have now surveyed the evidence for laterality effects in all major aspects of brain function, including sensory perception, long term memory, various aspects of motor function, short term memory, general alertness, selective attention, and emotional perception and expression. Each of these major functional categories was broken down into a variety of smaller subdivisions. With only a few significant exceptions, the laterality effects described could be predicted from or explained by the central hypothesis advanced in section 3.6. However, given the great diversity of the evidence reviewed, it is important to ask whether the many laterality effects described are aspects of a single unitary variable, which differs in monolithic fashion between hemispheres, or alternatively are relatively independent aspects of a multi-faceted manner of lateralization. Therefore it is necessary to consider the intercorrelations between different aspects of laterality, when studied in the same individuals.

There is another theoretical reason for considering correlation between different aspects of laterality, and their correlation with gender. When motor lateralization is considered, handedness is of course an important measure. However, this measure is especially useful because it has also been used as a variable in morphological studies of the brain, as mentioned in Chapter 4. In similar way, morphological asymmetries differ between genders. In both left handers and in females subjects, the morphological differences between the hemispheres are less than in males and right handers. Thus, if functional lateralization *is* based on anatomical differences between the hemispheres (including that proposed by our central hypothesis) one would expect that it would be less pronounced in left-handers and females than in right handed males.

Apart from the theoretical reasons behind the present book, there is an important practical reason for studying the intercorrelations between different laterality measures: In neurosurgical operations, where one lobe is to be removed to control epilepsy, it is important to know which is the language hemisphere. Specifically, it is crucial to detect the minority of cases where speech has a right hemisphere representation. The most rigorous way of testing this is by the use of the sodium amytal test (Wada and Rasmussen, 1960), where each hemisphere is selectively anaesthetised by injection of short acting barbiturate into the carotid artery, and the resulting speech and other impairments are assessed. However, this is a difficult procedure, with some hazard, and alternative, non-invasive tests are preferable if they have as good predictive value. Several of the studies described below have had this aim as their primary motive.

There are no studies which attempt to study very many aspects of lateralization together, in any comprehensive way in a population of subjects. However, there are a number of papers which attempt more limited intercorrelation of different measures of laterality. In the sections below we deal initially with evidence of intercorrelation between laterality effects for corresponding tests in different *perceptual* modalities (section 11.2). In similar way correlations between motor lateralization effects using *different effectors* (i.e. handedness *vs.* speech laterality) are dealt with in section 11.3. Thirdly, inter-correlations between the broader classes, of perceptual and motor lateralization, are considered (section 11.4). The present chapter is not intended to be a comprehensive review of the correlational evidence available. Nevertheless, it gives a clear indication that overall lateralization of function is the integral result of many other smaller factors, which need not co-vary exactly.

11.2 CORRELATIONS BETWEEN LATERALITY EFFECTS IN DIFFERENT SENSORY MODALITIES

Chapter 5 offered a series of predictions about perceptual aspects of laterality. These predictions were made in a general sense, rather than about specific sensory modalities. The evidence reviewed in Chapter 6 made it clear that these predictions did indeed apply to visual and auditory modalities, as well as to somatosensory stimuli in a few studies. This supports the conclusion from Chapter 4 that there be global differences between the hemispheres, quite independent of those found in specific language areas of the left hemisphere.

However, a few studies have investigated the correlation between asymmetries in visual and auditory tests of the types described in Chapter 6. Two studies (Zurif and Bryden, 1969; Hines and Satz, 1974) show that in the right handed majority, congruent left hemisphere advantages in visual and auditory tests occur in the majority of subjects. Nevertheless, statistical tests in two studies (Zurif and Bryden, 1969; Dagenbach, 1986) show that this congruence is not greater than would be expected by chance association of traits which separately occur in the large majority of subjects. Zurif and Bryden (1969) find that different tests within the same modality correlated across subjects, but tests with different modalities did not.

Some studies *do* report significant, albeit weak correlations between laterality effects in different modalities. Hines and Satz (1974) show a significant correlation between laterality scores in the auditory and visual modalities ($p < 0.05$). Piazza (1980) used two different dichotic tasks (words and environmental sounds) and two different tachistoscopic tasks (words and faces). Correlations between tests were generally congruent with the laterality scores on the individual tests, but were seldom statistically significant. Graves (1983) also found significant correlations between laterality scores using tachistoscopic presentation of words, and dichotic word presentations. There may also be a tendency for visual and auditory perceptual lateralization to be more strongly linked to each other in non-right handers (Dagenbach, 1986), though this was not found by Hines and Satz (1974). Bradshaw *et al.* (1981) used visual tests which are expected to give left hemisphere advantage (a verbal reaction time test) or right hemisphere advantage (a facial identity reaction time test). A strong negative correlation was found for the result in these two tests, such that subjects were either strongly right hemisphere dominant for faces and weakly left hemisphere dominant for words, or vice versa. A combination of strong advantages in each hemisphere for each of the tasks preferred by that hemisphere was not found. A study by Kim and Levine (1992) used three visual half-field tasks and three dichotic listening tasks. Within each modality the three tests correlated over individuals, so that characteristics of each subject accounted for about 50% of the total variance. In cross-modal statistical analyses both modality-specific and more general hemispheric effects were found. Tentatively the authors concluded that modality specific factors were more important than the general ones.

Apart from these studies in normal subjects, a study of brain-injured subjects supports a relation between lateralized visual function and receptive aphasia. Swisher and Hirsh (1972) presented two different-coloured stimuli in centre field and subjects had to tell which order they came in. The time difference needed to tell the order accurately was longer for fluent dysphasics (with presumed left temporal lobe damage) than for non-fluent dysphasics (with presumed left frontal damage), normal controls or right lesioned patients.

There is room for much further work about the detail of these correlations. However, any correlations between laterality effects in different modalities appear to be weak. Undoubtedly in individual subjects there is often a dissociation of the hemispheric advantage for one modality from that in another.

11.3 CORRELATIONS BETWEEN MOTOR LATERALITY EFFECTS FOR DIFFERENT EFFECTORS

Laterality effects in the motor domain concern mainly two sets of effectors: the vocal apparatus, and the hands and limbs. Study of the correlations between the lateralization for these two sets of effectors is possible in several ways: One can use handedness as a measure of manual motor laterality and either aphasia (after unilateral brain damage), or the results of the sodium amytal test as the measure of speech laterality. Alternatively, for cases of left brain damage, one can study the correlation between apraxia and aphasia. The amount of correlative data available for correlations between handedness and spoken language lateralization is greater than that between apraxia and aphasia. Thirdly, one can investigate correlations between the lateralization for different effectors demonstrated in normal subjects.

It is well known that, in right handed subjects, aphasia results almost entirely from left hemispheres lesions. Only 1–2% of cases have right hemisphere damage. Thus in right handers cerebral dominance for speech and manual praxis are both in the left hemisphere. When cerebral dominance for manual praxis shifts towards the right hemisphere — that is in left handers — one may ask if language lateralization is also shifted.

Goodglass and Quasfadel (1954) studied 123 cases of left handers with unilateral brain damage, half on the right side and half on the left. Regardless of the side of the damage, aphasia occurred in the vast majority of cases, and there was no difference in its incidence between right and left hemisphere lesions. Overall the incidence of aphasia in left handers (80–90%) was greater than in right handers (a finding confirmed by Hecaen *et al.*, 1981). These findings appear to point to a much greater involvement of the right hemisphere in language in left handers than in right handers, though the authors found that *strong* right hemisphere lateralization of language was very rare, even in left handers. The authors also concluded that in left handers, language was commonly bilaterally represented, so that damage to either side could cause impairment.

The idea of bilateral representation of language in left handers has been criticised by Hammond and Kaplan (1982). These authors argue that the earlier study overestimated the number of subjects with bilateral representation, since the clinical sample included those with brain damage and aphasia, but omitted those with brain damage but no aphasia. (Subjects of Goodglass and Quasfadel (1954) were selected by the presence of aphasia.) Assuming that only 50% of cases without aphasia are included, Hammond and Kaplan calculate that there need be no left handers with bilateral language representation, but rather a mixture of subjects with right and left sided unilateral language. However, other results (e.g. Zatorre, 1989) using the sodium amytal test, leave no doubt that language can have a bilateral representation: Out of 61 epileptic patients awaiting neurosurgery, 22 showed disruption of language with amytal injections on both left and right side. Regardless of this issue, it is still clear that left handers have a much higher rate of right hemisphere language, whether language as a whole was unilateral or bilateral.

The model of Hammond and Kaplan also had to conclude that left handers have a more diffuse language representation than in right handers, because aphasia for lesion of either hemisphere is more common for left handers than for right handers. Other findings point to the same conclusion: Goodglass and Quasfadel (1954) and others find that aphasia in left handers is more transient than in right handers. Brown and Hecaen (1976) find that left handers with aphasia have less clear-cut aphasic syndromes with localising value than do right handers.

The evidence from these studies is that left handedness does predispose towards right hemisphere language, but this appears not to be a very tight correlation. Brown and Hecaen (1976) give a figure of 20–30% of aphasic left handers as having right hemisphere lesions. More detail is given by Hecaen *et al.* (1981): When left handed patients have right hemisphere damage 33% or them (in either sex) suffer aphasia as opposed to 3% of right handed males or 13% or right handed females with right hemisphere damage. However, even in left handers, left hemisphere lesions are more likely to cause aphasia (in 69–89% of subjects according to Hecaen *et al.*, 1981) than are right hemisphere ones.

Similar conclusions can be drawn from a study comparing handedness and language lateralization as assessed by the sodium amytal test (Geffen and Caudrey, 1981). Handedness and language laterality were weakly but significantly correlated ($p < 0.05$). There were a enough exceptions to the correlation that handedness could not be used in practice as a measure of language lateralization.

In correlating aphasia and apraxia, it will be recalled from section 9.4.5, that the different tests of apraxia and communicative gesturing correlate for a given effector (oral effectors or hands). However, the syndrome for oral effectors is commonly *dis*sociated from that for hands. Goodglass and Kaplan (1963) found that aphasic patients might have an impairment in gesture and imitation with hands, but this deficit was not related to severity of aphasia, once auditory comprehension was controlled for. Kimura and Archibald (1974) and Selnes *et al.* (1991) also clearly describe cases where there is dissociation of manual apraxia from aphasia.

Overall one must conclude that the concurrence of left hemisphere control of both speech and hand movements is strongest in right handers. In left handers language *may* be switched to the other hemisphere along with hand control, but such concurrence is weaker than in right handers. Even in right handers it is clear that lateralization of motor function is not of a unitary nature, since cerebral lesions may differentially affect speech and manual praxis. Thus one may consider that there are parallel (and perhaps functionally analogous) pathways for oral and manual control, but with somewhat different locations within the fronto-temporo-parietal regions.

There is also a little evidence comparing motor laterality for different effectors in normal subjects. Results mentioned in section 9.4.6, of Chaurasia and Goswami (1975) will be recalled. They found that facial expressiveness was somewhat stronger in the left face of right handers, a probable indication that the right hemisphere can control more complex motor synergies than the left. They also found that in left handers, there was a marked majority of subjects (73%) for whom facial expression was stronger in the right face. Borod and Caron (1980) also found that the left facedness of right handers was not present in left handers. In a subsequent paper Borod *et al.* (1981) showed that right handed males were more expressive with the right face for expressing happy emotions, and this facial asymmetry was reverse in left handed males. Lomas and Kimura (1976) tested the interfering effect of speaking on a sequential tapping (sequence of fingers from index to little finger as fast as possible for 15 sec). As expected, speaking interfered with right but not left handed tapping in right handers. In left handers, speaking interfered with both hands. This result is consistent with the clinical data suggesting an increased bilateral representation of language in left handers.

Lastly, two studies report on eye movement laterality as a function of handedness. Pirozzolo and Rayner (1980) and Hutton and Palet (1986), as already mentioned, find that saccadic eye movements have a shorter latency in right handers when directed to the right side than to the left. While the exact reasons for this are unclear, both studies find that the asymmetry disappears in left handers.

11.4 CORRELATIONS BETWEEN MOTOR AND PERCEPTUAL LATERALITY EFFECTS

11.4.1 Dissociations of motor and sensory aphasia

According to the discussion of Chapter 8, and the evidence of section 9.2, the receptive and the productive aspects of language are closely intermingled, and to some extent make up a single conjoint function. Therefore, quite apart from the issue of whether all aspects of lateralization constitute a unitary functional specialisation, we should expect these two aspects of language lateralization to be closely correlated. In left handers, where there is a tendency for the right hemisphere to take over some of the roles normally carried out by the left, speech is sometimes represented bilaterally. In such cases, the theoretical arguments presented in Chapter 8 would suggest that the shift to right hemisphere or bilateral representation of language should apply equally to its productive and receptive aspects. The issue has not been widely studied. Efron (1963b) tested left head-injury patients in an auditory temporal order judgement. Non-aphasic head-injury patients (presumed to have a right hemisphere injury) were unimpaired. Aphasic patients (with presumed left hemisphere injury) were significantly impaired, but the expressive aphasic patients were more severely affected than the receptive aphasic ones (see section 6.6). This would not be expected if temporal analysis were more important for comprehension than expression of speech and also raises some doubts about the interdependence of the two aspects of language. Naeser and Borod (1986) describe two cases of left handers with large fronto-parieto-temporal lesions on the right side, which demonstrated Broca's aphasia but good comprehension. The authors suggested that in these subjects, Broca's area is on the right but Wernicke's area is on the left. Were this true, it would tend to undermine the theorising of Chapter 8, where time-structured modelling of speech was only possible when the delay lines from Broca's to Wernicke's area were direct ones. If the two areas were in opposite hemispheres, they could still be connected via delays lines, but it is less likely that they would be direct. However, the interpretation of their data given by Naeser and Borod may be incorrect: It is known that in right handers, the right hemisphere has a considerable capacity for comprehension, which is not associated with right hemisphere speech. The same may be true for the left hemisphere in left handers. It remains unresolved whether the productive and receptive aspects of language can be focused in opposite hemispheres.

11.4.2 Correlations of speech laterality and perceptual laterality

Three papers give detail about the correlation between language laterality as determined by the sodium amytal test, and lateralization in dichotic listening tasks requiring phonetic discrimination. Geffen and Caudrey (1981) found that reaction time data in the dichotic word test gave correct identification of language laterality in 30/37 subjects (including five with right hemisphere language). Data on accuracy of word identification gave correct prediction of laterality in 31 cases. Data from the dichotic tests were a better indication of language laterality than handedness data. Even better prediction of language lateralization could be obtained when data on both handedness and the dichotic tests were used in a discriminant function analysis: Correct prediction was then achieved in 35 out of 37 cases, including 2 correctly assigned as bilateral. Strauss *et al.* (1987) found that in a group of 63 epileptic patients awaiting surgery with left hemisphere language, a dichotic listening test gave a right ear advantage in 86%. For those with bilateral language the figure was 71%, while for those with right hemisphere language it was 50%. Zatorre (1989) tested 61 epileptic patients facing temporal lobectomy with the sodium amytal test and a dichotic word test. The dichotic word test predicted language laterality in 33/35 with left hemisphere language and 4/4 with right hemisphere language. However, there were also 22 patients with bilateral language. For these, the dichotic test results straddled the symmetry point, overlapping both the other groups. The contribution of handedness could not be assessed because this was a selected atypical group with more than normal bilateral/mixed cerebral dominance. Overall dichotic listening tests are better predictors of language laterality than is handedness, though dissociations are still quite common.

11.4.3 Correlations of handedness and perceptual laterality

An early study by Efron (1963a) studied judgements of simultaneity of two stimuli presented to opposite sides of the body (somaesthetic stimuli to opposite hands; opposite visual fields). Such judgements showed a consistent delay for the stimuli delivered to the left side of right handers, and to the right side of left handers. Additional data of Teuber (1962) using auditory stimuli were reanalysed to reach a similar conclusion. Efron concluded that conscious comparison of the time of occurrence of any two sensory stimuli requires the use of the hemisphere controlling the dominant hand, and that additional transcallosal conduction delays occur for the stimuli directed to the hemisphere controlling the non-dominant hand. A later study by Hammond (1981) showed that there was better temporal acuity for somatosensory stimuli presented to the preferred hand, and this generalisation included one strong left hander. Most work on the relation between handedness and perceptual lateralization concerns visual and auditory stimuli, and is reviewed below.

11.4.3.1 Visual tests

Visual measures of cerebral lateralization usually correlate with handedness. Amongst the tests where a right visual field advantage is expected in right handers, most studies report reduced lateralization in left handers. Zurif and Bryden (1969), employed three different tachistoscopic tests, and found that the majority of right handers (80%), and left handers without familial sinistrality (70%) showed a left hemisphere advantage. Left handers who had a family history of left handedness were equally divided (50% each) between left and right hemisphere dominance. Even in this selected group there was not a majority with right hemisphere dominance. Thus the correlation of handedness with visual lateralization seems to be similar to that with language functions: Left handedness predisposes toward the right hemisphere taking over some of the roles of the left, but this amounts to a reversed dominance in only a minority of subjects. Jeeves and Dixon (1972) found that the right hemisphere advantage in reaction time tests for visual stimuli is not seen in left handers. McKeever and Gill (1972a) used a masking paradigm in which two different letter stimuli were presented sequentially in the same field position (with 30 or 130 msec intervening between the onset of the two stimuli). Right handers showed an overall right field advantage for recognition of letters. Left handers showed a trend toward left field advantage, which became significant for recognition of the initial stimuli. In a subsequent paper using the same task (McKeever *et al.*, 1973) the right field advantage was seen in right handers without a family history of left handedness, but not in right handers with such a history, or in left handers. Davidoff (1975) tested subjects with rectangles of different brightness. 21/24 right handers saw the left side as lighter, while only 11/24 left handers saw left field as lighter. Schmuller and Goodman (1979) used high and low imagery words in a tachistoscopic tasks, and found that right handers and non-familial left handers reported more accurately from right fields, while familial left handers were more accurate with the left fields. Christman (1989b) gives detail of laterality as a function of both exposure time and handedness. He used a task involving temporal integration of digits, with partial images separated by different time intervals. Right handers showed a left field superiority at the shorter intervals (30 and 55 msec) which switched to the right field superiority at 80 msec, as discussed in Chapter 6. Left handers however, showed a left field advantage at both 55 and 80 msec separation. The reversal of lateralization as duration is increased did not appear as it did in right handers. One study fails to find the trend towards reversed laterality in left handers for tachistoscopic tests: Graves (1983) found that 80% of right handers and 77% of left handers showed a right field advantage. The discrepancy with other studies might be resolved if the left handers were non-familial, but this detail is not stated in their report.

Amongst the tasks where a left field advantage is to be expected, varying effects of left handedness have been reported, but they are usually in the direction of reversal of the lateralization expected in right handers. Kappauf and Yeatman (1970) found for right handers that the latency for detection of the start of a visual stimulus was longer for right field presentations than for left. In left handers, the reverse was found. Curiously, the latency for detection of stimulus cessation did not follow the same pattern: In right handers it was equal in either field, while for left handers the latency was longer in right fields. Gilbert and Bakan (1973) constructed chimeric faces either from the right side of a photographed face and its mirror image, or the left side and its mirror image. Subjects were asked to say which of the two composite faces was most like the original or the reversed original. The results showed that facial images projected to left fields had greater salience for right handers, but this asymmetry was lost in left handers. Likewise, Levy *et al.* (1983) found that right handers judged a chimeric face as happier if the left face was smiling than if the right face was smiling. Left handers showed some leftwards bias, but to a lesser degree than right handers. Strauss and Goldsmith (1987) asked subjects to make a visual matching to a target of laterally presented faces. A questionnaire was used to assess handedness, footedness, eyedness and earedness. None of these predicted laterality in perceptual tasks well, but eyedness was the best predictor. One study using visual stimuli compares lateralization in verbal and non-verbal tests in both right and left handers (Piazza, 1980): Negative correlations were sometimes seen (usually non-significant) between verbal and nonverbal laterality scores, as expected if they involve opposite hemispheric specialisations. However, this pattern was more typical of right handers than of left handers.

11.4.3.2 Auditory tests

In auditory tests of lateralization a variety of the relationships with handedness are reported. Several studies show, as with visual tests, that there is a reduced left hemisphere advantage in left handers. Zurif and Bryden (1969), using two different dichotic tests found that right handers, and left handers without familiar sinistrality had the expected left hemisphere advantage. Left handers with a family history of left handedness were equally divided between left and right hemisphere dominance. In other words, as with several results with visual stimuli, there was a shift to the right hemisphere but no overall majority of right hemisphere dominance even in this selected group. A similar result was obtained with accuracy of report as the dependent measures as when using the percentage of subjects with advantage to each hemisphere. Shankweiler and Studdert-Kennedy (1975) found that the degree of right ear advantage for CV syllable recognition correlated with measures of the degree of right handedness. Natale (1977) used a dichotic listening test involving short rhythms, with recognition of already-presented rhythms as the response measure. Male and female subjects were graded according to their degree of handedness. A right ear advantage was detected correlating in degree with degree of dexterity. Graves

(1983) and Iaccino and Sowa (1989) also found that right ear advantages in dichotic tests were reduced in left handers.

Strauss (1986) and Strauss and Goldsmith (1987) failed to confirm that lateralization of manual praxis was correlated with lateralization of auditory perception. In the earlier paper, neither hand preference nor eye preference (as determined by a questionnaire) predicted ear advantage in a dichotic task. Ear preference was the best predictor. In the later paper auditory tests were used which usually give right hemisphere dominance (matching the tone of voice between dichotic and binaural presentations). Neither handedness, footedness nor earedness was a significant predictor of laterality in the perceptual task. Eyedness was the best predictor.

Lake and Bryden (1976) obtained a result which was actually in conflict with the consensus view described above. They found that right handers are like left handers with familial sinistrality in having a high proportion with a right ear advantage in a dichotic task. Male left handers *without* family history were the group with the highest incidence of left ear advantage.

Another significant result is that of Bryden *et al.* (1991). Subjects were tested in discrimination of phonetic differences and affective tone in dichotically presented words. The right ear advantage for the phonetic task was slightly reduced in left handers. The left ear advantage in the identification of affective tone was slightly larger in left handers. In other words, whatever is determining lateralization, asymmetry in phonetic judgements is not simply the mirror of that in recognition of prosody. The two effects seem to have independent determinants, so that decreased leftwards asymmetry in one can occur at the same time as increased rightwards asymmetry in the other.

11.4.3.3 Hemispatial neglect

There is one paper comparing hemispatial neglect after left sided brain injury in right and left handers (Maeshima *et al.*, 1992). Amongst right handers this symptom was relatively transient (lasting less than six months), as described in section 10.3.1.1. Amongst non-right handers however, the symptom was usually longer lasting, as it is for right hemisphere lesions in right handers.

11.5 GENDER CORRELATIONS WITH LATERALITY

Many studies of functional lateralization report differences between male and female subjects. It is not necessary to review them thoroughly here, though there may be interesting ways in which the present psychobiological theory can be developed from such evidence. However, it is appropriate to give a few indications of the thrust of such evidence, using papers which have already been cited.

In an auditory study, Lake and Bryden (1976) found that females, regardless of handedness had an increased incidence of left ear advantages in a dichotic listening test for which a right ear advantage would be expected. Young and Ellis (1980) used monaurally presented words, with binaural white noise to make the task more difficult. In recognition tests, males showed a greater right ear advantage than females. Iaccino and Sowa (1989) found a right ear advantage provided attention was directed to the right ear, while in females this manoeuvre had no such effect. Not all auditory studies find interactions between laterality and gender. For instance Kelly and Orton (1979) found no gender effects on laterality for recall of successive dichotic word pairs.

In a visual study Bradshaw and Gates (1978) used tasks with either lexical decision or overt naming of tachistoscopically-presented words. For overt naming, males showed a right visual field, with both reaction time and accuracy as the dependent measure. In females in this task, a similar asymmetry was seen. However in the lexical decision task, males regularly showed a right field advantage, while females often showed a left field advantage. This suggests that spoken language is strongly lateralized to left hemisphere in both sexes, but perceptual lateralization has a stronger left hemisphere bias in males than in females. This may be an indication of the interdependence of perceptual and motor aspects of lateralization where speech production is concerned (as discussed in Chapter 8 and section 9.2), whereas for other functions the two aspects of laterality can exert their independence. Christman (1988) used a temporal integration task (over 55 msec) and found that females showed an overall left field advantage not seen across so many conditions in males.

Several visual studies of gender effects use faces as stimuli, for which a right hemisphere advantage would be expected. Ladavas *et al.* (1980) found a left field advantage for recognition of facial emotions in females, but in males there was no asymmetry. Rizzolatti and Buchtel (1977) found a left field reaction time advantage for recognising faces in males, but no asymmetry in females. Strauss and Moscovitch (1981) found a right field advantage for identifying facial emotions in males, but no difference between hemispheres in females. On the other hand, as with auditory tasks there are studies which find no interaction with gender. Diehl and McKeever (1987) found for both face recognition and object naming tasks that there was a significant left field advantage, but this did not interact with gender.

In a tactile task used by Witelson (1976) dichaptic recognition of nonsense shapes was required. Boys showed right hemisphere advantages as early as 6 years, girls had bilateral equality of skills as far as 13 years.

O'Boyle and Hoff (1987) required subjects to perform mirror tracing of random forms, a task with clear requirement for visuomotor guidance. Males showed significant advantage to the non-dominant hand and females show a non-significant trend in same direction.

McGlone (1980) reviewed the evidence on gender effects upon laterality, and concluded that overall females show less pronounced functional laterality effects than males. However, she writes: "when statistically analysed, sex-by-laterality interactions often tend to be weaker than the main effect of laterality, and they are easily altered by other factors, such as age, aetiology of the lesion, strategies and test procedures. Thus one must not overlook perhaps the most obvious conclusion, which is that basic patterns of male and female brain asymmetry seem to be more similar than they are different."

11.6 SUMMARY OF CORRELATIONS, AND DISCUSSION

In the paragraphs above, many correlations were described, mostly weak ones. These included correlations between different measures of perceptual laterality, between different measures of motor laterality, and between perceptual and motor laterality. Generally measures which in individual studies gave advantages to one hemisphere were positively correlated with others giving the same direction of advantage. However, in all cases there was plenty of room for exceptions in individual cases. Attempts to predict language lateralizations from non-invasive psychological tests are generally more successful using auditory tests than handedness, and still more successful if composite scores derived from several tests are used than if only a single test is employed (Geffen and Caudrey, 1981; Graves, 1983). Dagenbach (1986) found that congruent hemispheric advantages in dichotic and tachistoscopic tests (both using linguistic material) were more likely to be found in right handers than in left handers. Bryden *et al.* (1991) found that the reduction of left hemisphere superiority in one test, found in left handers, could not be used to predict degree of lateralization in left handers for another test for which right hemisphere superiority is expected. All this points clearly to the conclusion that *lateralization is not a unitary function, but is one made up of many individual components.*

This fits one aspect of the central hypothesis, that lateralization is a global result of the integrated effects of asymmetry in the calibre of axonal connections, repeated many times amongst different populations of cortico-cortical connections. Presumably different populations of axons, with different spectra of delay times mediate perceptual integration in visual and auditory senses or in time-structured motor modelling. Although all such populations may tend to contain more fine calibre or unmyelinated axons in the left than the right hemisphere, the characteristics of each population need not vary in exact correlation. With that as a general morphological premise, it follows easily that the degree of functional asymmetry may vary for different functions, and the side of advantage in different tests may not correlate over individuals.

In the situation of brain injury we already know that the dissociation of different aspects of laterality is likely to be based on regional differences: Apraxia and aphasia, while both motor disorders due to left hemisphere lesions, and possibly of analogous form, require damage in different regions. The correlational data reviewed above is consistent with the idea that many other aspects of lateralization, as studied in normal subjects, may also reflect regional differences in whatever factor underlies functional asymmetries. The fact that in left handers and females the morphological evidence indicates a lesser degree of asymmetry than in right handed males is in accord with the functional data that left handers and females have lesser degrees of functional asymmetry. Two reports give a few examples of the detailed individual anatomical variation in the brain which may underlie variants in functional lateralization. Foundas *et al.* (1994) found that, out of twelve patients subjected to both the sodium amytal test and MRI scanning, the one with right hemisphere language was also the only one with right hemisphere superiority in the size of the planum temporale. A later paper (Foundas *et al.*, 1995) extended this finding by showing a lack of asymmetry in superior temporal lobe structures in left handers. Overall the correlational data, and associated morphological data indicates clearly that morphological aspects of the brain are an important determinant of laterality, albeit not the sole determinants.

12 Summary, Synopsis of Predictions, and Concluding Remarks

12.1 SUMMARY

The argument of this book has covered a large area of research, spanning both anatomical and physiological data, as well as the psychological literature, which forms its largest part. The following points summarise the main thrust of the reasoning in previous chapters.

Chapter 1

1. If human language is to be given a scientific explanation, it is necessary to analyse language in terms of the same physical variables used in the rest of science. This means that neurolinguistics needs to describe its subject matter in terms of the variables used more widely in the anatomy and physiology of the brain, with especial attention to the physical characteristics of language segments considered as sensory stimuli and as motor responses. Psychological constructs, however useful, cannot substitute for this.
2. Given this, the argument in this book is developed from two clear facts about human language: (i) Most aspects of language are served by the left hemisphere (in the right handed majority); (ii) Both the perception of linguistic signals and their production involve processing of information with fine temporal structure, on the scale of about one tenth of a second. These two facts are taken as points of departure for an examination of mechanisms underlying cerebral laterality in a much wider sense than just language.

Chapter 2

3. In perception of elementary speech sounds, discriminations need to be made in duration and relative timing on a scale ranging from a few tens up to a few hundred

milliseconds. Production of speech sounds involves construction of motor command sequences with temporal structure on the same fine scale. These facts help to define what is lateralized in the human hemisphere, by focusing upon the representation of temporal structure in sensory stimuli and motor responses. However, because of the complexity of the speech articulators and their innervation and action, there can be no one-to-one correspondence between motor commands and movement, let alone between motor command and phonemes.

Chapter 3

4. Our best model for the representation of meaning in the cerebral cortex is the cell assembly concept of Donald Hebb (1949). Such assemblies are conceived to be distributed groups of neurones more strongly linked amongst themselves than with other surrounding neurones. The linkages are formed by a process of synaptic modification predicted by Hebb, and now known to exist in many structures of the brain, especially the cerebral cortex. Abeles has developed a version of the cell assembly concept, which he calls the "synfire chain" in which linked neurones fire with a temporal structure which is precisely measured over intervals up to a few hundred milliseconds. A body of experimental evidence demonstrates the existence of such temporal structure for firing patterns amongst neurones in various parts of the forebrain, especially the cerebral cortex. Such time-structured chains of neural activation in the cerebral cortex are likely to be the physical representations of time-structured information at the psychological level, whether in language perception and production, or for other categories of information.
5. The most likely neural process for representation of such precise temporal structure in the brain is axonal conduction time. However, in a network of interlinked neurones, where interactions are determined by convergence of excitatory post-synaptic potentials, and Hebbian synaptic change is dependent on co-operativity between co-active afferents, this process cannot register temporal structure to an accuracy much finer than 10 msec, the approximate time scale of these convergent interactions within each neurone. Therefore, use of axonal conduction time to represent temporal structure in sensory or motor signals will be possible only for networks where conduction delays are greater, perhaps much greater, than 10 msec.
6. If such long delay lines are used for representation of temporal structure, there will be an associated loss of amplification of signals transmitted along a temporal chain of neural activity. This will make such temporally-extended chains of activity vulnerable to transmission failure, unless activity is "boosted" at each step of the chain, by additional synaptic activations.
7. Axons in a network with conduction times of 10 msec or less may have a special role for maintenance of activity, and for ensuring security of transmission in temporally-extended chains of activity. They should also be important in representing spatial or "quasi-spatial" patterns, rather than temporal patterns.
8. Empirical evidence of cortico-cortical axonal connections using electrophysiological techniques is all biased, in a variety of ways, to give an undue impression of the importance of the rapidly conducting connections. However, the best data available from small-brained mammalian species, indicates, despite its biases, that conduction

times of several tens of milliseconds are common. If the biases are allowed for, and simple projections of the data are made for larger-brained animals, it is entirely plausible that in the human hemispheres there are abundant connections with conduction times up to several hundred milliseconds.

9. Given this, the central premise of this book arises simply, when one considers the lateralization to the left hemisphere of many language functions. *It is envisaged that (in the right handed majority) the cortico-cortical connections in the left hemisphere have a higher proportion and a richer repertoire of connections with long delay times (100 msec or greater) than those in the right hemisphere. The right hemisphere correspondingly may have a higher proportion of rapidly conducting axons than the left hemisphere.*

Chapter 4

10. From the central hypothesis a number of morphological consequences follow: The major prediction is that the left hemisphere, or lobes of it, should have a higher grey to white matter ratio than the right. Also, other things being equal, the right hemisphere and its lobes should be larger than those on the left.
11. Empirical data, using a variety of methods converges to suggest that the right frontal and temporal lobes are larger than their counterparts on the left, due mainly to smaller volumes of white matter. The right occipital lobe, and probably the right parietal lobe are smaller than their counterparts on the left, probably due mainly to greater volumes of grey matter on the left. Over all lobes, the grey-to-white matter ratio appears to be higher on the left side, in accordance with prediction. The right hemisphere as a whole is marginally larger than the left. To some extent lateral differences in frontal and temporal lobes counteract opposite lateral differences in parietal and occipital lobes.
12. The above large scale differences are quite independent of, and opposite to the better known asymmetries in the anterior and posterior language zones, which areas of grey matter are both larger on the left side. A few reports also give detail of cytoarchitectural differences in these areas from the corresponding regions in the right hemisphere.
13. A variety of other pieces of biological evidence relating to asymmetry support the central hypothesis: There is some evidence that long-latency components of evoked potentials have a longer latency in the left than the right hemisphere. Coherence of the EEG is greater in the right hemisphere than the left, while phase delays between EEG traces recorded at different sites in the same hemisphere may be greater on the left than the right. Transmission time across the corpus callosum (a likely indicator of conduction time in ipsilateral cortico-cortical connections), is greater from left to right than from right to left. Neurological signs after cortical damage can give a more precise localization in the left than the right hemisphere, for lesions of similar size. All these observations can be predicted from the central hypothesis.

Chapter 5

14. In the discipline of perceptual psychophysics it is well known that for brief stimuli (up to 100–150 msec) perceived intensity is determined not solely by objective

stimulus intensity, but by a roughly linear integration of intensity over time. This integration has been attributed, at least in part, to processes within the forebrain, rather than just within the sense organs. A model is proposed in which this time/intensity integration is a function of the temporal convergence possible within a network of cortico-cortical connections where a wide spectrum of delay times exists amongst the population of connections.

15. The central premise of this book, that there is a richer spectrum of long axonal conduction delays in the left than the right hemisphere, leads to a number of specific predictions in the light of this model of perceptual processes. The central premise suggests that there should be a reversal of hemispheric superiority from right to left hemispheres as stimulus duration is extended. The critical duration at which hemispheric advantage reverses should be somewhere in the range 50–150 msec. This reversal should be detectable in reaction time experiments, and in measures of the rate of encoding information. Under certain circumstances, the model also predicts that there will be reversals in superiority from right to left hemispheres as intensity is increased (duration held constant). Cessation of a constant stimulus should be accompanied by longer persistence of a subjective trace in the left than the right hemisphere. The left hemisphere should be superior to the right in abstraction of information from a stimulus about its temporal structure, and, unlike some other measures where there is a reversal, this superiority should hold regardless of stimulus duration. The right hemisphere, however, should be superior to the left in detecting stimuli of large size, or high complexity, and should be able to deal with stimuli of low contrast or definition better than the left.

Chapter 6

16. From a review of a large body of evidence on lateralization of perceptual processes, all the predictions mentioned in item 15 receive support, and sometimes this is a striking confirmation of the model presented in Chapter 5. The critical duration for reversal of hemispheric advantages is about 100 msec, as detected in relative timing or reaction time experiments. Evidence about rate of information coding, though not so strong, points in the same direction. A very large amount of evidence shows that the left hemisphere is superior to the right in abstraction of information about temporal structure, and, in accordance with prediction, there is negligible evidence for a reversal of laterality for short duration stimuli. A limited amount of evidence shows a reversal of hemispheric advantage from right to left as stimulus intensity is increased. From the model of Chapter 5, this implies that the left hemisphere has a faster gradient of time/intensity integration than the right instead of (or as well as) a longer duration before saturation is reached. A limited amount of data also suggests that for stimuli of supra-optimal duration in both hemispheres, the left hemisphere has greater absolute sensitivity than the right. The right hemisphere is superior to the left in perception when definition is low, or when stimuli are spatially large (for visual stimuli) or are complex.
17. The morphological implications of these conclusions would seem to be that, for cortico-cortical pathways mediating perceptual integration, there is indeed a larger proportion of connections with delays in the range 20–150 msec in the left than the

right hemisphere; while the right hemisphere should have an excess of those less than 20 msec, compared with the left. The left hemisphere may also have a greater total number of connections (i.e. synapses) than the right. These morphological conclusions are of course dependent on the validity of the model for perceptual integration put forward in Chapter 5. There may be alternative models which can deal equally well (or better) with the available psychophysical evidence. Despite the apparent success of the model, there can be no substitute for more direct evidence.

Chapter 7

18. The differences between hemispheres in perceptual processes, discussed in Chapters 5 and 6 are reflected in the sort of information stored in each hemisphere, at least as far as this can be probed with word stimuli. The left hemisphere is superior in representation of phonemic structure, while the right is superior in representing names of objects perceived visually. The latter specialisation is often referred to as a superiority in representation of "concrete" or "imageable" words, as opposed to abstract ones. However, the more fundamental advantage of the right hemisphere, from which this is derived, is in representation of named objects which are spatially complex.

Chapter 8

19. The theory underlying lateralization of motor function is complex. In this work it is built on a body of theory, advanced in previous works, for instrumental actions in general. This is briefly summarised: The logical structure of instrumental conditioning implies a form of synaptic modification more complex than Hebb's rule, which has been called the "three-factor rule". It has been suggested that the various components of the three factor rule correspond to important aspects of structure and function of the mammalian striatum, although information about instrumental programs acquired there is envisaged to achieve its permanent representation in the cortex. If the cortex has delay lines adequate to represent temporal structure of an instrumental task, this also can be laid down in the cortex, as a consequence of cortico-striatal interaction.
20. An additional component needed for the construction of the schemata for advanced motor performance are the neurones, known to exist in the parietal cortex of experimental animals, which can represent the association between motor programs (or just the preparatory stages of such programs) and their sensory consequences. In humans, it is assumed, on the basis of suggestive cytoarchitectural evidence, but without direct proof, that *such motor-sensory links exist for vocal/auditory association* as well as for association of limb movements with consequent kinesthetic feedback.
21. As an extension of the central hypothesis of this book, it is also suggested that such vocal/auditory links (i.e. from Broca's to Wernicke's area) provide a rich repertoire of delay lines in the left hemisphere, not available in the right hemisphere.
22. It has previously been suggested (Miller, 1981) that combinations of activation of such motor-sensory links can allow the modelling of possible consequences of combinations of actions, in advance of their actually being emitted. Thus it becomes possible to prepare motor programs which have never before been emitted, without first subjecting them to trial and error in the real environment.

23. The final premise needed to account for lateralization of motor function is that certain motor effectors (the primate arm and hand under many circumstances, and the vocal apparatus) have mechanical properties which allow themselves to be used in a manner where each motor command dictates unconditionally the sensory consequences (in the auditory or kinesthetic modalities), as well as their exact timing in relation to the motor commands.
24. Given these premises individually, they may be combined in a way which permits radically new forms of motor co-ordination to emerge. If the motor-sensory delay lines in the left hemisphere can "bridge the gap" between the activation of a motor program (e.g. for speech articulation) and the reception of the consequent sensory feedback (i.e. the speech sound), *time-structured modelling* of motor programs can be constructed so that a motor program is emitted with a definite intention of the speech sound it will achieve. This is a necessary condition for spoken language. Time structured modelling elaborated by the left hemisphere is predicted to apply to other forms of motor control by that hemisphere (hands, arms, eye movements) as well as to speech.
25. These arguments have implications for the exact form of the superiority of the left hemisphere for its motor control. Both for speech, manual praxis and other examples of motor control by the left hemisphere, the superiority should be in co-ordinating rapid movements which require accurate advance planning over the next few hundreds of milliseconds. The right hemisphere however, may have an advantage in motor control (with any effector) which requires rapid moment-by-moment sensory guidance, and also in co-ordinating complex synergies of many effectors
26. Given that any connections which exist between Broca's and Wernicke's areas are likely to be reciprocal, cell assemblies incorporating neurones in both areas may develop which form a unitary representation of speech sounds as abstractions, where both perception and production are combined.

Chapter 9

27. The evidence referring to these predictions is reviewed. Vocal babbling reveals a very lengthy learning process for production of adequate speech sounds, and similar manual babbling occurs in acquisition of sign language. Such stages appear to be the developmental phases in which time-structured models of rapid motor sequences are acquired.
28. Functional brain imaging in normal subjects shows that premotor and sensory association areas (*viz:* Broca's and Wernicke's areas) come to have rather similar roles, indicating, as predicted, that assemblies may be spread across both areas.
29. Available evidence strongly supports the view that the right hand has an advantage in rapid aiming movements, and that this advantage is based on more accurate open-loop programs, rather than on more accurate visual guidance. Evidence from brain-damaged subjects, especially those suffering from apraxia, though not so strong, points mainly to the same conclusion. A little clinical evidence about lateralization of saccadic eye control supports the idea that the left hemisphere is superior in open-loop control. More evidence about this aspect of lateralization in normals in needed. Evidence using the delayed auditory feedback paradigm gives some support to pre-

dictions from Chapter 8, that matching of an internal and a feedback representation of auditory consequences of an action is important for preplanned performance, especially of speech. However the effects shown are capricious.

30. Study of aphasias, by both clinical and experimental methods, can be interpreted in terms of motor-sensory cortico-cortical delay lines being the substrate for preplanning of speech programs. Some of the evidence using delayed auditory feedback in aphasic subjects is particularly striking in this respect.
31. The right hemisphere, in accordance with tentative suggestions made in Chapter 8, appears to be superior in motor acts co-ordinated by sensory guidance, using the visual sense for eye- or hand-movement co-ordination. However there are also some negative findings. Clinical evidence tends to support this role for the right hemisphere. This evidence includes a clinical report about abnormal prosody after right hemisphere lesions, the detail of which is suggestive of loss of auditory-motor guidance.

Chapter 10

32. There is no evidence that the process of short term memory differs between the hemispheres either in duration or capacity. Likewise, selective attention in itself does not appear to show asymmetry. These conclusions are parsimonious in the present context: Both of these functional constructs are probably based on cortico-striatal interaction in maintaining selected cell assemblies in a state of heightened excitability (as discussed in Chapters 8 and 10), this mechanism being quite independent of the central hypothesis about axonal conduction time.
33. Although the process of short-term memory does not differ between hemispheres, the work accomplished by it does, because the raw material upon which it works differs between left and right hemispheres. Short term memory in the right hemisphere can represent slowly changing aspects of sensory input (prosody, "contour" in music) as well as non-temporal integrations of the more complex semantic attributes of sentences. Short term memory in the left hemisphere allows the brief (but precisely represented) temporal patterns of left hemisphere perception to be joined together, to form extended temporal patterns, such as complete melodies or sentences. The left hemisphere is probably also superior in comprehension of grammar and syntax, aspects of language where precise details of sequence are often critical. Differences between the hemispheres similar to these may exist for non-linguistic patterns in the visual sense, but these have been less thoroughly investigated than auditory patterns or linguistic ones.
34. Although the process of selective attention does not differ between hemispheres, selective attention can influence the degree and the direction of laterality effects. The data on this subject is very complex, with many separate effects having been described. Certainly functional as well as structural influences determine many examples of overtly-lateralized function. However, a more solid explanation of these effects may be found by looking for physiological measures rather than by relying on the rather malleable concept of "attention". For instance, neuronal membrane potential, averaged over whole populations of neurones, as the determinant of cell assembly "ignition", may be the physical basis for many so-called attentional effects.

35. Apart from *selective* attention, there is a considerable amount of mutually consistent evidence, both experimental and clinical, for asymmetries in tests of *general arousal, alertness and vigilance*. Using simple stimuli, the left hemisphere is more vulnerable to attentional lapses (loss of vigilance) than the right. This well-documented lateral difference is easily predicted from the central hypothesis (section 12.1: item 9; see also item 6). Usually these differences are minimised by interhemispheric interactions. The inferior left hemisphere performance in vigilance tasks can also be brought up to that of the right hemisphere, but this requires special circumstances (manoeuvres for creating transitory states of "heightened alertness").
36. Much evidence purports to show asymmetry between the hemispheres in representation of emotions. However, the predominance of the right hemisphere for representation of emotional attributes of perception appear to be determined by aspects of the stimuli used other than emotion *per se*. Likewise, lateralization of emotional expression does not go beyond the principles of lateralization already analysed, for motor performance. On the other hand, there is consistent evidence that enduring emotional states (e.g. "moods") are differentially lateralized, those with positive valence to the left hemisphere, those with negative valence to the right. These differences, shown mainly with physiological measures of hemispheric activation are found mainly in the frontal lobes. These conclusions can be derived indirectly from the central hypothesis of this book, if it is assumed that the most important instrumental behaviours of human (successful performance of which produces positive affective states), are those connected with speech and dextrous manual praxis, for which the left hemisphere is superior. Many aspects of the evidence on lateralization of emotion and emotional states are inconclusive, and require further work. In particular, lateralized effects of induced emotional states on perception have been described. They are difficult to relate, either directly or indirectly to the central hypothesis.
37. A variety of papers report on intercorrelations between different measures of laterality. In perceptual tests, hemispheric superiority in tests with one sensory modality generally show correlations with analogous tests in another sensory modality, but these are not strong ones. Likewise, in motor tests, hemispheric superiorities for tests with different effectors (e.g. those indicated by handedness *vs.* speech laterality) are weakly correlated. Perceptual and motor laterality effects are also weakly correlated in the direction to be expected by their individual hemispheric superiorities. The fact that these correlations are all weak ones suggests that overall cerebral laterality is the sum of many independent effects, rather than arising from a single unitary factor. This conclusion is readily compatible with the hypothesis that interhemispheric differences in axonal conduction times, in many different populations of cortico-cortical axons, underlie cerebral specialisation of function.

12.2 UNTESTED PREDICTIONS

This section is not concerned with the majority of the predictions made in the course of Chapters 1 to 11, because the evidence related to them has already been discussed in detail, and in very many cases gives the predictions good support. However, at various stages in the argument, predictions were made for which evidence is not yet available, or for which the evidence is equivocal. Therefore, it is appropriate to summarise all these "loose ends" to guide further work, which may either give further support to the main hypothesis, or refute it.

In Chapter 3, before the central hypothesis was explicitly formulated, a prediction was made, without implications for lateralization, that conduction delays in cerebral white matter should be substantially longer than is generally recognised. A variety of biasing factors, it was suggested, give a false impression of the importance of the rapidly conducting axons. Therefore, it is important to use electrophysiological methods, especially antidromic recording of responses relayed in cortico-cortical axons, to obtain unbiased information about latencies, from long trajectory axons, in large-brained animals. Signal averaging methods should be used along with suitable microelectrodes, for detecting the smallest consistent-latency action potentials detectable in such experiments.

The most direct prediction from the central hypothesis is of course that, when subcortical white matter from human hemispheres is studied by electronmicroscopy, there should be a higher numerical fraction of fine calibre or unmyelinated axons in the left hemisphere than in the right. Such a morphological analysis is not possible in humans (see Chapter 4), because of the ethical and technical problems of electron microscopy in the human brain.

However, there is a possibility that the functional differences between hemispheres documented in humans may apply in some sense to many laboratory animals. This may seem a startling conjecture, for those who think human cerebral laterality is closely related to the distinctively human function of language. However, it has been suggested many times in the previous chapters that human hemispheric specialisation concerns a much more fundamental dichotomy in styles of information processing. According to this view, those aspects of language function which are centred on the left hemisphere are just some of the more special derivatives of this fundamental difference. Seen in this context, it is not so startling to suggest that subhuman mammalian species should also show hemispheric specialisation. Indeed there is some evidence that this is the case:

In some subhuman primates there is evidence for lateralization of function along similar lines to that found in humans. It has already been mentioned in Chapter 4 that chimpanzees show asymmetry of the Sylvian fissure similar to that in most humans (Yeni-Komshian and Benson, 1976). This asymmetry however, is likely to be related to the small zones specialised for speech (or equivalent function), rather than to the larger-scale asymmetry which is the central hypothesis postulated here. More relevant is data from a species of monkey (*Macaca fuscata*) where there appears to be a left-hemisphere dominance for recognition of the species' own vocalisations (Petersen *et al.*, 1978).

Since 1978, quite a variety of examples of hemispheric morphological asymmetry (or related functional asymmetry) in several mammalian species have been reported. Bradshaw (1991) reviewed this subject. In rodents, the right hemisphere appears to be larger than the left (as in humans), and similar asymmetry, though less well documented has been found in mouse, rabbit, cat (Kolb *et al.*, 1982) as well as in dogs and dolphins (review of Bradshaw, 1991). In rats the cortex is thicker on the right in most regions, the length, height and wet weight of the hemispheres is also larger on the right, although subcortical width is not asymmetrical (Kolb *et al.*, 1982). In primates the right frontal lobe and the left occipital lobe appear to be larger than on the opposite sides (as in humans). Emotional responses appear to be dealt with more in the right than the left hemisphere of rats (as in humans). Split-brain monkeys can discriminate faces of monkeys better with the right than the left hemisphere (corresponding to the asymmetry found in humans). Monkeys have a left hand superiority in discriminating tactile shapes (as in humans). Several species of monkey are known to favour the right hand in complex manipulations (perhaps corresponding to human right handedness). In rhesus monkeys, the left face begins to display facial expression at an earlier age than the right face (Hauser, 1993).

The Russian neurobiologist, V.L. Bianki, has performed many physiological, and behavioural experiments some in cats, more in rats, which demonstrate functional asymmetry between the hemispheres (Bianki, 1988). Many of Bianki's experiments involved selective inactivation of right or left hemisphere of the rat by the method of cortical spreading depression. This is a test somewhat analogous to the sodium amytal test of Wada and Rasmussen (1960) used in humans. It is interesting to mention a few of his results. In a task involving discrimination of visual shapes, degraded by visual noise, inactivation of the right hemisphere led to greater performance deficits than that of the left (p. 122). In a task involving time estimation (similar to the "fixed-interval schedule" of Skinnerian psychology), the left hemisphere was superior to the right (p. 315). In tasks involving recognition of complex visual stimuli all of whose components were present simultaneously, right hemisphere inactivation led to greater impairment than left hemisphere inactivation. In tasks where the components were presented successively, the reverse was found (p. 325). In a rat version of a reaction time task for discrimination of visual stimuli, performance was impaired more by right than left hemisphere inactivation (p. 329). In a superficial way, many of Bianki's results (such as those just mentioned) are reminiscent of laterality effects in somewhat comparable tests in humans. Of course, one cannot claim this in a strong way, without exploring each behavioural paradigm in some depth, to investigate just what features of each test give rise to the asymmetrical performance. Nevertheless, there is now a case to made that cerebral asymmetry is not confined to the human brain but exists in a number of other mammalian species.

This being the case, it seems reasonable to suggest that the central premise of the present book can be investigated in some species of experimental animal, including both rats and some experimental primates. In these species, it is possible to prepare electronmicrographs of cerebral white matter of the highest quality, and then to make the appropriate counts of myelinated and unmyelinated axons of different calibres, in carefully matched regions of the left and right hemisphere. Regional differences in calibre spectra may also be important, as they appear to be in the corpus callosum (Lamantia and Rakic, 1990a).

Using neurochemical measures, a less direct approach to the central hypothesis can be made, with the advantage that it is applicable in humans. If the central hypothesis is true, one would expect there to be a slightly increased concentration of markers for axoplasm (e.g. neurofilament protein) in white matter from the left hemisphere, and an increased volume fraction of myelin in that from the right.

Fine calibre axons are likely to have smaller cell bodies than large calibre ones. Assuming equal cell numbers in the two hemispheres, the volume of both cortico-cortical cell bodies and associated neuropil would be less in the left than the right hemisphere, and total volume of grey matter should be less. One would therefore expect a higher neuronal density (neurones per unit volume) in the left than the right hemisphere. Data from the older German literature are mentioned in Chapter 4. Such studies need to be repeated. The difference may be most marked in cortical laminae II and III, in which the largest proportion of projection neurones provide the origin of cortico-cortical axons.

Electrophysiological methods may be capable of giving new information of the manner of asymmetry between the hemisphere. In particular consideration needs to be given to making further estimates of wave velocity for the EEG, comparing left with right hemispheres. Apart from this the central hypothesis makes predictions about the integration time for conscious experience when the surface of the cortex is stimulated electrically. Libet (1982) has shown that this occurs over a period of about 0.5 sec., and it is suggested in Chapter 4 that this represents integration by temporal convergence along connections with a variety of conduction delays. It is predicted that electrical stimulation of the brain on the left side should exhibit a longer integration time than that of the right.

As mentioned above, the predictions made in Chapter 5 for lateralization of perceptual processes was largely supported by the evidence reviewed in Chapter 6. However, the important prediction that there be reversal of dominance in some tasks, from right to left, as stimulus duration is increased, was evaluated (and confirmed) mainly by a meta-analytic approach, piecing together a pattern from many different studies with different methodologies. A more rigorous test of this prediction would involve systematic comparison, within a single study, of time/intensity integration in each hemisphere separately, using a wide range of stimulus durations. Evidence of this sort is needed particularly in the auditory modality.

Lateralization of the representation of words and their meanings (discussed in Chapter 7) is largely what would be predicted. However, there is one significant aspect where evidence available so far is equivocal with regard to a prediction. The right hemisphere should have a capacity for richer semantic associations than the left. Some evidence supports this view, but the conclusion cannot be made rigorously because word-types are often used for which the right hemisphere has an advantage, regardless of that hemisphere's innate capacity for association. More rigorous studies are required in which differences in associability in each hemisphere are sought using abstract as well as visually imageable words as probes, which (when using both hemispheres together) have comparable associability. If, given such matching, there is still an interaction between associability and hemisphere of presentation, one could conclude that the right hemisphere *does* indeed have a richer semantic network than the left.

Chapter 8 extended the basic morphological hypothesis of this book, to account for the motor aspects of hemispheric specialisation. For speech lateralization it was suggested that there are projections from the inferior frontal gyrus to the superior temporal lobe, which could subserve vocal/auditory associations in the same manner that projections from premotor areas to parietal lobe in subhuman species subserve motor/kinesthetic or motor/visual association. More specifically, it was suggested that in the left hemisphere there should be connections from Broca's area to the region in and near Wernicke's area in humans, which are more numerous and/or slower conducting than the corresponding connections on the right side. These predictions can only be addressed by studies in the human brain. The first, less specific prediction (that there should be pathways from Broca's to Wernicke's area) is accessible to study using anatomical methods developed recently. It has been found that silver impregnation methods for demonstrating degenerating axons can detect such axons a considerable time after the lesion which caused the degeneration (Mesulam, 1979; Grafe and Leonard, 1980). Such methods can thus be used to show connections from regions of brain tissue damaged by cerebro-vascular accidents (see for instance, Clarke and Miklossy, 1990). If brains can be obtained in which, by chance, small cerebral infarcts have occurred confined as far as possible to the inferior frontal gyrus, existence of the basic connection between Broca's and Wernicke's areas could be confirmed. However, to evaluate the second, more specific prediction, that such a connection constituted a richer repertoire of delay lines on the left than the right, is not possible using anatomical techniques available for use in the human brain. Nevertheless, clinical electrophysiology might be able to address this question (by comparing phase delays for EEG records between Broca's and Wernicke's with that between the corresponding regions in the right hemisphere.)

The evidence of motor lateralization, reviewed in Chapter 9 largely supports the major predictions made in Chapter 8. However, there are a number of minor points where further evidence is needed to complete the picture. It may be possible to strengthen the small body of evidence for lateralization of eye movement control, the major prediction being that the left hemisphere is better for controlling saccadic eye movements, the right hemisphere for smooth pursuit movements. Future experimental work may also be able to clarify the involvement of each hemisphere in both open- and closed-loop varieties of auditory-motor control. Abnormal prosody after right hemisphere lesions may reflect in part at least, defective auditory-motor sensory guidance. Experimental study of this is required.

An important prediction was also made in Chapter 8 about motor behaviour, on which there is as yet very little evidence. This is that the right hemisphere should be responsible for the construction of the more complex motor synergies than the left. A few clinical anecdotes support this role, but there are no experimental studies on the subject. These are much needed.

In conduction aphasia, it was predicted in Chapter 9 that the usual activation during spontaneous speech of anterior regions such as Broca's area should still take place, but the activation of Wernicke's area, described for normal subjects in section 9.2.1 should not occur.

Several loose ends in the evidence about motor lateralization concern the possible analogy between manual praxis and speech production. Detailed developmental studies may be able to show that the acquisition of dextrous control of right hand in infants has some formal similarities to vocal or manual babbling, using the language effectors. Likewise there are suggestions that there are formal parallels between some forms of apraxia ("afferent apraxia" according to Luria (1973)) and receptive (Wernicke's) aphasia. This requires further clinical and experimental investigation, and if the parallel exists, the locality of lesions required to produce "afferent apraxia" needs to be defined. Thirdly, in PET studies, it may be possible to show that imagined limb movements activate the parietal cortex, in a manner similar to the activation of Wernicke's area by imagined speech.

It is also becoming possible to investigate some of these issues with EEG methodology. Asymmetries of the EEG during processing of visually presented word *vs.* non-word stimuli have recently been reported (Lutzenberger *et al.*, 1994). Similar methods might be applicable during speech production, to supplement and amplify data using PET.

The diverse and complex evidence reviewed in Chapter 10 was often equivocal. A few more specific predictions required experimental investigation. There are two aspects of the lateralization of the products of short term memory which may be investigable: It was predicted that there be alternative ways of perceiving contrapuntal music, corresponding to left and right hemispheric specialisations. Experimental investigation of the serial order effect for brief sequences of visual stimuli deserves further study, with the particular aim of determining whether information about order in sequences lasting seconds can be recalled better for right field presentations.

Some specific aspects of the lateralization of emotional representation deserve further study. The emotional attributes of words requires further examination. If right hemisphere lateralization of all emotions (in faces, words etc.) is based on the fact that emotional stimuli have generally higher associability than similar stimuli without emotional attributes, it is predicted that emotional words should *not* show a right lateralization compared with non-emotional words matched for associability. Apart from this, there is a major prediction about lateralization of emotional states: It was suggested in Chapter 10 that the interaction between affective valence and hemisphere is based on the affective state habitually produced by the most important types of instrumental behaviours controlled by each hemisphere. The data reviewed was compatible with the view that left hemisphere instrumental tasks tend to generate positive affective states in that hemisphere, with negative ones occurring in the right. If this reasoning is correct, it might be possible to devise experiments where transient euphoria is produced by stimuli directed to the right hemisphere, and to show greater cerebral activation by induced euphoria in the right hemisphere. The conditions in which this might occur would be those associated with instrumental responses which use sensory guidance (apparently a right hemisphere function).

12.3 CLOSING REMARKS

Figure 12.1 depicts the overall structure of the theory of cerebral lateralization offered in this book. It will be seen that the central hypothesis, and its extensions are more specifically directed to explaining the special functions of the left hemisphere. Given this, the predictions which follow from the basic morphological postulate about the left hemisphere are supported in almost all areas from existing evidence. For the right hemisphere, the central hypothesis is not a radical departure from assumptions made hitherto. However, there are a number of predictions about functional specialisations of the right hemisphere which are made from the reasoning offered here, where existing evidence is mainly lacking. These issues are thus tests of the finer details implied by the proposed morphological difference between hemispheres. As for the credibility of the argument as a whole, readers must form their own judgement. However, although the central morphological premise is not directly accessible to experiment in humans, there are enough indirect ways of evaluating it that the theory as a whole can be refuted or verified by future work.

One broad issue deserves further comment. As discussed in Chapter 8, Bradshaw and Nettleton (1983) believe that motor lateralization is more fundamental that perceptual lateralization. The outcome of the present work is similar, but not quite congruent, with this view. The hypothesised large scale differences between cortico-cortical connections in left and right hemispheres are envisaged to underlie most examples of cerebral lateralization, including perceptual ones. Motor lateralization, especially for speech production,

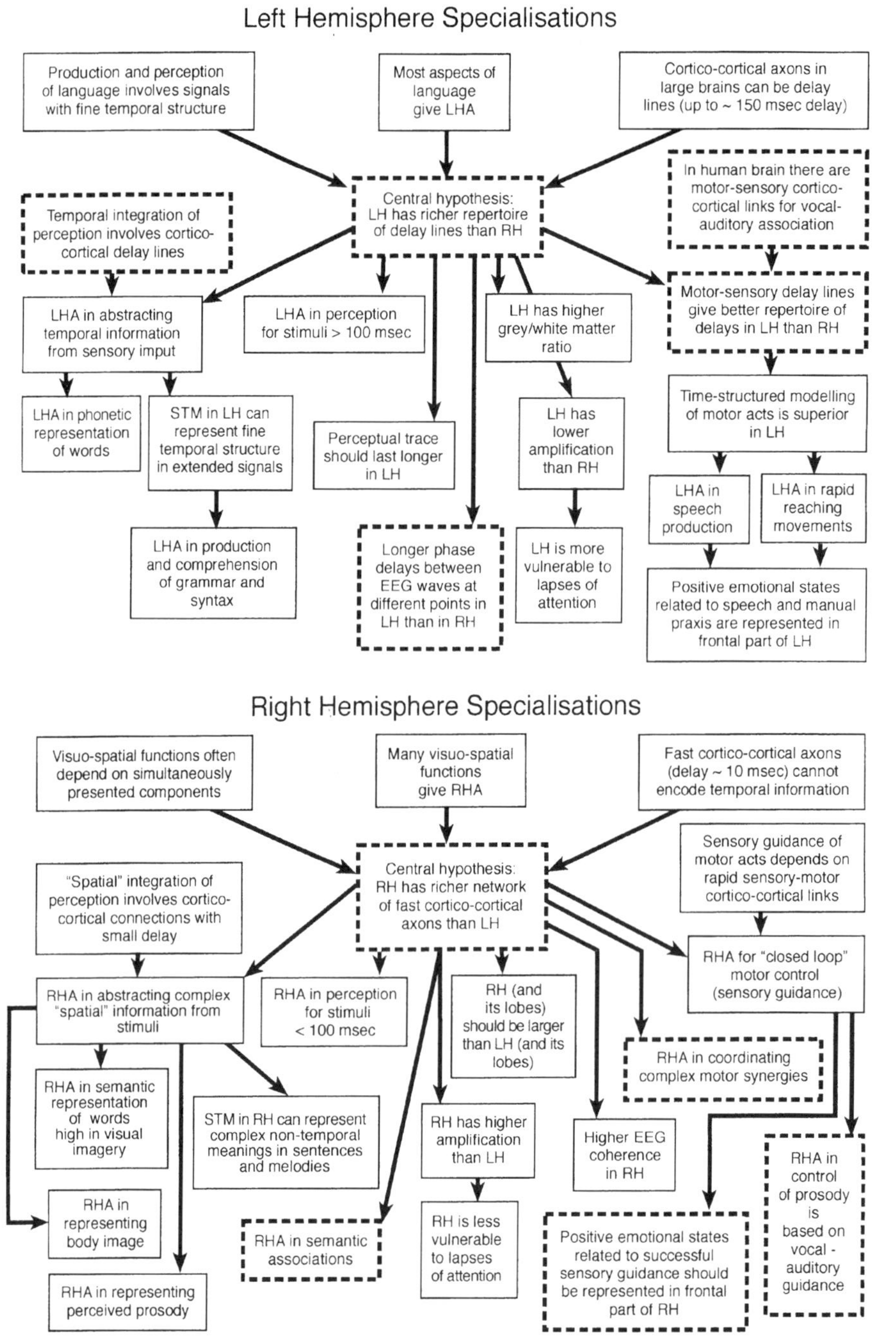

Figure 12.1 Diagrammatic Summaries of the Specialisations of Left and Right Hemispheres, and their Proposed Derivations from the Central Hypothesis. **A:** Left hemisphere specialisations. **B:** Right hemisphere specialisations. LH, RH: Left hemisphere, right hemisphere; LHA,RHA: Left hemisphere advantage, right hemisphere advantage. Boxes enclosed by continuous lines: Plausible (uncontentious) assumptions, or conclusions well supported by existing evidence. Boxes enclosed by broken lines: Central or supporting hypotheses, which await definitive evidence, or predictions which at present have not been properly tested.

needs, in addition, a somewhat more specific morphological specialisation, that is, the existence of a particular cortico-cortical pathway between Broca's and Wernicke's area. This is probably combined with the enlargement of these special speech areas, compared with their counterparts in the right hemisphere. These specialisations could not allow speech production were it not for the global difference, that the left hemisphere also has a range of longer time delays of cortico-cortical conduction than does the right. This difference, it is suggested, applies to the specific pathway between Broca's and Wernicke's areas, as to many other channels of information flow. In this sense, motor lateralization may be a more specific development than perceptual lateralization, but relies on another specialisation which could be thought equally fundamental. However, in view of the preceding discussion about lateralization in non-human animals, the specialisation for speech production may be the one which distinguishes humans from non-humans, though this is mere conjecture at present.

In previous works (e.g. Miller, 1981; 1991; Miller and Wickens, 1991) the theory of the function of the mammalian forebrain has been addressed from a variety of perspectives. There is a common thread running through these previous essays: To develop a theory of the mammalian forebrain it is necessary to address in turn the cybernetic function performed by each of the major entities of forebrain, such as the striatum and associated subcortical structures, the hippocampus, the thalamus, and of course the neocortex. However, to take any of these structures by itself severely limits the power of the theory which can be constructed, because any one of these displays its proper large-scale dynamic functions and computations only when it is connected to and, is interacting with, the others. Thus it is necessary to explore and unravel the mode of the dynamic interaction of these structures.

In particular, it is necessary to examine how the largest of these structures — the cerebral neocortex — has its functional repertoire both constrained and extended by its interplay with each of the other structures of the forebrain. We have many good reasons to suspect that the cerebral neocortex is a large "mixing" device (Braitenberg and Schüz, 1991), where a wide range of associations can be registered between signals which occur in contiguity. However, the cerebral neocortex in the intact forebrain is more than that: It can abstract patterns whose components occur widely separated in time. It can plan and control skilled bodily actions. It can model the world in both space and time.

Previous works have considered the interplay between striatum and neocortex, and between neocortex and hippocampus. The present work is a continuation of the same strategy of theoretical research. It considers the dynamic behaviour of two radically different types of cortical network, such as appear to exist in the human brain. One is specialised for perceptual integration and motor planning in the temporal domain, immediately for short intervals of much less than one second, but by extension to longer time frames as well. The other is specialised for perceptual integration in a "spatial" sense, and probably for those aspects of motor planning and control requiring immediate feedback and immediate co-ordination of many effectors. The possible interaction of each type of cortical network with the basal ganglia was considered in some detail, as well as the mutual interaction between the two sorts of hemisphere.

The picture of the forebrain is not yet complete. For instance, we need also to consider in detail the three-way interplay of neocortex, striatum and hippocampus. It is also necessary to examine the different sort of interaction between neocortex and hippocampus which can occur when two quite different types of dynamic behaviour can occur in the two hemispheres. The data base for understanding these interactions is much smaller than for the issues dealt with so far, and may need to wait for a later generation of theoretically-orientated researchers.

One of the explicit guiding principles in this book has been to use the descriptive language of sensory and motor physiology, and, implicitly, the language used more widely throughout science, for unravelling human brain processes. Along the way, the language of psychology also provided some very useful signposts in the analysis of cerebral laterality. Following such lines of thought there are a variety of formulations of the differences between right and left hemisphere. These include the ideas that the right hemisphere is better capable of "imagery", or "holistic processing", or "emotional representation", or "analogue processing", as opposed to the left hemisphere which is better at dealing with "logical", "linear", "symbolic" and "propositional" information. Most of these ideas are difficult to define precisely. Some of them have been discussed in previous chapters. Overall such ideas have a distinctly metaphysical tone to them. For each of them, careful analysis is needed to define what these psychological or quasi-metaphysical concepts might mean in terms of the physical variables used in the rest of science. The thrust of the present chapter is that when that careful analysis is done, the resulting differences between the hemispheres is in no sense metaphysical, but is a direct derivation from the differences in information processing which we have discussed in previous chapters. Those differences in turn were related to a simple hypothesis about interhemispheric differences which is in principle very solid and physical, a difference in microscopic structure of the connections linking different parts of each hemisphere. Thus, although these psychological concepts are useful guides to an explanation of observed aspects of lateral specialisation, in the end they should be converted to the language of the physical variables used in neurobiology and in the rest of science. Thus, the metaphysical or quasi-metaphysical implications of many psychological terms, so common when dealing with higher functions of the human brain, have been avoided in this work as the *primary* constructs for framing explanations.

As a result, a view of distinctively human psychological processes has been developed which some readers may find disturbingly deterministic. No apology is offered for this. The language of science is the language of determinism. Science is based on the hunch that nature does behave consistently, and if we are to continue scientific endeavours we have to adopt this paradigm of thought. However science is not everything. From a different perspective one may wish to view human nature as something more than a piece of clockwork mechanism, however complex. This other perspective has been advocated in a recent work (Miller, 1995b). It is a perspective that is different from the scientific one in a truly metaphysical way. However, as far as the study of the brain in a strictly scientific sense goes, our understanding is best developed by a remorselessly deterministic programme of theory development, whether or not that is the metaphysical perspective which really commands our allegiance.

References

Abbs, J.H. and Smith, K.U. (1970) Laterality differences in the auditory feedback control of speech. *J. Speech Hear. Res.*, **13**, 298–303.

Abeles, M. (1982a) Role of the cortical neuron: integrator or coincidence detector? *Isr. J. Med. Sci.*, **18**, 83–92.

Abeles, M. (1982b) Local cortical circuits. *Studies in Brain Function No. 6.* Berlin, Heidelberg: Springer.

Aboitiz, F., Schiebel, A.B., Fisher, R.S. and Zaidel, E. (1992) Fiber composition of human corpus callosum. *Brain Res.*, **598**, 143–153.

Adams, J.A. (1971) A closed-loop theory of motor learning. *J. Motor Behav.*, **3**, 111–149.

Ahern, G.L. and Schwartz, G.E. (1979) Differential lateralization for positive versus negative emotion. *Neuropsychologia*, **17**, 693–698.

Ahern, G.L. and Schwartz, G.E. (1985) Differential lateralization for positive and negative emotion in the human brain: EEG spectral analysis. *Neuropsychologia*, **23**, 745–755.

Ahern, G.L., Herring, A.M., Tackenberg, J.N., Schwartz, G.E., Seeger, J.F., Labiner, D.M., Weinand, M.E. and Oommen, K.J. (1994) Affective self-report during the intracarotid sodium amobarbital test. *J. Clin. Exp. Neuropsychol.*, **16**, 372–376.

Aiba, T.S. and Stevens, S.S. (1964) Relation of brightness to duration under light and dark adaptation. *Vision Res.*, **4**, 391–401.

Alajouanine, T., Lhermitte, F., Ledoux, M., Renaud, D. and Vignolo, L.A. (1964) Les composantes phonemiques et semantiques de la jargonaphasie. *Rev. Neurol. (Paris),* **110**, 5–20.

Albert, M.L. (1973) A simple test of visual neglect. *Neurology*, **23**, 658–664.

Alexander, G.E., DeLong, M.R. and Strick, P.L. (1986) Parallel organization of functionally segregated circuits linking basal ganglia and cortex. *Annu. Rev. Neurosci.,* **9**, 357–381.

Andersen, R.A., Asanuma, C., Essick, G. and Siegel, R.M. (1990) Cortico-cortical connections of anatomically and physiologically defined subdivisions within the inferior parietal lobule. *J. Comp. Neurol.*, **296**, 65–113.

Annett, M. (1970) The growth of manual performance and speed. *Br. J. Psychol.*, **61**, 545–548.

Annett, J., Annett, M., Hudson, P.T.W. and Turner, A. (1979) The control of movement in the preferred and non-preferred hand. *Q. J. Exp. Psychol.*, **31**, 641–652.

Arieli, A., Shoham, D., Hildesheim, R. and Grinvald, A. (1995) Coherent spatiotemporal patterns of ongoing activity revealed by real-time optical imaging coupled with single unit recording in the cat visual cortex. *J. Neurophysiol.* **73**, 2072–2093.

Asbjornsen, A., Hugdahl, K. and Bryden, M.P. (1992) Manipulation of subjects' level of arousal in dichotic listening. *Brain Cogn.*, **19**, 183–194.

Bakker, D.J. (1968) Ear asymmetry with monoaural stimulation. *Psychon. Sci.*, 12, 62.

Bakker, D.J. and van der Kleig, P.C.M. (1978) Development of lateral asymmetry in the perception of sequentially touched fingers. *Acta Psychol. (Amst.)*, **42**, 357–365.

Balfour, C.A., Clark, C.R. and Geffen, G.M., (1991) Interhemispheric differences in the planning and execution of sequences of skilled finger movement. *Neuropsychologia*, **29**, 889–905.

Ban, T., Shiwa, T. and Kawamura, K. (1991) Cortico-cortical projections from the prefrontal cortex to the superior temporal sulcal area (STs) in the monkey studied by means of HRP method. *Arch. Ital. Biol.*, **129**, 259–272.

Barta, P.E., Pearlson, G.D., Powers, R.E., Richards, S.S. and Tune, L.E. (1990) Auditory hallucinations and smaller superior temporal gyral volume in schizophrenia. *Am. J. Psychiatry*, **147**, 1457–1462.

Bartlett, E.J., Brown, J.W., Wolf, A.P. and Brodie, J.D. (1987) Correlations between glucose metabolic rates in brain regions of healthy male adults at rest and during language stimulation. *Brain Lang.*, **32**, 1–18.

Barto, A.G., Sutton, R.S. and Anderson, C.W. (1983) Neuronlike adaptive elements that can solve difficult learning control problems. *IEE Transaction on Systems, Man and Cybernetics*, SMC-13, 834–846.

Battersley, W.S. and Defabaugh, G.L. (1969) Neural limitations on visual excitability: after effects of subliminal stimulation. *Vision Res.*, **9**, 757–768.

Baum, S.R., Blumstein, S.E. Naeser, M.A. and Palumbo, C.L. (1990) Temporal dimensions of consonant and vowel production: an acoustic and CT analysis of aphasic speech. *Brain Lang.*, **39**, 33–56.

Bear, D.M. (1983) Hemispheric specialization and the neurology of emotion. *Arch. Neurol.*, **40**, 195–202.

Bear, D., Schiff, D., Saver, J., Greenberg, M. and Freeman, R. (1986) Quantitative analysis of cerebral asymmetries: Frontal-occipital correlation, sex dimorphism and association with handedness. *Arch. Neurol.*, **43**, 598–603.

Beck, E. (1955) Typologie des Gehirns am Biespiel des dorsalen menschlichen Schlafenlappens nebst weiteren Beitrage zurfrage des Links-Rechts hirnigheit. *Deutsche Zeitschrift fur Nervenheilkunde*, **173**, 267–306.

Begg, A. and Wickens, J.R. (1993) A long-lasting effect of dopamine on synaptic transmission in the corticostriatal pathway in vitro. Paper presented at: *11th International Australasian Conference on Brain Research, Queenstown, New Zealand, August, 1993.*

Behrens, S.J. (1985) The perception of stress and lateralization of prosody. *Brain Lang.*, **26**, 332–348.

Beisteiner, R., Hollinger, P., Lindiger, G., Lang, W. and Berthoz, A. (1995) Mental representation of movements. Brain potentials associated with imagination of hand movements. *Electroencephalogr. Clin. Neurophysiol.*, **96**, 183–195.

Bellugi, U., Poizner, H. and Klima, E.S. (1983) Brain organization for language: clues from sign aphasia. *Hum. Neurobiol.*, **2**, 155–170.

Bellugi, U., Poizner, H. and Klima, E.S. (1989) Language, modality and the brain. *Trends Neurosci.*, **12**, 380–388.

Belmore, S.M. (1981) The fate of ear asymmetries in short term memory. *Brain Lang.*, **12**, 101–115.

Bendat, J. (1986) *Random data*. New York: Wiley-Interscience.

Beninger, R.J. (1983) The role of dopamine in locomotor activity and learning. *Brain Res.*, 287 (Brain Res. Rev. 6), 173–196.

Benson, D.F. and Barton, M.I. (1970) Disturbances in constructional ability. *Cortex*, **6**, 19–46.

Benson, D.F. (1979) *Aphasia, alexia and agraphia*. New York: Churchill-Livingstone.

Bentin, S. and Feinsod, M. (1983) Hemispheric asymmetry for word perception: behavioural and ERP evidence. *Psychophysiology*, **20**, 489–497.

Benton, A. and Tranel, D. (1993) Visuoperceptual visuospatial and visuoconstructive disorders. In: Heilman, K.M. and Valenstein, E. (eds) *Clinical Neuropsychology, Third Edn.* Oxford: Oxford University Press, pp. 165–213.

Berbel, P. and Innocenti, G.M. (1988) The development of the corpus callosum in cats: light- and electron-microscopic study. *J. Comp. Neurol.*, **276**, 132–156.

Bergego, C., Delocke, C., Pradat-Diehl, P., Robineau, F., Lauriot-Prevost, M.C. (1993) Visual recognition in right brain-damaged patients: evidence from a tachistoscopic confrontation naming task. *Brain Lang.*, **44**, 181–190.

Berger, H. (1921) Untersuchungen uber den Zellgehalt der menschlichen Grosshirnrinde. *Z. Gesamte Neurol. Psychiatr.*, **69**, 46–60.

Bernstein, N.A. (1967) Biodynamics of locomotion. In: *The coordination and regulation of movements (selected papers of N.A. Bernstein).* Oxford: Pergamon Press, pp. 60–113.

Bertelson, P. and Tissegre, F. (1972) Lateral asymmetry in the perceived sequence of speech and non-speech stimuli. *Percept. Psychophysics*, **11**, 356–362.

Besner, D., Grimsell, D. and Davis, R. (1979) The mind's eye and the comparative judgement of number. *Neuropsychologia*, **17**, 373–380.

Bever, T. (1971) The nature of cerebral dominance in speech behaviour of the child and adult. In: Huxley, R. and Ingram, E. (eds) *Language acquisition: models and methods.* New York: Academic Press, pp. 231–255.

Bever, T.G., Lackner, J. and Kirk, R. (1969a) The underlying structure of sentences are the primary units of immediate speech processing. *Percept. Psychophysics*, **5**, 225–234.

Bever, T.G., Lackner, J. and Stolz, W. (1969b) Transitional probability is not a general mechanism for the segmentation of speech. *J. Exp. Psychol.*, **79**, 387–394.

Bever, T.G. and Chiarello, R.J. (1974) Cerebral dominance in musicians and non-musicians. *Science*, **185**, 537–539.

Bianki, V. (1988) *The right and left hemispheres of the animal brain. Cerebral lateralization of function.* New York, London, Paris: Gordon and Breach.

Binder, J., Marshall, R., Lazar, R. and Benjamin, J. (1992) Distinct syndromes of hemineglect. *Arch. Neurol.*, **49**, 1187–1194.

Bisiacchi, P., Marzi, C.A., Nicoletti, R., Carena, G., Mucignat, C. and Tomaiulo, F. (1994) Left-right asymmetry of callosal transfer in normal human subjects. *Behav. Brain Res.*, **64**, 173–178.

Blackburn, J.R., Phillips, A.G., Jakubovic, A. and Fibiger, H.C. (1986) Increased dopamine metabolism in the nucleus accumbens and striatum following consumption of a nutritive meal but not palatable non-nutritive saccharin. *Pharmacol. Biochem. Behav.*, **25**, 1095–1100.

Blackwood, D.H.R., Young, A.H., McQueen, J.K., Martin, M.J., Roxborough, H.M., Muir, W.J., St Clair, D.M. and Kean, D.M. (1991) Magnetic resonance imaging in schizophrenia: altered brain morphology associated with P300 abnormalities and eye tracking dysfunction. *Biol. Psychiatry*, **30**, 753–769.

Blauenstein, U.W., Halsen, J.H., Wilson, J.H., Wills, E.L. and Risberg, J. (1977) 133-Xenon inhalation method. Analysis of reproducibility: Some of its physiological implications. *Stroke*, **8**, 92–102.

Blinkov, S.M. and Glezer, I.I. (1968) *Das Zentralnervensystem in Zahlen und Tabellen.* Jena: Gustav Fischer Verlag.

Blonder, L.X., Bowers, D. and Heilman, D.K. (1991) The role of the right hemisphere in emotional communication. *Brain*, **114**, 1115–1127.

Blumenthal, A.L. (1977) *The process of cognition.* Englewood Cliffs, New Jersey: Prentice-Hall.

Blumstein, S.E., Bakes, E. and Goodglass, H. (1977) Phonological factors in auditory comprehension in aphasia. *Neuropsychologia*, **15**, 19–30.

Bogousslavsky, J. and Regli, F. (1986) Pursuit gaze defects in acute and chronic unilateral parieto-occipital lesions. *Eur. Neurol.*, **25**, 10–18.

Boivin, M.J., Giordani, B., Berent, S., Amato, D.A., Lehtinen, S., Koeppe, R.A., Buchtel, H.A., Foster, N.L. and Kuhl, D.E. (1992) Verbal fluency and positron emission tomographic mapping of regional cerebral glucose metabolism. *Cortex*, **28**, 231–239.

Boller, F., Vrtunski, P.B., Kim, Y. & Mack, J.L. (1978) Delayed auditory feedback and aphasia. *Cortex*, **14**, 212–216.

Bondareff, W., Baldy, R., and Levy, R. (1981) Quantitative computed tomography in senile dementia. *Arch. Gen. Psychiatry*, **38**, 1365–1368.

Borod, J.C. and Caron, H.S. (1980) Facedness and emotion related to lateral dominance, sex and expression type. *Neuropsychologia*, **18**, 237–241.

Borod, J.C., Caron, H.S. and Koff, E. (1981) Asymmetry in positive and negative facial expressions: sex difference. *Neuropsychologia*, **19**, 819–824.

Borod, J.C., Koff, E., Lorch, M.P. and Nicholas, M. (1985) Channels of emotional expression in patients with unilateral brain damage. *Arch. Neurol.*, **42**, 345–348.

Borod, J.C., Koff, E., Lorch, M.P. and Nicholas, M. (1986) The expression and perception of facial emotion in brain damaged patients. *Neuropsychologia*, **24**, 169–180.

Bottini, G., Cappa, S.F., Sterzi, R., and Vignolo, L.A. (1995) Intramodal somaesthetic recognition disorders following right and left hemisphere injury. *Brain*, **118**, 395–400.

Bowen, F.P., Hoehn, M.H. and Yahr, M.D. (1972) Cerebral dominance in relation to tracking and tapping performance in patients with Parkinsonsim. *Neurology*, **22**, 32–39.

Bowen, R.W., Pola, J. and Martin, L. (1974) Visual persistence: effects of flash luminance, duration and energy. *Vision Res.*, **14**, 295–303.

Bowers, D., Bauer, R.M., Coslett, H.B. and Heilman, K.M. (1985) Processing of faces by patients with unilateral hemispheric lesions. I. Dissociation between judgments for facial affect and facial identity. *Brain Cogn.*, **4**, 258–272.

Boysson-Bardies, B. de, Sagart, L. and Bacri, N. (1981) Phonetic analysis of late babbling: a case study of a French child. *J. Child Lang.*, **8**, 511–524.

Bracewell, R.M., Husain, M. and Stein, J.F. (1990) Specialization of the right hemisphere for visuomotor control. *Neuropsychologia*, **28**, 763–775.

Bradshaw, J.L. (1991) Animal asymmetry and human hereditary: dextrality, tool use and language in evolution. *Br. J. Psychol.*, **82**, 39–59.

Bradshaw, J.L., Nettleton, N.C. and Geffen, G. (1971) Ear differences and delayed auditory feedback. Effects on a speech and a music task. *J. Exp. Psychol.*, **91**, 85–92.

Bradshaw, J.L., Nettleton, N.C. and Geffen, G. (1972) Ear asymmetry and delayed auditory feedback: effects of task requirements and competitive stimulation. *J. Exp. Psychol.*, **94**, 267–275.

Bradshaw, J.L. and Gates, E.A. (1978) Visual field differences in verbal tasks: effects of task familiarity and sex of subject. *Brain Lang.*, **5**, 166–187.

Bradshaw, J.L., Nettleton, N.C. & Taylor, M.J. (1981) Right hemisphere language and cognitive deficit in sinistrals. *Neuropsychologia*, **19**, 113–132

Bradshaw, J.L. and Nettleton, N.C. (1983) *Human cerebral asymmetry.* Englewood Cliffs, New Jersey: Prentice Hall.

Bradshaw, J.L., Bradshaw, J.A. and Nettleton, N.C. (1988) Movement initiation and control: abduction, adduction and locus of limb. *Neuropsychologia*, **26**, 701–709.

Bradshaw, J.L., Bradshaw, J.A. and Nettleton, N.C. (1990) Abduction, adduction and hand differences in simple and serial movements. *Neuropsychologia*, **28**, 917–931.

Braitenberg, V. (1977) *On the texture of brains.* Heidelberg: Springer Verlag.

Braitenberg, V. (1978) Cell assemblies in the cerebral cortex. In: R. Heim and G. Palm (eds.) *Lecture notes on Biomathematics.,* Heidelberg: Springer Verlag, pp. 171–188.

Braitenberg, V. and Schuz, A. (1991) Anatomy of the cortex: statistics and geometry. *Studies on Brain Function No. 18.* Heidelberg, Berlin: Springer Verlag.

Brown, J.W. (1974) *Aphasia, apraxia and agnosia. Clinical and theoretical aspects.* C.C. Thomas, Springfield, Illinois, p. 153.

Brown, J.W. and Hecaen, H. (1976) Lateralization and language representation. *Neurology*, **26**, 183–189.

Brown, T.H., Kairiss, E.W. and Keenan, C.L. (1990) Hebbian synapses: Biophysical mechanisms and algorithms. *Annu. Rev. Neurosci.,* **13**, 475–511.

Brown, W.S., Barsz, K., Sprecker, E. and Emmerich, D.S. (1984) Laterality effects in response to the onset of visual stimuli. *Neuropsychologia*, **22**, 627–629.

Brownell, H.H., Michel, D., Pavelson, J. and Gardner, H. (1983) Surprise but not coherence: sensitivity to verbal humour in right hemisphere patients. *Brain Lang.,* **18**, 20–27.

Brownell, H.H., Potter, H.H. and Bihrle, A.M. (1986) Inference deficits in right brain-damaged patients. *Brain Lang.,* **27**, 310–321.

Bryden, M.P., Ley, R.G. and Sugarman, J.H. (1982) A left-ear advantage for identifying the emotional quality of tonal sequences. *Neuropsychologia*, **20**, 83–97.

Bryden, M.P., Free, T., Gagne, S, and Goff, P. (1991) Handedness effects in the detection of dichotically-presented words and emotions. *Cortex*, **27**, 229–235.

Bryson, S.E., McLaren, J., Wadden, N.P. and MacLean, M. (1991) Differential asymmetries for positive and negative emotion: hemisphere or stimulus effects. *Cortex*, **27**, 359–365.

Buchtel, H., Campari, F., de Risio, C. and Romanella, R. (1978) Hemispheric difference in discriminative reaction time to facial expressions. *Ital. J. Psychol.,* **5**, 159–169.

Bullier, J., McCourt, M.E. and Henry, G.H. (1988) Physiological studies on the feedback connections to the striate cortex from cortical areas 18 and 19 of the cat. *Exp. Brain Res.,* **70**, 90–98.

Bulman-Fleming, M.B. and Bryden, M.P. (1994) Simultaneous verbal and affective laterality effects. *Neuropsychologia*, **32**, 787–797.

Burton, A., Morton, N. and Abbess, S. (1989) Mode of processing and hemisphere differences in the judgement of musical stimuli. *Br. J. Psychol.,* **80**, 169–180.

Campain, R. and Minckler, J. (1976) A note on the gross configuration of the human auditory cortex. *Brain Lang.,* **3**, 318–323.

Cancelliere, A.E.B. and Kertesz, A. (1990) Lesion localization in acquired deficits of emotional expression and comprehension. *Brain Cogn.,* **13**, 133–147.

Carmon, A. (1978) Spatial and temporal factors in visual perception of patients with unilateral cerebral lesions. In: Kinsbourne, M. (ed.) *Asymmetrical functions of the brain.* Cambridge (UK), Cambridge University Press, pp. 86–98.

Carmon, A. and Nachshon, I. (1971) Effect of unilateral brain damage on perception of temporal order. *Cortex*, **7**, 411–418.

Carmon, A. and Nachshon, I. (1973) Ear asymmetry in perception of emotional non-verbal stimuli. *Acta Psychol. (Amst.),* **37**, 351–357.

Carpenter, K., Berti, A., Oxbury, S., Molyneux, A.J., Bisiach, E. and Oxbury, J.M. (1995) Awareness of and memory for arm weakness during intracarotid sodium amytal testing. *Brain*, **118**, 243–251.

Carr, C.E. and Konishi, M. (1988) Axonal delay lines for time measurement in the owl's brainstem. *Proc. Natl. Acad. Sci. USA,* **85**, 8311–8315.

Carson, R.G., Elliott, D., Goodman, D. and Dickinson, J. (1990a) Manual asymmetries in the reproduction of a 3-Dimensional spatial location. *Neuropsychologia*, **28**, 99–103.

Carson, R.G., Chua, R., Elliott, D. and Goodman, D. (1990b) The contribution of vision to asymmetries in manual aiming. *Neuropsychologia*, **28**, 1215–1220.

Carson, R.G., Goodman, D. and Elliott, D. (1992) Asymmetries in the discrete and pseudocontinuous regulation of visually guided reaching. *Brain Cogn.,* **18**, 169–191.

Cavada, C. and Goldman-Rakic, P.S. (1989) Posterior parietal cortex in rhesus monkey: evidence for segregated corticocortical networks linking sensory and limbic areas with the frontal lobe. *J. Comp. Neurol.,* **287**, 442–445.

Chapin, C., Blumstein, S.E., Meissner, B. and Boller, F. (1981) Speech production mechanisms in aphasia: a delayed auditory feedback study. *Brain Lang.*, **14**, 106–113.

Charbonneau, G. and Risset, J.-C. (1975) Difference entre oreille droite et oreille gauche pour la perception de la hauteur des sons. *C.R. Acad. Sci. Paris (ser. D)*, **281**, 163–166.

Chaurasia, B.D. and Goswani, H.K. (1975) Functional asymmetry in the face. *Acta Anat. (Basel)*, **91**, 154–160.

Chedru, F., Bastard, V. and Efron, R. (1978) Auditory micropattern discrimination in brain damaged subjects. *Neuropsychologia*, **16**, 141–149.

Chentanez, T., Keatsuwan, W., Akaraphan, A., Chaunchiiyakul, R., Lechanavanich, C., Hiranrati, S., Chaiwatcharaporu, C. and Glinsukonu, T. (1988) Reaction time, impulse speed, overall synaptic delay and number of synapses in tactile reaction neuronal circuits of normal subjects and thinner sniffers. *Physiol. Behav.*, **42**, 423–431.

Chi, J.G., Dooling, E.C. and Gilles, F.H. (1977a) Left-right asymmetries of the temporal speech areas of the human fetus. *Arch. Neurol.*, **34**, 346–348.

Chi, J.G., Dooling, E.C. and Gilles, F.H. (1977b) Gyral development of the human brain. *Ann. Neurol.*, **1**, 86–93.

Chiarello, C. (1985) Hemispheric dynamics in lexical access: automatic and controlled priming. *Brain Lang.*, **26**, 146–172.

Chiarello, C., Knight, R. and Mandel, M. (1982) Aphasia in a prelingually deaf woman. *Brain*, **105**, 29–51.

Chiarello, C. and Church, K.L. (1986) Lexical judgments after right- or left-hemisphere injury. *Neuropsychologia*, **24**, 623–630.

Chiarello, C., Seneki, J. and Nuding S. (1987) Semantic priming with abstract and concrete words: differential asymmetry may be post-lexical. *Brain Lang*, **31**, 43–60.

Christman, S. (1987) Effects of perceptual quality on hemispheric asymmetries in visible persistence. *Percept. Psychophysics*, **41**, 367–374.

Christman, S. (1988) Effects of luminance and blur on hemispheric asymmetries in temporal integration. *Neuropsychologia*, **28**, 361–374.

Christman, S. (1989a) Perceptual characteristics in visual laterality research. *Brain Cogn.*, **11**, 238–257.

Christman, S. (1989b) Temporal integration of form as a function of subject handedness and retinal locus of presentation. *Neuropsychologia*, **27**, 1373–1383.

Chua, R., Carson, R.G. and Goodman, D. (1992) Asymmetries in the spatial localization of transformed targets. *Brain Cogn.*, **20**, 227–235.

Chui, H.C. and Damasio, A.R. (1980) Human cerebral asymmetry evaluated by computed tomography. *J. Neurol. Neurosurg. Psychiatry*, **43**, 873–878.

Clarke, J.M., Lufkin, R.B. and Zaidel, E. (1993) Corpus callosum morphometry and dichotic listening performance: individual differences in functional interhemispheric inhibition? *Neuropsychologia*, **31**, 547–557.

Clarke, S. and Miklossy, J. (1990) Occipital cortex in man: Organization of callosal connections, related myelo- and cytoachitecture and putative boundaries of functional visual areas. *J. Comp. Neurol.*, **298**, 188–214.

Clarke, S., Assal, G. and de Tribolet, N. (1993) Left hemisphere strategies in visual recognition, topographical orientation and time planning. *Neuropsychologia*, **31**, 99–113.

Coffmann, J.A. and Bloch, S. (1984) Interhemispheric differences in regional density of the normal brain. *J. Psychiatr. Res.*, **18**, 269–275.

Coffmann, J.A., Andreasen, N.C. and Nasrallah, H.A. (1984) Left hemisphere density deficits in chronic schizophrenia. *Biol. Psychiatry*, **19**, 1237–1247.

Cohen, G. (1972) Hemispheric differences in a letter classification task. *Percept. Psychophysic*, **11**, 139–142.

Cohen, G. (1976) Components of the laterality effect in letter recognition: asymmetries in iconic storage. *Q. J. Exp. Psychol.*, **28**, 105–114.

Conolly, C. (1950) *External morphology of primate brain*. Illinois: Thomas Springfield.

Cook, J.D. and Diggles, V.A. (1984) Rapid error correction during human arm movements: evidence for central monitoring. *J. Motor Behav.*, **16**, 348–363.

Cook, N.D. (1986) *The brain code: mechanisms of information transfer and the role of the corpus callosum*. London, New York: Methuen.

Cook, N.D. and Beech, A.R. (1990) The cerebral hemispheres and bilateral neural nets. *Int. J. Neurosci.*, **52**, 201–210.

Cook, N.D., Brugger, P., Regard, M. and Landis, T. (1990) On the role of the corpus callosum in cerebral laterality: a comment on Lassonde, Bryden and Demers. *Brain Lang.*, **39**, 471–474.

Cook, N.D., Fruh, H., Mehr, A., Regard, M. and Landis, T. (1994) Hemispheric cooperation in visuospatial rotations: evidence for a manipulation role for the left hemisphere and a reference role for the right hemisphere. *Brain Cogn.*, **25**, 240–249.

Corina, D.P., Vaid, J., Bellugi, U. (1992a) The linguistic basis of left hemisphere specialization. *Science*, **255**, 1258–1260.

Corina, D.P., Poizner, H., Bellugi, U., Feinberg, T., Dowd, D. & O'Grady-Batch, L. (1992b) Dissociation between linguistic and non-linguistic gestural systems: a case for compositionality. *Brain Lang.*, **43**, 414–447.

Coslett, H.B., Bowers, D. and Heilman, K.M. (1987) Reduction in cerebral activation after right hemisphere stroke. *Neurology*, **37**, 957–962.

Cremonini, W., De Renzi, E. and Faglioni, P. (1980) Contrasting performance of right and left-hemisphere patients on short-term and long-term sequential visual memory. *Neuropsychologia*, **18**, 1–9.

Crichton-Browne, J. (1880) On the weight of the brain and its component parts in the insane. *Brain*, **2**, 42–67.

Crowne, D.O., Yeo, C.H. and Russell, I.S. (1981) The effects of unilateral frontal eye fields in the monkey: visual motor guidance and avoidance behaviour. *Behav. Brain Res.*, **2**, 165–187.

Curtis, J.N. and Foster, W.S. (1917) Size vs. intensity as a determinant of attention. *Am. J. Psychol.*, **28**, 293–296.

Cutting, J.E. (1974) Two left hemisphere mechanisms in speech perception. *Percept. Psychophysics*, **16**, 601–612.

Dagenbach, D. (1986) Subject variable effects in correlations between auditory and visual language processing asymmetries. *Brain Lang.*, **28**, 169–177.

Dallenbach, K.M. (1923) Position vs. intensity as a determinant of clearness. *Am. J. Psychol.*, **34**, 282–286.

Dalrymple-Alford, E.C. (1972) Associative facilitation and interference in the Stroop colour-word task. *Percept. Psychophysics*, **11**, 274–276.

Daniel, D.G., Kim, E., Kostianovsky, D., Goldberg, T.E., Casanova, M.F., Pickar, D., Kleinman, J.E. and Weinberger, D.R. (1992) Computed tomography measurements of brain density in schizophrenia. *Biol. Psychiatry*, **29**, 745–756.

Dauphinais, I.D., DeLisi, L.E., Crow, T.J., Alexandropoulos, K., Colter, N., Tuma, I. and Gershon, E.S. (1990) Reduction in temporal lobe size in siblings with schizophrenia: a magnetic resonance imaging study. *Psychiatr. Res.*, **35**, 137–147.

Davidoff, J. (1975) Hemispheric differences in perception of lightness. *Neuropsychologia*, **13**, 121–124.

Davidoff, J. (1976) Hemispheric sensitivity differences in the perception of colour. *Q. J. Exp. Psychol.*, **28**, 387–394.

Davidoff, J. (1977) Hemispheric difference in dot detection. *Cortex*, **13**, 434–444.

Davidson, R.J. (1988) EEG measures of cerebral asymmetry: conceptual and methodological issues. *Int. J. Neurosci.*, **39**, 71–89.

Davidson, R.J. and Schwartz, G.E. (1976) Patterns of cerebral lateralization during cardiac biofeedback versus the self-regulation of emotion: sex differences. *Psychophysiology*, **13**, 62–68.

Davidson, R.J., Schwartz, G.E., Saron, C., Bennett, J. and Coleman, D.J. (1979) Frontal versus parietal EEG asymmetry during positive and negative affect. *Psychophysiology*, **16**, 202–203.

Davidson, R.J., Leslie, S.C. and Saron, C. (1990a) Reaction time measures of interhemispheric transfer time in reading disabled and normal children. *Neuropsychologia*, **28**, 471–485.

Davidson, R.J., Ekman, P., Saron, C.D., Senulis, J.A. and Friesen, W.F. (1990b) Approach-withdrawal and cerebral asymmetry: emotional expression and brain physiology. *J. Pers. Soc. Psychol.*, **58**, 330–341.

Davis, R. and Schmit, V. (1973) Visual and verbal coding in the interhemispheric transfer of information. *Acta Psychol. (Amst.)*, **37**, 229–240.

Day, J. (1977) Right-hemisphere language processing in normal right handers. *J. Exp. Psychol. Hum. Percept. Perform.*, **3**, 518–528.

Day, J. (1979) Visual half-field word recognition as a function of syntactic class and imageability. *Neuropsychologia*, **17**, 515–519.

Day, R.S. and Vigorito, J.M. (1973) A parallel between degree of encodedness and the ear advantage: evidence from a temporal order judgement. *J. Acoust. Soc. Am.*, 53, 368A.

Dayhoff, J.E. and Gerstein, G.L. (1983) Favoured patterns in spike trains. II. Application. *J. Neurophysiol.*, **49**, 1349–1363.

De Ajuriaguerra, J. and Tissot, R. (1969) The apraxias. In: Vinken, P.J. and Bruyn, G.W. (eds) *Handbook of Clinical Neurology Vol. 4.* Amsterdam: North Holland, pp. 48–66.

Deacon, T.W. (1992) Cortical connections of the inferior arcuate sulcus cortex in the macaque brain. *Brain Res.*, **573**, 8–26.

Dee, H.L. and Fontenot, D.J. (1973) Cerebral dominance and lateral differences in perception and memory. *Neuropsychologia*, **11**, 167–173.

Deecke, L., Scheid, P. and Kornhuber, H.H. (1969) Distribution of readiness potenital, pre-motion positivity, and motor potential of the human cerebral cortex preceding voluntary finger movements. *Exp. Brain Res.*, **7**, 158–168.

De Kosky, S.T., Heilman, K.M., Bowers, D. and Valenstein, E. (1980) Recognition and discrimination of emotional faces and pictures. *Brain Lang.*, **9**, 206–214.

de Lacoste-Utamsing, C. and Woodward, D.J. (1982) Intra- and interhemispheric asymmetries of the human brain. *Soc. Neurosci. Abstr.*, 8, 212.

de Lacoste, M.C., Adesanya, T. and Woodward, D.J. (1990) Measures of gender differences in the human brain and their relationship to brain weight. *Biol. Psychiatry*, **28**, 931–942.

de Lacoste, M.-C., Horvath, D.S. and Woodward, D.J. (1991) Possible sex differences in the developing human fetal brain. *J. Clin. Exp. Neuropsychol.*, **13**, 831–846.

De Lisi, L., Stritzke, P., Riordan, H., Holan, V., Boccio, A., Kushner, M., McClelland, J., van Eyl, O. and Anand, A. (1992) The timing of brain morphological changes in schizophrenia and their clinical outcome. *Biol. Psychiatry*, **31**, 241–54.

Dennis, M. and Whitaker, H.A. (1976) Language acquisition following hemidecortication: linguistic superiority of the left over the right hemisphere. *Brain Lang.*, **3**, 404–433.

De Renzi, R.K., Pieczuro, A. and Vignolo, L.A. (1966) Oral apraxia and aphasia. *Cortex*, **2**, 50–73.

De Renzi, E. and Nichelli, P. (1975) Verbal and non-verbal short-term memory impairment following hemispheric damage. *Cortex*, **11**, 341–354.

Derryberry, D. (1989) Hemispheric consequences of success-related emotional states: roles of arousal and attention. *Brain Cogn.*, **11**, 258–274.

Derryberry, D. (1990) Right hemisphere sensitivity to feedback. *Neuropsychologia*, **28**, 1261–1271.

Deuel, R.K. (1977) Loss of motor habits after cortical lesions. *Neuropsychologia*, **15**, 205–213.

Deutsch, D. and Roll, P.L. (1976) Separate "what" and "where" decision mechanisms in processing a dichotic tonal sequence. *J. Exp. Psychol. Hum. Percept. Perform.*, **2**, 23–29.

Dewan, M.J., Pandurangi, A.K., Lee, S.H., Ramachandran, T., Levy, B., Boucher, M., Yozawitz, A. and Major, L. (1986) A comprehensive study of chronic schizophrenic patients. I. Quantitative computed tomography: cerebral density, ventricle and sulcal measures. *Acta Psychiatr. Scand.*, **73**, 152–160.

Diaz, G. and Guiard, Y. (1984) Directional and amplitude error in open-loop aimed movements: laterality effects. *Behav. Brain Res.*, 12, 183.

Diehl, J.A. and McKeever, W.F. (1987) Absence of exposure time influence on lateralized face recognition and object naming latency. *Brain Cogn.*, **6**, 347–357.

Di Lollo, V. (1981) Hemispheric symmetry in duration of visual persistence. *Percept. Psychophysics*, **29**, 21–25.

Dimond, S.J. (1979) Tactual and auditory vigilance in split brain man. *J. Neurol. Neurosurg. Psychiatry*, **42**, 70–74.

Dimond, S.J. and Beaumont, J.G. (1972) A right hemisphere basis for calculation in the human brain. *Psychon Sci.*, **26**, 137–138,

Dimond, S.J. and Beaumont, J.G. (1973) Difference in the vigilance performance of the right and left hemispheres. *Cortex*, **9**, 259–265.

Dimond, S.J., Farrington, L. and Johnson, P. (1976) Differing emotional response from right and left hemisphere. *Nature*, **261**, 690–692.

Dimond, S.J. and Farrington, L. (1977) Emotional response to films shown to right and left hemisphere of the brain measured by heart rate. *Acta Psychol. (Amst.)*, **41**, 255–260.

Dinopoulos, A., Dori, I., Davies, S.W. and Parnavelas, J.G. (1989) Neurochemical heterogeneity among corticofugal and callosal projections. *Exp. Neurol.*, **105**, 36–44.

Divenyi, P.L. and Efron, R. (1979) Spectral versus temporal features in dichotic listening. *Brain Lang.*, **7**, 375–386.

Doehring, D.G. and Ling, D. (1971) Matching to sample of three-tone simultaneous and successive sounds by musical and non-musical subjects. *Psychon. Sci.*, **25**, 103–105.

Dowling, W.J. (1978) Dichotic recognition of musical canons: effects of leading ear and time lag between ears. *Percept. Psychophysics*, **23**, 321–325.

Duysens, J., Gulyas, B. & Maes, H. (1991) Temporal integration in cat visual cortex: a test of Bloch's law. *Vision Res.*, **9**, 1517–1528.

Efron, R. (1963a) The effect of handedness on the perception of simultaneity and temporal order. *Brain,* **86**, 261–284.

Efron, R. (1963b) Temporal perception, aphasia and deja vu. *Brain*, **86**, 403–424.

Eidelberg, D. and Galaburda, A.M. (1984) Inferior parietal lobule. Divergent architectonic asymmetries in the human brain. *Arch. Neurol.,* **41**, 843–852.

Eilers, R.E. and Oller, D.K. (1994) Infant vocalizations and the early diagnosis of severe hearing impairment. *J. Pediatrics*, **124**, 199–203.

Elias, J.W., Yairi, E., Wright, L., Adams, L.A. and Villescas, R. (1977) The use of delayed auditory feedback in the identification of the left cerebral hemisphere as a temporal/duration processor. *J. Audit. Res.,* **17**, 155–160.

Elias, J.W., Marshall, P.H., Wright, J. and Hutton, J.T. (1988) Parkinson's disease and motor error detection and correction. *J. Hum. Movement Studies*, **15**, 51–59.

Eling, P. (1983) Consistency of ear advantage: an improvement due to increase in presentation rate. *Neuropsychologia*, **21**, 419–423.

Ellis, H.C. and Shepherd, J.W. (1974) Recognition of abstract and concrete words presented in left and right visual fields. *J. Exp. Psychol.,* **103**, 1035–1036.

Elman, J.L. (1983) Oral vs. manual tapping with delayed auditory feedback as measures of cerebral dominance. *J. Speech Hear. Res.,* **26**, 106–110.

Emmerich, D.S. and Pitchford, L.J. (1976) Ear differences in reaction time to onset and offsets of monaurally presented signals. *J. Acoust. Soc. Am.*, **59**, S3.

Emmerich, D.S., Pitchford, L.J. and Becker, C.A. (1976) Reaction time to tones in tonal backgrounds and a comparison of reaction time to signal onset and offset. *Percept. Psychophys*, **20**, 210–214.

Emmerich, D.S., Pitchford, L.J., Joyce, C.M. and Kopell, S.M. (1981) Laterality effects in response to offsets of tonal stimuli. *Neuropsychologia*, **19**, 227–234.

Emmerich, D.S., Harris, J., Brown, W.S. and Springer, S.P. (1988) The relationship between auditory sensitivity and ear asymmetry on a dichotic listening task. *Neuropsychologia*, **26**, 133–143.

Emmorey, K. and Corina, D. (1990) Lexical recognition in sign language: effects of phonetic structure and morphology. *Percept. Mot. Skills*, **71**, 1227–1252.

Emmorey, K. and Corina, D. (1993) Hemispheric specialization for ASL signs and English words: differences between imageable and abstract forms. *Neuropsychologia*, **31**, 645–653.

Ericksen, C.W. and Schultz, D.W. (1977) Retinal locus and acuity in visual information processing. *Bull. Psychon. Sci.*, **9**, 81–84.

Etcoff, N.L. (1984) Selective attention to facial identity and facial emotion. *Neuropsychologia*, **22**, 281–295.

Falkai, P., Schneider, T., Greve, B., Klieser, E., and Bogerts, B. (1995) Reduced frontal and occipital lobe asymmetry on the CT-scans of schizophrenic patients. Its specificity and clinical significance. *J. Neural Transm. Gen. Sect.*, **99**, 63–77.

Falzi, G., Perrone, P. and Vignolo, L.A. (1982) Right-left asymmetry in anterior speech region. *Arch. Neurol.*, **39**, 239–240.

Farah, M.J. (1984) The neurological basis of mental imagery: a componential analysis. *Cognition*, **18**, 245–272.

Farah, M.J., Gazzaniga, M.S., Holzman, J.D. and Kosslyn, S.M. (1985) A left hemisphere basis for mental imagery. *Neuropsychologia*, **23**, 115–118.

Fennell, E.B., Satz, P. and Morris, R. (1983) The development of handedness and dichotic ear listening asymmetries in relation to school achievement: a longitudinal study. *J. Exp. Child. Psychol.*, **35**, 248–262.

Ferraya Moyano, H. and Molina, J.C. (1980) Axonal projections and conduction properties of olfactory peduncle neurons in the rat. *Exp. Brain Res.*, **39**, 241–248.

Fisk, J.D. and Goodale, M.A. (1988) The effects of unilateral brain damage on visually guided reaching: hemispheric differences in the nature of the deficit. *Exp. Brain Res.*, **72**, 425–435.

Fitts, P.M. (1954) The information capacity of the human motor system in controlling the amplitude of movement. *J. Exp. Psychol.*, **47**, 381–391.

Fleischhauer, K. and Wartenberg, H. (1967) Elektronenmikroscopische Untersuchungen uber das Wachstum der Nervenfasern und uber das Auftreten von Markscheiden in Corpus callosum der Katze. *Z. Zellforsch.*, **83**, 568–581.

Flowers, K. (1975) Handedness and controlled movement. *Br. J. Psychol.*, **66**, 39–52.

Fodor, J. and Bever, T.G. (1965) The psychological reality of linguistic segments. *J. Verb. Learn. Verb. Behav.*, **4**, 414–420.

Fontenot, D.J. (1973) Visual field differences in the recognition of verbal and non-verbal stimuli in man. *J. Comp. Physiol. Psychol.*, **85**, 564–569.

Foundas, A.L., Leonard, C.M., Gilmore, R., Fennell, E. and Heilman, K.M. (1994) Planum temporale asymmetry and language dominance. *Neuropsychologia*, **32**, 1225–1232.

Foundas, A.L., Leonard, C.M. and Heilman, K.M. (1995) Morphologic asymmetries and handedness. The pars triangularis and planum temporale. *Arch. Neurol.*, **52**, 501–508.

Frankfurter, A. and Honeck, R.P. (1973) Ear differences in the recall of monaurally presented sentences. *Q. J. Exp. Psychol.*, **25**, 138–146.

Freund, T.F., Powell, J.F. and Smith, A.D. (1984) Tyrosine hydroxylase-immunoreactive boutons in synaptic contact with identified striatonigral neurons, with particular reference to dendritic spines. *Neuroscience*, **13**, 1189–1215.

Friston, K.J., Frith, C.D., Liddle, P.F. and Frackowiak, R.S.J. (1991) Investigating a network model of word generation with positron emission tomography. *Proc. Roy. Soc. Lond. Biol. Sci.*, **244**, 101–106.

Frith, C.D., Bloxham, C.A. and Carpenter, K.N. (1986) Impairments in the learning and performance of a new manual skill in patients with Parkinson's disease. *J. Neurol. Neurosurg. Psychiatry*, **49**, 661–668.

Frostig, R.D., Frysinger, R.C. and Harper, R.M. (1990) Recurring discharge patterns in multiple spike trains. II. Application in forebrain areas related to cardiac and respiratory control during different sleep-waking states. *Biol. Cybern.* **62**, 495–502.

Gado, M., Danziger, W.L., Chi, D., Hughes, C.P., Coben, L.A. (1983) Brain parenchymal density measurements by CT in demented subjects and normal controls. *Radiology*, **147**, 703–710.

Gaede, S.E., Parsons, O. and Bertera, J.H. (1978) Hemispheric difference in music perception: aptitude vs. experience. *Neuropsychologia*, **16**, 369–373.

Gainotti, G., Caltagirone, C., Miceli, G. and Masullo, C. (1981) Selective semantic-lexical impairment of language comprehension in right brain-damaged patients. *Brain Lang.*, **13**, 201–211.

Galaburda, A.M. (1980) La region de Broca: Observations anatomiques faites un siecle apres la mort de son decoureur. *Rev. Neurol. (Paris)*, **136**, 609–616.

Galaburda, A.M., Le May, M., Kemper, T.L. and Geschwind, N. (1978a) Right-left asymmetries in the brain. *Science*, **199**, 852–856.

Galaburda, A.M., Sanides, F. and Geschwind, N. (1978b) Human brain: cytoarchitectonic left-right asymmetries in the temporal speech regions. *Arch. Neurol.*, **35**, 812–817.

Galaburda, A.M., Corsiglia, J., Rosen, G.D. and Sherman, G.F. (1987) Planum temporale asymmetry, reappraisal since Geschwind and Levitsky. *Neuropsychologia*, **25**, 853–868.

Gardner, E.B. and Branski, D.M. (1976) Unilateral cerebral activation and perception of gaps: a signal detection analysis. *Neuropsychologia*, **14**, 43–53.

Gardner, H., Ling, K., Flam, L. and Silverman, J. (1975) Comprehension and appreciation of humorous material in brain damaged patients. *Brain*, **98**, 399–412.

Gardner, H., Silverman, J., Wapner, W. and Zurif, E. (1978) The appreciation of antonymic contrasts in aphasia. *Brain Lang.*, **6**, 301–317.

Gassanov, V.G., Merzhanova, G.Kh. and Galashina, A.G. (1985) Interneuronal relations within and between cortical areas during conditioning in cats. *Behav. Brain Res.*, **15**, 137–146.

Gates, E.A., Bradshaw, J.L. and Nettleton, N.C. (1974) Effect of different delayed auditory feedback intervals on a music performance task. *Percept. Psychopysics*, **15**, 21–25.

Gates, E.A. and Bradshaw, J.L. (1977) Music perception and cerebral asymmetries. *Cortex*, **13**, 396–401.

Gazzaniga, M.S. and Sperry, R.W. (1967) Language after section of the cerebral commissures. *rain*, **90**, 131–148.

Gaymard, B., Rivaud, S. and Pierrot-Deseilligny, C. (1993) Role of the left and right supplementary motor areas in memory-guided saccade sequences. *Ann. Neurol.*, **34**, 404–406.

Geffen, G. (1978) The development of the right ear advantage in dichotic listening with focused attention. *Cortex*, **14**, 169–177.

Geffen, G. (1988) The effects of orientation and maintenance of attention on hemispheric asymmetry for speech perception. *Cortex*, **24**, 255–265.

Geffen, G., Bradshaw, J.L. and Nettleton, N.C. (1972) Hemispheric asymmetry: verbal and spatial encoding of visual stimuli. *J. Exp. Psychol.*, **95**, 25–31.

Geffen, G. and Caudrey, D. (1981) Reliability and validity of the dichotic monitoring test for language laterality. *Neuropsychologia*, **19**, 413–421.

George, M.S., Ketter, T.A., Parekh, P.I., Horwitz, B., Herscovitch, P. and Post, R.M. (1995) Brain activity during transient sadness and happiness in healthy women. *Am. J. Psychiatry*, **152**, 341–351.

Geschwind, N. (1965) Disconnection syndromes in animals and man. Part I. *Brain*, **88**, 237–294.

Geschwind, N. (1975) The apraxias: neuromechanisms of disorders of learned movement. *Am. Scientist* **63**, 188–195.

Geschwind, N. and Levitsky, W. (1968) Human brain: left right asymmetries in the temporal speech region. *Science*, **161**, 186–187.

Gilbert, C. and Bakan, P. (1973) Visual asymmetry in perception of faces. *Neuropsychologia*, **11**, 355–362.

Glass, C., Bradshaw, J.L., Day, R.H. & Umilta, C. (1985) Familiarity, spatial frequency and task determinants in processing laterally presented representations of faces. *Cortex*, **21**, 513–531.

Glosser, G. and Goodglass, H. (1991) Idiosyncratic word association following right hemisphere damage. *J. Clin. Exp. Neuropsychol.*, **13**, 703–711.

Golden, C.J., Graber, B., Coffman, J., Berg, R.A., Newin, D.B. and Block, S. (1981) Structural brain deficits in schizophrenia. Identification by computed tomographic scan density measurements. *Arch. Gen. Psychiatry*, **38**, 1014–1017.

Goldman, P.S., Lodge, A., Hammer, L.R., Semmes, J. and Mishkin, M. (1968) Critical flicker frequency after unilateral temporal lobectomy in man. *Neuropsychologia*, **6**, 355–363.

Goodale, M.A., Pelisson, D. and Prablanc, C. (1986) Large adjustments in visually guided reaching do not depend on vision of the hand or perception of target displacement. *Nature*, **320**, 748–750.

Goodglass, H. and Quasfadel, F.A. (1954) Language laterality in left-handed aphasics. *Brain*, **77**, 521–548.

Goodglass, H. and Kaplan, E. (1963) Disturbance of gesture and pantomime in aphasia. *Brain*, **86**, 703–720.

Gordon, H.W. (1970) Hemispheric asymmetries in the perception of musical chords. *Cortex*, **6**, 387–397.

Gordon, H.W. (1978a) Hemispheric asymmetry for dichotically presented chords in musicians and non-musicians. *Acta Psychol. (Amst.)*, **42**, 383–395.

Gordon, H.W. (1978b) Left hemisphere dominance for rhythmic elements in dichotically-presented melodies. *Cortex*, **14**, 58–70.

Gott, P.S. (1973) Language after dominant hemispherectomy. *J. Neurol. Nerurosurg. Psychiatry*, **36**, 1083–1088.

Gracco, V.L. and Abbs, J.H. (1986) Variant and invariant characteristics of speech movements. *Exp. Brain Res.*, **65**, 156–166.

Gracco, V.L. and Abbs, J.H. (1988) Central patterning of speech movements. *Exp. Brain Res.*, **71**, 515–526.

Grafe, M.R. and Leonard, C.M. (1980) Successful silver impregnation of degenerating axons after long survivial in the human brain. *J. Neuropathol. Exp. Neurol.*, **39**, 555–574.

Graves, R. (1983) Mouth asymmetry, dichotic ear advantage, and tachistoscopic visual field advantage as measures of language lateralization. *Neuropsychologia*, **21**, 641–649.

Gregory, A.H. (1978) Perception of clicks in music. *Percept. Psychophysics*, **24**, 171–174.

Guariglia, C. and Antonucci, G. (1992) Personal and extrapersonal space: a case of neglect dissociation. *Neuropsychologia*, **30**, 1001–1009.

Guariglia, C., Padovani, A., Pantano, P. and Pizzimiglio, L. (1993) Unilateral neglect restricted to visual imagery. *Nature*, **364**, 235–237.

Gudmundsson, E. (1993) Lateral preference of preschool and primary school children. *Percept. Mot. Skills*, **77**, 819–823.

Guiard, Y., Diaz, G. and Beaubaton, D. (1983) Left-hand advantage in right handers for spatial constant error: preliminary evidence in a unimanual ballistic aimed movement. *Neuropsychologia*, **21**, 111–115.

Gur, R.C., Reivitch, M., Obrist, W.D., Amarnek, W.S. and Sackheim, H.A. (1980) Differences in the distribution of grey and white matter in human cerebral hemispheres. *Science*, **207**, 1226–1228.

Gur, R.E., Mosley, D., Resnick, S.M., Shtasel, D., Kohn, M., Zimmerman, R., Herman, G., Atlas, S., Grossman, R., Erwin, R., and Gur, R.C. (1991) Magnetic resonance imaging in schizophrenia. I. Volumetric analysis of brain and cerebrospinal fluid. *Arch. Gen. Psychiatry*, **48**, 407–412.

Gur, R.C., Skolnick, B.E., Gur, R.E. (1994) Effects of emotional discrimination tasks on cerebral blood flow: regional activation and its relation to performance. *Brain Cogn.*, **25**, 271–286.

Gustafsson, B. and Wigstrom, H. (1986) Hippocampal long-lasting potentiation produced by pairing single volleys and brief conditioning tetani in separate afferents. *J. Neurosci.*, 6, 1575–1582.

Haaland, K.Y., Porch, B.E. and Delaney, H.D. (1980). Limb apraxia and motor performance. *Brain Lang.*, **9**, 315–322.

Haaland, K.Y. and Harrington, D. (1989a) The role of the hemispheres in closed loop movements. *Brain Cogn.*, **9**, 158–180.

Haaland, K.Y. and Harrington, D. (1989b) Hemispheric control of the initial and corrective components of aiming movements. *Neuropsychologia*, **27**, 961–969.

Haaland, K.Y. and Harrington, D. (1994) Limb-sequencing deficits after left but not right hemispheric damage. *Brain Cogn.*, **24**, 104–122.

Haber, R.N. and Standing, L.G. (1969) Direct measurment of short-term visual storage. *Q. J. Exp. Psychol.*, **21**, 43–54.

Haggard, M.P. and Parkinson, A.M. (1971) Stimulus and task factors as determinants of ear advantages. *Q. J. Exp. Psychol.*, **23**, 168–177.

Halligan, P.H. and Marshall, J.C. (1991) Left neglect for near but not far space in man. *Nature*, **350**, 498–500.

Halperin, Y., Nachson, I. and Carmon, A. (1973) Shift of ear superiority in dichotic listening to temporally patterned non-verbal stimuli. *J. Acoust. Soc. Am.*, **53**, 46–50.

Hammond, G.R. (1981) Fine temporal acuity for stimuli applied to the preferred hand. *Neuropsychologia*, **19**, 325–329.

Hammond, G.R. and Kaplan, R.J. (1982) Language lateralization and handedness: estimates based on clinical data. *Brain Lang.*, **16**, 348–351.

Hannay, H.J. and Malone, D.R. (1976) Visual field effects and short-term memory for verbal material. *Neuropsychologia*, **14**, 203–209.

Hannay, H.J., Rogers, J.P. and Durant, R.F. (1976) Complexity as a determinant of visual field effects for random forms. *Acta Psychol. (Amst.)*, **40**, 29–34.

Hannay, H.J. and Smith, A.C. (1979) Dichhaptic perception of forms by normal adults. *Percept. Mot. Skills*, **49**, 991–1000.

Harmon, D.W. and Ray, W.J. (1977) Hemispheric activity during affective verbal stimuli: an EEG study. *Neuropsychologia*, **15**, 457–460.

Harvey, A.R. (1980) A physiological analysis of subcortical and commissural projections of areas 17 and 18 of the cat. *J. Physiol. (Lond.)*, **302**, 507–534.

Harvey, I., Ron, M.A., Du Boulay, G., Wicks, D., Lewis, S.W. and Murray, R.M. (1993) Reduction of cortical volume in schizophrenia on magnetic resonance imaging. *Psychol. Med.*, **23**, 591–604.

Hashimoto, T., Toyama, M., Mori, K., Fujino, K., Miyazaki, M. and Kuroda, Y. (1989) Magnetic resonance imaging in autism: preliminary report. *Neuropaediatrics*, **20**, 142–146.

Hashemzadeh-Gargari, H., Colbert, C.M. and Levy, W.B. (1991) Refining the temporal definition of an association at the neuronal level using long-term potentiation and long-term depression in the dentate gyrus. *Neurosci. Lett.*, **126**, 37–40.

Hass, E.J. and Whipple, J.L. (1985) Effects of a concurrent memory task on hemispheric asymmetries in categorization. *Brain Cogn.*, **4**, 13–26.

Hauser, M.D. (1993) Right hemisphere dominance for the production of facial expression in monkeys. *Science*, **261**, 475–477.

Hayes, T.L. and Lewis, D.A. (1993) Hemispheric differences in layer III pyramidal neurones of the anterior language area. *Arch. Neurol.*, **50**, 501–505.

Hebb, D.O. (1949) *The organization of behavior: a neuropsychological theory.* New York: Wiley.

Hecaen, H. (1969) Aphasia, apraxia and agnosic syndromes in right and left hemispheric lesions. In: Vinken, P.J. and Bruyn, G.W. (eds). *Handbook. Clin. Neurology. Volume 4.* Amsterdam: North Holland, pp. 291–311.

Hecaen, H. and Albert, M.L. (1978) *Human neuropsychology.* New York: Wiley.

Hecaen, H., DeAgostini, M. and Monzon-Montes, A. (1981) Cerebral organization in left handers. *Brain Lang.*, **12**, 261–284.

Heckers, S., Heinsen, H., Heinsen, Y. and Beckmann, H. (1991) Cortex, white matter and basal ganglia in schizophrenia. *Biol. Psychiatry*, **29**, 556–566.

Heffner, T.G., Luttinger, D., Hartman, J.A. and Sieden, L.S. (1981) Regional changes in brain catecholamine turnover in the rat during performance on fixed ratio and variable interval schedules of reinforcement. *Brain Res.*, **214**, 215–218.

Heilman, K.M., Pandya, D.N. and Geschwind, N. (1970) Trimodal inattention following parietal lobe lesions. *Trans. Am. Neurol. Ass.*, **95**, 259–261.

Heilman, K.M., Schwartz, H.D. and Watson, R.T. (1978) Hypoarousal in patients with the neglect syndrome and emotional indifference. *Neurology*, **28**, 229–232.

Heilman, K.M. and van den Abell, T. (1979) Right hemispheric dominance for mediating cerebral activation. *Neuropsychologia*, **17**, 315–321.

Heilman, K.M. and Valenstein, E. (1979) Mechanisms underlying hemispatial neglect. *Ann. Neurol.*, **5**, 166–170.

Heilman, K.M. and van den Abell, T. (1980) Right hemisphere dominance for attention mechanisms underlying hemispheric asymmetries of inattention (neglect). *Neurology*, **30**, 327–330.

Heilman, K.M., Bowers, D., Speedie, L. and Coslett, H.B. (1983) The comprehension of emotional and non-emotional prosody. *Neurology*, 33, 241.

Heilman, K.M., Bowers, D., Speedie, L. and Coslett, H.R. (1984) Comprehension of affective and non-affective prosody. *Neurology*, **34**, 917–921.

Heilman, K.M. and Rothi, J.G. (1985) Apraxia. In: Heilman, K.M. and Valenstein, E. *Clinical Neuropsychology.* Oxford: Oxford University Press, pp. 131–150.

Heller, W. and Levy, J. (1981) Perception and expression of emotion in right handers and left handers. *Neuropsychologia*, **19**, 263–272.

Hellige, J.B. (1976) Changes in same-different laterality patterns as a function of practice and stimulus quality. *Percept. Psychophysics*, **20**, 267–273.

Hellige, J.B. and Cox, P.J. (1976) Effects of concurrent verbal memory on recognition of stimuli from the left and right visual fields. *J. Exp. Psychol. Hum. Percept. Perform.*, **2**, 210–221.

Hellige, J.B., Cox, P.J. and Litvac, L. (1979) Information processing in the cerebral hemispheres: selective hemispheric activation and capacity limitation. *J. Exp. Psychol. Gen.*, **108**, 251–279.

Hellige, J.B. and Webster, R. (1981) Case effect in letter-name matching: a qualitative visual field difference. *Bull. Psychon. Sci.*, **17**, 179–182.

Henderson, V.W. (1986) Paul Broca's less heralded contributions to aphasia research. Historical perspective and contemporary relevance. *Arch. Neurol.*, **43**, 609–612.

Hermelin, B. and O'Connor, N. (1971) Functional asymmetry in the reading of Braille. *Neuropsychologia*, **9**, 431–435.

Hilgard, E.R. and Marquis, D.G. (1964) *Conditioning and learning.* London: Methuen.

Hill, W.C.O. (1970) *Primates: comparative anatomy and taxonomy.* Edinburgh University Press.

Hines, D. (1976) Recognition of verbs, abstract nouns and concrete nouns from the left and right visual half fields. *Neuropsychologia*, **14**, 211–216.

Hines, D., Satz, P., Schell, B. and Schmidlin, S. (1969) Differential recall of digits in the left and right visual half-fields under free and fixed order of recall. *Neuropsychologia*, **7**, 13–22.

Hines, D. and Satz, P. (1971) Superiority of right visual half-fields in right handers for recall of digits presented at varying rates. *Neuropsychologia*, **9**, 21–25.

Hines, D., Satz, P. and Clementino, T. (1973) Perceptual and memory components of the superior recall of letters from the right visual half-fields. *Neuropsychologia*, **11**, 175–180.

Hines, D. and Satz, P. (1974) Cross-modal asymmetries in perception, related to asymmetry in cerebral function. *Neuropsychologia*, **12**, 239–247.

Hines, D., Sutker, L.W., Satz, P. and Altmann, A. (1976) Recall of letters from the left and right visual half-fields under unilateral and bilateral presentation. *Percept. Mot. Skills*, **42**, 531–539.

Hirata, K. and Bryden, M. (1976) Right visual field superiority for letter recognition with partial report. *Can. J. Psychol.*, **30**, 134–139.

Hirsh, I.J. and Sherrick, C.E. (1961) Perceived order in different sense modalities. *J. Exp. Psychol.*, **62**, 423–432.

Howes, D. and Boller, F. (1975) Simple reaction time: evidence for focal impairment from lesions of the right hemisphere. *Brain*, **98**, 317–332.

Hughes, C.M. and Peters, A. (1992) Symmetric synapses formed by callosal afferents in rat visual cortex. *Brain Res.,* **583**, 271–278.

Hutton, J.T. and Palet, J. (1986) Lateral saccadic latencies and handedness. *Neuropsychologia*, **24**, 449–451.

Iaccino, J.F. and Sowa, S.J. (1989) Asymmetrical processing of dichotic inputs in undergraduates across sex, handedness, ear-side and experimental instructions. *Percept. Mot. Skills*, **68**, 1003–1010.

Iwasaki, S., Nakagama, H., Fukushimi, A., Kichikawa, K., Kitamura, K., Otsuji, H., Uchida, H., Ohishi, H., Yaguchi, K., Sumie, H and Kuru, Y. (1990) Left right asymmetry of the temporal and parietal regions in children: based on the medullary pattern of cerebral white matter. *Surgical Radiol. Anat.,* **12**, 209–214.

Jack, C.R., Gehring, D.G., Sharborough, F.W., Felmlee, J.P., Forbes, G., Hench, V.S. and Zinsmeister, A.R. (1988) Temporal lobe measurements from MR images: Accuracy and left-right asymmetry in normal persons. *J. Comput. Assist. Tomogr.,* **12**, 21–29.

Jack, C.R., Twomey, C.K., Zinsmeister, A.R., Sharbrough, F.W., Peterson, R.C. and Cascino, G.D. (1989) Anterior temporal lobes and hippocampal formation: normative volumetric measurements from MR images in young adults. *Radiology*, **172**, 549–554.

Jackman, M.K. (1985) The recognition of tachistoscopically presented words, varying in imagery, part of speech and word frequency in left and right visual fields. *Br. J. Psychol.,* **76**, 59–74.

Jarvella, R.J. and Herman, S.J. (1973) Speed and accuracy of sentence recall: effects of ear of presentation, semantics and grammar. *J. Exp. Psychol.,* **97**, 108–110.

Jason, G.W. (1983a) Hemispheric asymmetries in motor function: I. Left-hemisphere specialization for memory but not performance. *Neuropsychologia*, **21**, 35–42.

Jason, G.W. (1983b) Hemispheric asymmetries in motor function: II. Ordering does not contribute to left hemisphere specialization. *Neuropsychologia*, **21**, 47–58.

Jeannerod, M. (1986) Are corrections in accurate arm movements corrective? *Prog. Brain Res.,* **64**, 353–360.

Jeeves, M. and Dixon, N. (1970) Hemispheric differences in response rates to visual stimuli. *Psychon. Sci.,* **20**, 249–251.

Jeeves, M. (1972) Hemispheric differences in response rates to visual stimulus in children. *Psychon. Sci.,* **27**, 201–203.

Jernigan, T.L., Zatz, L.M., Moses, J.A. and Cardellino, J.P. (1982) Computed tomography in schizophrenics and normal volunteers. *Arch. Gen. Psychiatry*, **39**, 771–773.

Johnson, D., Sato, S., Quarles, R.H., Inuzuka, T., Brady, R.O. and Tortelotte, W.W. (1986) Quantification of the myelin-associated glycoprotein in human nervous tissue from control and multiple sclerosis patients. *J. Neurochem.,* **46**, 1089–1093.

Jones, E.G. and Powell, T.P.S. (1970) An anatomical study of converging sensory pathways within the cerebral cortex of monkeys. *Brain*, **93**, 793–820.

Jones-Gotman, M. and Milner, B. (1978) Right temporal-lobe contribution to image-mediated verbal learning. *Neuropsychologia*, **16**, 61–71.

Jonsson, J. and Hellige, J. (1986) Lateralized effects of blurring: a test of the visual spatial frequency model of cerebral asymmetry. *Neuropsychologia*, **24**, 351–362.

Juraska, J.M. and Kopcik, J.R. (1988) Sex and environmental influences on the size and ultrastructure of the rat corpus callosum. *Brain Res.,* **450**, 1–8.

Kakeshita, T. (1925) Zur Anatomie der Operkularen temporal region (Vergleichende Untersuchung der rechten und Linken Seite) *Arbeiten aus dem Neurologischen Institut an der Weiner Universitat*, **27**, 292–326.

Karlin, R., Weinapple, M., Rochford, J. and Goldstein, L. (1979) Quantitated EEG features of negative affective states: report of some hypnotic studies. *Res. Commun. Psychol. Psychiatry. Behav.* **4**, 397–413.

Kallman, H.J. and Corballis, M.C. (1975) Ear asymmetry in reaction time to musical sounds. *Percept. Psychophysics*, **17**, 368–370.

Kappauf, W.E. and Yeatman, F.R. (1970) Visual on- and off- latencies and handedness. *Percept. Psychophysics*, **8**, 46–50.

Katz, A.N. (1980) Cognitive arithmetic: evidence for right hemispheric mediation in an elementary component stage. *Q. J. Exp. Psychol.,* **32**, 69–84.

Katz, W.F. (1988) Anticipatory coarticulation in aphasia: acoustic and perceptual data. *Brain Lang.,* **35**, 340–368.

Katz, W., Machetanz, J., Orth, U. and Shonle, P. (1990) A kinematic analysis of anticipatory coarticulation in the speech of anterior aphasic subjects using electromagentic articulography. *Brain Lang.,* **37**, 172–176.

Kellar, L.A. and Bever, T.G. (1980) Hemispheric asymmetries in the perception of musical intervals as a function of musical experience and family handedness background. *Brain Lang.,* **10**, 24–38.

Kelly, R.R. and Orton, K.D. (1979) Dichotic perception of word-pairs with mixed image values. *Neuropsychologia*, **17**, 363–371.

Kelsoe, J.R., Cadet, J.L., Pickar, D. and Weinberger, D.R. (1988) Quantitative neuroanatomy in schizophrenia. A controlled magnetic resonance imaging study. *Arch. Gen. Psychiatry*, **45**, 533–541.

Kent, R.D., Osberger, M.J., Netsell, R., Goldschmidt-Hustedde, C. (1987) Phonetic development in identical twins differing in auditory function. *J. Speech Hear. Disord.,* **52**, 64–75.

Kertesz, A., Polk, M., Black, S.E., Howell, J. (1990). Sex, handedness, and the morphometry of cerebral asymmetry on magnetic resonance imaging. (1990) *Brain Res.*, **530**, 40–48.

Kester, D.B., Saykin, A.J., Sperling, M.R., O'Connor, M.J., Robinson, L.J. and Gur, R.C. (1991) Acute effects of anterior temporal lobectomy on musical processing. *Neuropsychologia*, **29**, 703–708.

Kim, C.-W. (1978) Experimental phonetics. In: W.O. Dingwall (ed.) *A survey of linguistic science*. Stamford, Connecticut: Greylock Publishers, pp. 159–201.

Kim, H. and Levine, S.C. (1992) Variations in characteristic perceptual asymmetry: modality specific and modality general components. *Brain Cogn.*, **19**, 21–47.

Kimura, D. (1977) Acquisition of a motor skill after left-hemisphere damage. *Brain*, **100**, 527–542.

Kimura, D. (1982) Left hemisphere control of oral and brachial movements and their relation to communication. *Phil. Trans. Roy. Soc. Lond. (Biol.)*, **298**, 135–149.

Kimura, D. and Archibald, Y. (1974) Motor functions of the left hemisphere. *Brain*, **97**, 337–350.

Kimura, D., Battison, R. and Lubert, B. (1976) Impairment of non-linguistic hand movement in a deaf aphasic. *Brain Lang.*, **3**, 566–571.

Kinsbourne, M. (1970) The cerebral basis of lateral asymmetries in attention. *Acta Psychol. (Amst.)*, **33**, 193–201.

Kinsbourne, M. (1971) The minor cerebral hemisphere as a source of aphasic speech. *Arch. Neurol.*, **25**, 302–306.

Kinsbourne, M. (1973) The control of attention by interaction between the cerebral hemispheres. In: S. Kornblum (ed.), *Attention and performance Vol. IV* New York and London: Academic Press, pp. 239–258.

Kinsbourne, M. and Byrd, M. (1985) Word load and visual hemifield shape recognition: priming and interference effects. In: M.I. Posner and O.S.M. Marin, (eds), *Attention and Performance Vol. XI*. New York: Lawrence Erlbaum, pp. 529–546.

Kitterle, F.L. and Kaye, R.S. (1985) Hemispheric symmetry in contrast and orientation sensitivity. *Percept. Psychophysics*, **37**, 391–396.

Klatt, D.H. (1976) Linguistic uses of segmental duration in English: Acoustic and perceptual evidence. *J. Acoust. Soc. Am.*, **59**, 1208–1221.

Kleist, K. (1934) *Gehirnpathologie*. Barth, Leipzig.

Koff, E., Naeser, M., Pieniadz, J.M., Foundas, A.L. and Levine, H.L. (1986) Computed tomographic scan hemispheric asymmetries in right- and left-handed male and female subjects. *Arch. Neurol.*, **43**, 487–491.

Kohler, W. (1927) *The mentality of apes*. New York: Harcourt Brace. (Translated by E. Winter).

Kolb, B. and Milner, B. (1981a) Performance of complex arm and facial movements after focal brain lesions. *Neuropsychologia*, **19**, 491–503.

Kolb, B. and Milner, B. (1981b) Observations on spontaneous facial expression after focal cerebral excisions and after intracarotid injection of sodium amytal. *Neuropsychologia*, **19**, 505–514.

Kolb, B., Sutherland, R.J., Nonneman, A.J. and Whishaw, I.Q. (1982) Asymmetry in the cerebral hemisphere of the rat, mouse, rabbit and cat: the right hemisphere is larger. *Exp. Neurol.*, **78**, 348–359.

Kolk, H.H.J. (1978) Judgement of sentence structure in Broca's aphasia. *Neuropsychologia*, **16**, 617–625.

Kopp, N., Michel, F., Carrier, H., Biron, A. and Duvillard, P. (1977) Etude de certaines asymmetries hemispherique du cerveau humain. *J. Neurol. Sci.*, **34**, 349–363.

Koppel, H. and Innocenti, G.M. (1983) Is there a genuine exuberancy of callosal projections in development? A quantitative electron microscopic study in the cat. *Neurosci. Lett.*, **41**, 33–40.

Kornhuber, H.H. and Deecke, L. (1965) Hirnpotentialanderungen bei Wilkurbewegungen und passiven Bewegungen des Meschen: Bereitschaftspotential und reafferente Potential. *Pflugers Arch.*, **284**, 1–17.

Kosslyn, S.M., Holtzman, J.D., Farah, M.J. and Gazzaniga, M.S. (1985) A computational analysis of mental image generation: evidence from functional dissociations in split-brain patients. *J. Exp. Psychol. Gen.*, **114**, 311–341.

Kubota, Y., Inagaki, S., Shimeda, S., Kito, S. and Wu, J.-Y. (1987) Glutamate decarboxylase-like immunoreactive neurons in the rat caudate putamen. *Brain Res. Bull.*, **18**, 687–697.

Kumar, S (1977) Short-term memory for a non-verbal task after cerebral commissurotomy. *Cortex*, **13**, 55–61.

Lackner, J.R. (1982) Alterations in resolution of temporal order after cerebral injury in man. *Exp. Neurol.*, **75**, 501–509.

Ladavas, E., Umilta, C. and Ricci-Bitti, P.E. (1980) Evidence for sex differences in right-hemisphere dominance for emotions. *Neuropsychologia*, **18**, 361–366.

Ladavas, E., Nicoletti, R., Umilta, C. and Rizzolatti, G. (1984) Right hemisphere interference during negative affect: a reaction time study. *Neuropsychologia*, **22**, 479–485.

Ladavas, E., Del Pesce, M. and Provinciali, L. (1989) Unilateral attention deficits and hemispheric asymmetry in the control of visual attention. *Neuropsychologia*, **27**, 353–366.

Ladavas, E., Del Pesce, M., Mangun, G.R. and Gazzaniga, M.S. (1994) Variations in attentional bias of the disconnected cerebral hemispheres. *Cogn. Neuropsychol.*, **11**, 57–74.

Lake, D.A. and Bryden, M.P. (1976) Handedness and sex differences in hemispheric asymmetry. *Brain Lang.*, **3**, 266–282.

Lalande, S., Braun, C.M.J., Charlebois, N. and Whitaker, H.A. (1992) Effects of right and left hemisphere cerebrovascular lesions on discrimination of prosodic and semantic aspects of affect in sentences. *Brain Lang.*, **42**, 165–186.

Lamantia, A.S. and Rakic, P. (1990a) Cytological and quantitative characteristics of four cerebral commissures in the rhesus monkey. *J. Comp. Neurol.*, **291**, 520–537.

Lamantia, A.S. and Rakic, P. (1990b) Axon overproduction and elimination in the corpus callosum of the developing rhesus monkey. *J. Neurosci.*, **10**, 2156–2175.

Lambert, A. and Voot, N. (1993) A left visual field bias for semantic encoding of unattended words. *Neuropsychologia*, **31**, 67–73.

Lambert, A.J. (1993) Attentional interaction in the split brain: evidence from negative priming. *Neuropsychologia*, **31**, 313–324.

Lambert, A.J. and Beaumont, J.G. (1983) Imageability does not interact with visual field lateral word recognition with oral report. *Brain Lang.*, **20**, 115–142.

Landsell, H. (1968) Effect of extent of temporal lobe ablations on two lateralized deficits. *Physiol. Behav.*, **3**, 271–273.

Lansing, R., Schwartz, E. and Lindsley, P.B. (1959) Reaction time and EEG activation under alerted and nonalerted conditions. *J. Exp. Psychol.*, **58**, 1–7.

Largen, J.W., Calderon, M. and Smith, R.C. (1983) Asymmetries in the densities of white and grey matter in the brains of schizophrenic patients. *Am. J. Psychiatry*, **140**, 1060–1062.

Larmande, P., Dongmo, L., Belin, C. and Limodin, J. (1985) Cerebral activation during pursuit eye movements: localization by the dichotic listening test. *J. Fr. Ophtalmol.*, **8**, 813–817.

Larsen, B., Skinkoj, E. and Lassen, N.A. (1978) Variations in regional cortical blood flow in the right and left hemispheres during automatic speech. *Brain*, **101**, 193–209.

Lashley, K.S. (1951) The problem of serial order in behaviour. In: L.A. Jeffress (ed.) *Cerebral mechanisms in Behaviour.* New York: Wiley, pp. 112–136.

Leandersen, R., Persson, A. and Ohman, S. (1970) Electromyographic studies of the function of the facial muscles in dysarthria. *Acta Otolaryng. (Stokh.)*, **263**, 89–94.

Ledlow, A., Swanson, J.M. and Kinsbourne, M. (1978) Reaction times and evoked potentials as indicators of hemispheric differences for laterally presented name and physical matches. *J. Exp. Psychol. Hum. Percept. Perform.*, **4**, 440–454.

Lee, G.P., Loring, D.W., Meador, K.J. and Brooks, B.B. (1990) Hemispheric specialization for emotional expression: a reexamination of results from intracarotid administration of sodium amobarbital. *Brain Cogn.*, **12**, 267–280.

Lee, K.H., Chung, K., Chung, J.M. and Coggeshall, R.E. (1986) Correlation of cell body size, axon size and signal conduction velocity for individually labelled dorsal root ganglion cells in the cat. *J. Comp. Neurol.*, **243**, 335–346.

Lefton, L.A. and Haber, R.N. (1974) Information extraction from different retinal locations. *J. Exp. Psychol.*, **102**, 975–980.

Leichnetz, G.R. (1989) Inferior frontal eye field projections to the pursuit-related dorsolateral pontine nucleus and middle temporal area (MT) in the monkey. *Visual Neurosci.*, **3**, 171–180.

Le May, M. (1976) Morphological cerebral asymmetries of modern man, fossil man and non-human primates. *Ann. N.Y. Acad. Sci.*, **280**, 349–366.

Le May, M. and Kido, D.K. (1978) Asymmetries of the cerebral hemispheres on computed tomograms. *J. Comput. Assist. Tomogr.*, **2**, 471–476.

Lencz, T., McCarthy, G., Bronen, R.A., Scott, T.M., Inserni, J.A., Sass, K.J., Novelly, R.A., Kim, J.H. and Spencer, D.D. (1992) Quantitative magnetic resonance imaging in temporal lobe epilepsy. *Ann. Neurol.*, **31**, 629–637.

Leonard, C.M., Voeller, K.S., Lombardino, L.J., Morris, M.K., Hynd, G.W., Alexander, A.W., Anderson, H.G., Garofalakis, M., Honeyman, J.C., Mao, J., Agee, O.F. and Stoab, E.K. (1993) Anomalous cerebral structure in dyslexia revealed with nuclear magnetic imaging. *Arch. Neurol.*, **50**, 461–469.

Leonard, G. and Milner, B. (1991) Contribution of the right frontal lobe to the encoding and recall of kinesthetic distance information. *Neuropsychologia*, **29**, 47–58.

Lesser, R. (1974) Verbal comprehension in aphasia: an English version of three Italian tests. *Cortex*, **10**, 247–263.

Leuchter, A.F., Dunkin, J.J., Lufkin, R.B., Anzai, Y., Cook, I.A. and Newton, T.F. (1994) Effect of white matter disease onfunctional correlation in the aging brain. *J. Neurol. Neurosurg. Psychiatry*, **57**, 1347–1354.

Levin, H.S. and Spiers, P.A. (1985) Acalculia. In: Heilman, K.M. and Valenstein, E. *Clinical neuropsychology, 2nd edition.* Oxford: Oxford University Press, pp. 97–114.

Levy, J. and Trevarthan, C. (1977) Perceptual, semantic and phonetic aspects of elementary language processes in split-brain patients. *Brain*, **100**, 105–118.

Levy, J., Heller, W., Banich, M.T. and Burton, L.A. (1983) Asymmetry of perception in free viewing of chimeric faces. *Brain Cogn.*, **2**, 404–419.

Levy, W.B. and Steward, O. (1983) Temporal contiguity requirements for long-term associative potentiation/depression in the hippocampus. *Neuroscience*, **8**, 791–797.

Lewinsohn, P.M., Zieler, R.E., Libet, J., Eyeberg, S. and Nielson, G. (1972) Short term memory: a comparison between frontal and non-frontal right- and left-hemisphere brain damaged patients. *J. Comp. Physiol. Psychol.*, **81**, 248–255.

Ley, R.G. and Bryden, M.P. (1979) Hemispheric differences in processing emotions and faces. *Brain Lang.*, **7**, 127–138.

Ley, R.G. and Bryden, M.P. (1982) A dissociation of right and left hemisphere effects for recognizing emotional tone and verbal content. *Brain Cogn.*, **1**, 3–9.

Liberman, A.M., Cooper, F.S., Shankweiler, D.F. and Studdert-Kennedy, M. (1967) Perception of the speech code. *Psychol. Rev.*, **74**, 431–461.

Libet, B. (1982) Brain stimulation in the study of neuronal functions for conscious sensory experiences. *Hum. Neurobiol.*, **1**, 235–242.

Liepmann, H. (1900) Der weitere Krankheitsverlauf bie dem einseitig Apraktischen und der Gehirnbefund auf Grund von Serienschnitten. *Monatsschrift fur Psychiatrie und Neurologie*, **19**, 217–243.

Liu, G.T., Bolton, A.K., Price, B.H. and Weintraub, S. (1992) Dissociated perceptual-sensory and exploratory-motor neglect. *J. Neurol. Neurosurg. Psychiatry*, **55**, 701–706.

Locke, J.L. and Pearson, D.M. (1990) Linguistic significance of babbling: evidence from a tracheostomized infant. *J. Child Lang.*, **17**, 1–6

Lofqvist, A., McGarre, N.S. and Honda, K. (1984) Laryngeal muscles and articulatory control. *J. Acoust. Soc. Am.*, **76**, 951–954.

Lomas, J. (1980) Competition within the left hemisphere between speaking and unimanual tasks performed without visual guidance. *Neuropsychologia*, **18**, 141–149.

Lomas, J. and Kimura, D. (1976) Intrahemispheric interaction between speaking and sequential manual activity. *Neuropsychologia*, **14**, 23–33.

Loring, D.W., Meador, K.J. and Lee, G.P. (1989) Differential-handed response to verbal and visual spatial stimuli: evidence of specialized hemispheric processing following callosotomy. *Neuropsychologia*, **27**, 811–827.

Lupien, S., LaFrance, M.S. and Lecours, A.R. (1993) On the interdependency between attentional biases and visual field effects with bilateral presentations of stimuli. *Brain Cogn.*, **21**, 153–162.

Luria, A.R. (1973) *The working brain: An introduction to neuropsychology.* London: Penguin Books.

Lutzenberger, W., Pulvermueller, F. and Birbaumer, N. (1994) Words and pseudowords elicit distinct patterns of 30-Hz EEG response in humans. *Neurosci. Lett.*, **176**, 115–118.

Maeshima, S., Shigeno, K., Dohi, N., Kajiwara, T. amd Kowai, N. (1992) A study of right unilateral spatial neglect in left hemispheric lesions: the difference between right handed and non-right handed post stroke patients. *Acta Neurol. Scand.*, **85**, 418–424.

MacNeilage, P.F. and Davis, B. (1990) Acquisition of speech production: frames, then content. In: Jeannerod, M. (Ed.) *Attention and Human Performance XIII.* Hillsdale, NJ: Lawrence Erlbaum Associates, pp. 453–476.

Malone, D.R. and Hannay, H.J. (1978) Hemispheric dominance and normal colour memory. *Neuropsychologia*, **16**, 51–59.

Malone, M.J. and Szoke, M.C. (1985) Neurochemical changes in white matter. Aged human brain and Alzheimer's disease. *Arch. Neurol.*, **42**, 1063–1066.

Marcel, A.J. and Patterson, K.E. (1978) Word recognition and production: reciprocity in clinical and normal studies. In: J. Requin (ed.), *Attention and Performance VII.* Hillsdale, N.J. : Lawrence Erlbaum and Associates, pp. 209–226.

Markowitsch, H.J., Emmans, D., Irle, E., Streicher, M. and Preilowski, B. (1985) Cortical and subcortical afferent connections of the primate temporal pole: a study of rhesus monkeys, squirrel monkeys and marmosets. *J. Comp. Neurol.*, **242**, 425–428.

Marshall, J.C. and Newcombe, F. (1973) Patterns of paralexia: a psycholinguistic approach. *J. Psycholinguist. Res.*, **2**, 175–199.

Martin, M. (1978) Hemispheric asymmetry for physical and semantic selection of visually presented words. *Neuropsychologia*, **16**, 717–724.

Martin, M. (1979) Hemispheric specialization for local and global processing. *Neuropsychologia*, **17**, 33–40.

Marzi, C.A., Stefano, M. Di, Tassarini, G. and Crea, F. (1979) Iconic storage in the two hemispheres. *J. Exp. Psychol. Hum. Percept. Perform.*, **5**, 31–41.

Marzi, C.A., Bisiacchi, P. and Nicoletti, R. (1991) Is interhemispheric transfer of visuomotor information asymmetric? Evidence from a meta-analysis. *Neuropsychologia*, **29**, 1163–1177.

McFarland, K. and Ashton, R. (1978) The influence of brain lateralization on function in a manual skill. *Cortex*, **14**, 102–111

McFarland, K., McFarland, M.L., Bain, J.D. and Ashton, R. (1978) Ear differences of abstract and concrete word recognition. *Neuropsychologia*, **16**, 555–561.

McGlone, J. (1980) Sex differences in human brain asymmetry: a critical survey. *Behav. Brain Sci.*, **3**, 215–263.

McHenry, L.C., Merory, J., Bass, E., Stump, D.A., Williams, R. Witcofski, R., Howard, G. and Toole, J.F. (1978) Xenon-133 inhalation method for regional cerebral blood flow measurements: normal values and test-retest results. *Stroke*, **9**, 396–399.

McKeever, W.F. and Gill, K.M. (1972a) Visual half-field differences in masking effects for sequential letter stimuli in the right and left handed. *Neuropsychologia*, **10**, 111–117.

McKeever, W.F. and Gill, K.M. (1972b) Interhemispheric transfer time for visual stimulus information varies as a function of the retinal locus of stimulation. *Psychon. Sci.*, **26**, 308–310.

McKeever, W.F., van Deventer, A. and Suberi, M. (1973) Avowed, assessed and familial handedness and differential hemispheric processing of brief sequential and non-sequential visual stimuli. *Neuropsychologia*, **11**, 235–238.

McKeever, W.F. and Suberi, M. (1974) Parallel but temporally displaced visual half-field meta contrast functions. *Q. J. Exp. Psychol.*, **26**, 258–265.

McKeever, W.F., Hoenemann, H.W., Florian, V.A. and Van Deventer, A.D. (1976) Evidence of minimal cerebral asymmetries for the processing of English words and American sign language in the congenitally deaf. *Neuropsychologia*, **14**, 413–423.

McKeever, W.F. and Dixon, M.S. (1981) Right hemisphere superiority for discriminating memorized from non-memorized faces: affective imagery, sex and perceived emotionality of faces. *Brain Lang.*, **12**, 246–260.

Mellus, L.E. (1911) A contribution to the study of the cerebral cortex in man. *Anat. Rec.*, **5**, 473–482.

Mennemeier, M., Wertman, E. and Heilman, K.M. (1992) Neglect of near peripersonal space. Evidence for multidirectional attentional systems in humans. *Brain*, **115**, 37–50.

Merrin, E.L., Floyd, T.C. and Fein, G. (1989) EEG coherence in unmedicated schizophrenic patients. *Biol. Psychiatry*, **25**, 60–66.

Messerli, P., Pegna, A. and Sordet, N. (1995) Hemispheric dominance for melody recognition in musicians and non-musicians. *Neuropsychologia*, **33**, 395–405.

Mesulam, M.-M. (1979) Tracing neural connections of human brain with selective silver impregnation. Observations on geniculo-calcarine, spinothalamic and entorhinal pathways. *Arch. Neurol.*, **36**, 814–818.

Meyer, B.-U., Roricht, S., von Einsiedel, H.G., Kruggel, F. and Weindl, A. (1995) Inhibitory and excitatory interhemispheric transfers between motor cortical areas in normal humans and patients with abnormalities of the corpus callosum. *Brain*, **118**, 429–440.

Meyer, J.S., Ishihara, N., Desmukh, V.D., Naritomi, H., Saki, F., Hsu, M.C. and Pollack, P. (1978) Improved method for noninvasive measurement of regional cerebral blood flow by 133-Xenon inhalation. Part I. Description of method and normal values obtained in healthy volunteers. *Stroke*, **9**, 195–204.

Michimata, C. and Hellige, J. (1987) Effects of blurring and stimulus size on the lateralized processing of non-verbal stimuli. *Neuropsychologia*, **25**, 397–407.

Milberg, W., Blumstein, S.E. and Dworetzky, B. (1987) Processing of lexical ambiguities in aphasia. *Brain Lang.*, **31**, 138–150.

Miller, A.K.H., Alston, R.L. and Corsellis, J.A.N. (1980) Variation with age in the volumes of grey and white matter in the cerebral hemispheres of man: measurements with an image analyser. *Neuropathol. Appl. Neurobiol.*, **6**, 119–132.

Miller, R. (1975) Distribution and properties of commissural and other neurons in the cat sensorimotor cortex. *J. Comp. Neurol.*, **164**, 361–374.

Miller, R. (1981) *Meaning and purpose in the intact brain.* Oxford: Oxford University Press.

Miller, R (1988) Cortico-striatal and cortico-limbic circuits: a two tiered model of learning and memory functions. In: H.J. Markowitsch (ed), *Information processing by the brain: views and hypotheses from a cognitive-physiological perspective.* Bern: Hans Huber Press, pp. 179–198.

Miller, R. (1989) Cortico-hippocampal interplay: self-organizing phase-locked loops for indexing memory. *Psychobiology*, **17**, 115–128.

Miller, R. (1991) Cortico-hippocampal interplay and the representation of contexts in the brain. *Studies of Brain Function No. 17.* Heidelberg, New York: Springer Verlag.

Miller, R. (1994) An interpretation, based on cell assembly theory, of the psychological impairments following lesions of the hippocampus and related structures. In: J. Delacour (ed.). *The memory system of the brain.* Singapore: World Scientific, pp. 659–712.

Miller, R. (1995a) What is the contribution of axonal conduction delay to temporal structure in brain dynamics? In: C. Pantev, Th. Elbert and B. Lutkenhoener (eds), *Oscillatory event related brain dynamics, NATO-ASI series.* New York: Plenum: In Press.

Miller, R. (1995b) *Arguments against secular culture.* London: SCM Press.

Miller, R., Wickens, J.R. and Beninger, R.J. (1990) Dopamine D-1 and D-2 receptors in relation to reward and performance: a case for the D-1 receptor as a primary site of therapeutic action of neuroleptic drugs. *Prog. Neurobiol.*, **34**, 143–183.

Miller, R. and Wickens, J.W. (1991) Corticostriatal cell assemblies in selective attention and in representation of predictable and controllable events. *Concepts in Neuroscience*, **2**, 65–95.

Mills, L. and Rollman, G.B. (1980) Hemispheric asymmetry for auditory perception of temporal order. *Neuropsychologia*, **18**, 41–47.

Milner, B. and Taylor, L. (1972) Right hemisphere superiority in tactile pattern-recognition after cerebral commissurotomy: evidence for non-verbal memory. *Neuropsychologia*, **10**, 1–15.

Mirenowicz, J. and Schultz, W. (1994) Importance of unpredictability for reward response in primate dopamine neurones. *J. Neurophysiol.*, **72**, 1024–1027.

Moll, L. and Kuypers, H.G.J.M. (1977) Premotor coertical ablations in monkeys: contralateral changes in visually guided reaching behavior. *Science*, **198**, 317–319.

Mondor, T.A. and Bryden, M.P. (1991) The influence of attention on the dichotic REA. *Neuropsychologia*, **29**, 1179–1190.

Mondor, T.A. and Bryden, M.P. (1992) On the relation between auditory spatial attention and auditory perceptual asymmetries. *Percept. Psychophysics*, **52**, 393–402.

Morrell, F. and Jasper, H.H. (1956) Electrographic studies of the formation of temporary connections in the brain. *Electroencephalogr. Clin. Neurophysiol.*, **8**, 201–215.

Mountcastle, V.B., Lynch, J.C., Georgopoulos, A., Sakata, H. and Acuna, C. (1975) Posterior parietal association cortex of the monkey: command functions for operations within extrapersonal space. *J. Neurophysiol.*, **38**, 871–908.

Mountcastle, V.B. (1978) An organizing principle for cerebral function: the unit module and the distributed system. In: Edelman, G.M. and Mountcastle, V.B. (eds), *The Mindful Brain.* Cambridge, Mass: MIT Press.

Mullan, S. and Penfield, W. (1959) Illusions of comparative interpretation and emotion. *Arch. Neurol. Psychiatr. (Chic.)*, **81**, 269–284.

Muller-Preuss, P. and Ploog, D. (1981) Inhibition of auditory cortical neurons during phonation. *Brain Res.*, **215**, 61–76.

Murphy, D.G.M., De Carli, C., Daly, E., Haxby, J.V., Allen, G., White, B.J., McIntosh, A.R., Powell, C.M., Horwitz, B., Rapoport, S.I. and Schapiro, M.B. (1993) X-chromosome effects on female brain: a magnetic resonance imaging study of Turner's syndrome. *Lancet*, **342**, 1197–1200

Nagafuchi, M. (1980) The effects of delayed auditory feedback (DAF) on patients with unilateral hemispheric damage. Nippon *Jibiinkoka Gakkai Kaiho* (Japanese Journal for Ear, Nose and Throat), **83**, 1042–1051.

Naeser, M.A. and Borod, J.C. (1986) Aphasia in left handers: lesion site, lesion side and hemispheric asymmetries on CT. *Neurology*, **36**, 471–488.

Naito, H., Miyakawa, F. and Ito, N. (1971) Diameters of callosal fibres interconnecting cat sensorimotor cortex. *Brain Res.*, **27**, 369–372.

Natale, M. (1977) Perception of non-linguistic auditory rhythms by the speech hemisphere. *Brain Lang.*, **4**, 32–44.

Natale, M., Gur, R.E. and Gur, R.C. (1983) Hemispheric asymmetries in processing emotional expressions. *Neuropsychologia*, **21**, 555–565.

Neal, J.W., Pearson, R.C.A. and Powell, T.P.S. (1990) The ipsilateral cortico-cortical connections of area 7 with the frontal lobe in monkeys. *Brain Res.*, **509**, 31–40.

Neville, H.J. and Lawson, D. (1987) Attention to central and peripheral visual space in a movement detection task. III. Separate effects of auditory deprivation and acquisition of a visual language. *Brain Res.*, **405**, 284–294.

Newcombe, F. and Marshall, J.C. (1967) Immediate recall of "sentences" by subjects with unilateral cerebral lesions. *Neuropsychologia*, **5**, 329–334.

Nicholls, M.E.R. (1994a) Hemispheric asymmetries for temporal resolution: a signal detection analysis of threshold and bias. *Q. J. Exp. Psychol. A*, **47**, 291–310.

Nicholls, M.E.R. (1994b) The non-contribution of attentional biases to visual field asymmetries in temporal discrimination. *Neuropsychologia*, **32**, 209–220.

Nicholls, M.E.R. and Cooper, C.J. (1991) Hemispheric differences in the rates of information processing for simple non-verbal stimuli. *Neuropsychologia*, **29**, 677–684.

Nicholls, M.E.R. and Atkinson, J. (1993) Hemispheric asymmetries for an inspection time task: a general left hemisphere temporal advantage. *Neuropsychologia*, **31**, 1181–1190.

Nunez, P. (1989) *Electric fields of the brain: the neurophysics of the EEG.* Oxford: Oxford University Press.

O'Boyle, M. and Hoff, E.J. (1987) Gender and handedness difference in mirror-tracing random forms. *Neuropsychologia*, **25**, 977–982.

Ojemann, G. and Mateer, C. (1979) Human language cortex: localization of memory, syntax and sequential motor-phoneme identification systems. *Science*, **205**, 1401–1403.

Oke, A., Keller, R., Mefford, I. and Adams, R.N. (1978) Lateralization of norepinephrine in human thalamus. *Science*, **200**, 1411–1413.

Oller, D.K. and Eilers, R.E. (1988) The role of audition in infant babbling. *Child Dev.*, **59**, 441–449.

Orenstein, H.B. and Meighan, W.B. (1976) Recognition of bilaterally presented words varying in concreteness and frequency: lateral dominance or sequential processing? *Bull. Psychon. Soc.*, **7**, 179–180.

Paivo, A., Yuille, J.C. and Madigan, S.A. (1968) Concreteness, imagery and meaningfulness values for 925 nouns. *J. Exp. Psychol. Monogr. Suppl.*, **76**, 1–25.

Palm, G. (1982) *Neural assemblies*. Heidelberg: Springer.

Pandya, D.N. and Kuypers, H.G.J.M. (1969) Cortico-cortical connections in the rhesus monkey. *Brain Res.*, **13**, 13–36.

Pandya, D.N. and Vignolo, L.A. (1971) Intra- and interhemispheric projections of the precentral, premotor and arcuate areas in the rhesus monkey. *Brain Res.*, **26**, 217–333.

Pandya, D.N., Dye, P. and Butters, N. (1971) Efferent cortico-cortical projections of the prefrontal cortex in the rhesus monkey. *Brain Res.*, **31**, 35–46.

Pandya, D.N. and Yeterian, E.H. (1985) Architecture and connections of cortical association areas. In: Peters, A. and Jones, E.G. (eds), *Cerebral Cortex, Volume 4.* New York and London: Plenum Press, pp. 3–61.

Papanicolou, A.C., Levin, H.S., Eisenberg, H.M. and Moore, B.D. (1983) Evoked potential indices of selective hemispheric engagement in affective and phonetic tasks. *Neuropsychologia*, **21**, 401–405.

Papcun, G., Krashun, S., Terbeck, D., Remington, R and Hershman, R. (1974) Is the left hemisphere specialized for speech, language (or something else)? *J. Acoust. Soc. Am.*, **55**, 319–327.

Pardo, J.V., Fox, P.T. and Raichle, M.E. (1991) Localization of a human system for sustained attention by positron emission tomography. *Nature*, **349**, 61–64.

Patterson, K.E. and Marcel, A.J. (1977) Aphasia, dyslexia and the phonological coding of written words. *Q. J. Exp. Psychol.*, **29**, 307–318.

Peretz, I. (1990) Processing of local and global musical information by unilateral brain-damaged patients. *Brain*, **113**, 1185–1205.

Peretz, I. and Morais , J. (1979) A left-ear advantage for chords in non-musicians. *Percept. Mot. Skills*, **49**, 957–958.

Peretz, I. and Morais, J. (1980) Modes of processing melodies and ear asymmetries in non-musicians. *Neuropsychologia*, **18**, 477–489.

Peretz, I. and Morais, J. (1987) Analytic processing in the classification of melodies as same or different. *Neuropsychologia*, **25**, 645–652.

Peretz, I., Morais, J. and Bertelson, P. (1987) Shifting ear differences in melody recognition through strategy inducement. *Brain Cogn.*, **6**, 202–215.

Peretz, I and Babai, M. (1992) The role of contour and intervals in the recognition of melody parts: evidence from cerebral asymmetries in musicians. *Neuropsychologia*, **30**, 277–292.

Peters, A. (1987) Numbers of neurons and synapses in primary visual cortex. In Jones, E.G. and Peters, A. (eds). *Cerebral Cortex, vol 6, (Further aspects of cortical function including hippocampus).* New York: Plenum, pp. 267–294.

Peters, M. (1976) Prolonged practice of a simple motor task by preferred and non-preferred hands. *Percept. Mot. Skills*, **43**, 447–450.

Peters, M (1983) Reaction time to tactile stimuli presented ipsi and contralaterally to the responding hand. *Q. J. Exp. Psychol. A*, **35**, 397–410.

Petersen, M.R., Beecher, M.D., Zoloth, S.R., Moody, D.B. and Stebbins, W.C. (1978) Neural lateralization of species specific vocalizations by Japanese macaques (*Macaca fuscata*). *Science*, **202**, 324–327.

Petersen, S.E., Fox, P.T., Posner, M.I., Mintun, M. and Raichle, M.E. (1988) Positron emission tomographic studies of the cortical anatomy of single-word processing. *Nature*, **331**, 585–589.

Petito, L.A. and Martenette, P.F. (1991) Babbling in the manual mode: evidence for the ontogeny of language. *Science*, **251**, 1493–1496.

Pfaus, J.G., Dammsa, G., Nomikos, G.G., Wenkstern, D.G., Blaha, C.D., Phillips, A.G. and Fibiger, H.C. (1990) Sexual behavior enhances central dopamine transmission in the male rat. *Brain Res.*, **530**, 345–348.

Phelps, E.A. and Gazzaniga, M.S. (1992) Hemispheric differences in mnemonic processing: the effects of left hemisphere interpretation. *Neuropsychologia*, **30**, 293–297.

Piaget, J. and Inhelder, B. (1969) *The psychology of the child.* London: Routledge and Kegan Paul.

Piazza, D.M. (1980) The influence of sex and handedness in the hemispheric specialization of verbal and non-verbal tasks. *Neuropsychologia*, **18**, 163–176.

Pieniadz, J.M. and Naeser, D.S. (1984) Computed tomographic scan cerebral asymmetries and morphological brain asymmetries. *Arch. Neurol.*, **41**, 403–409.

Pillai, K. (1989) *The role of the dopaminergic signal to the striatum in the acquisition of motor control.* University of Otago, B. Med. Sci. Thesis, 1989.

Pirozzolo, F.J. and Rayner, K. (1980) Handedness, hemispheric specialization and saccadic eye movement latencies. *Neuropsychologia*, **18**, 225–229.

Pleim, E.T., Matochik, J.A., Barfield, R.J. and Auerbach, S.B. (1990) Correlation of dopamine release in the nucelus accumbens with masculine sexual behaviour in rats. *Brain Res.*, **524**, 160–163.

Podbros, L.Z. and Wyke, M.A. (1988) Laterality differences for speed but not for control in sequential finger tapping. *Percept. Mot. Skills*, **67**, 927–933.

Poeck, K. (1986) The clinical examination of apraxia. *Neuropsychologia*, **24**, 129–134.

Pointer, J.S. and Hess, R.F. (1989) The contrast sensitivity gradient across the human visual field: with emphasis on the low spatial frequency range. *Vision Res.*, **29**, 1133–1151.

Poizner, H. and Lane, H. (1979) Cerebral asymmetry in the perception of American sign language. *Brain Lang.*, **7**, 210–226.

Poizner, H., Battison, R. & Lane, H. (1979) Cerebral asymmetry for American sign language: the effects of moving stimuli. *Brain Lang.*, **7**, 351–362.

Poizner, H., Mack, L., Verfaillie, M., Gonzalez-Rothi, L.G. and Heilman, K.M. (1990) Three-dimensional computergraphic analysis of apraxia. Neural representations of learned movement. *Brain*, **113**, 85–101.

Poizner, H., Clark, M.A., Meians, A.S., Macaulay, B., Gonzalez-Rothi, L.J. and Heilman, K.M. (1995). Joint coordination deficits in limb apraxia. *Brain*, **118**, 227–242.

Polich, J.M. (1978) Hemispheric differences in stimulus identification. *Percept. Psychophysics*, **24**, 49–57.

Polich, J.M. (1982) Hemispheric memory for random forms revisited. *Brain Cogn.*, **1**, 360–370.

Polzella, D.L., DaPolito, F. and Hinsman, M.C. (1977) Cerebral asymmetry in time perception. *Percept. Psychophysics*, **21**, 187–192.

Posner, M.I., Inhoff, A.W. and Cohen, W. (1987) Isolating attentional systems: a cognitive anatomical analysis. *Psychobiology*, **15**, 107–121.

Posner, M.I. and Petersen, S.E. (1990) The attention system of the human brain. *Annu. Rev. Neurosci.*, **13**, 25–42.

Preisler, A.-M., Gallasch, E. and Schulter, G. (1989) Hemispheric asymmetry and the processing of harmonies in music. *Int. J. Neurosci.*, **47**, 131–140.

Previc, F.H. (1982) Visual pattern recognition in the cerebral hemispheres: the role of spatial filtering. *Percept. Mot. Skills*, **55**, 1319–1326.

Pring, T. (1981) The effect of stimulus size and exposure duration on visual field asymmetries. *Cortex*, **17**, 227–240.

Provins, K.A. and Glencross, D.J. (1968) Handwriting, typewriting and handedness. *Q. J. Exp. Psychol.*, **20**, 282–289.

Pulvermuller, F., Lutzenberger, W. and Birbaumer, N. (1995) Electrocortical distinction of vocabulary types. *Electroencephalogr. Clin. Neurophysiol.* **94**, 357–370.

Qureshi, R. and Dimond, S.J. (1979) Calculation and the right hemisphere. *Lancet*, (i) 322–333.

Radhadishun, F.S., van Ree, J.M. and Westerinck, B.H.C. (1988) Scheduled eating increases dopamine release in the nucleus accumbens of food-deprived rats as assessed with on-line brain dialysis. *Neurosci. Lett.*, **85**, 351–356.

Rains, J.D. (1963) Signal luminance and position effects in human reaction time. *Vision Res.*, **3**, 239–251.

Rasmussen, T and Milner, B. (1977) The role of early left-brain injury in determining lateralization of cerebral speech functions. *Ann. N.Y. Acad. Sci.*, **299**, 355–369.

Rastatter, M.P. and Gallaher, A.J. (1982) Reaction-times of normal subjects to monoaurally presented verbal and tonal stimuli. *Neuropsychologia*, **20**, 465–473.

Rausch, R. (1981) Lateralization of temporal lobe dysfunction and verbal encoding. *Brain and Lang.*, **12**, 92–100.

Regard, M., Cook, N.D., Wieser, H.G. and Landis, T. (1994) The dynamics of cerebral dominance during unilateral limbic seizures. *Brain*, **117**, 91–104.

Reuter-Lorenz, P. and Davidson, R.J. (1981) Differential contribution of the two cerebral hemispheres to the perception of happy and sad faces. *Neuropsychologia*, **19**, 609–613.

Reuter-Lorenz, P., Givis, R.P. and Moscovitch, M. (1983) Hemispheric specialization and the perception of emotion: evidence from right handers and from inverted and non-inverted left handers. *Neuropsychologia*, **21**, 687–692.

Reveley, M.A., Reveley, A.M. and Baldy, R. (1987) Left cerebral hemisphere hypodensity in discordant schizophrenic twins. *Arch. Gen. Psychiatry*, **44**, 625–632.

Rizzolatti, G., Umilta, C. & Berlucchi, G. (1971) Opposite superiorities of the right and left cerebral hemispheres in discriminative reaction time to physiognomic and alphabetic material. *Brain*, **94**, 431–442.

Rizzolatti, G. and Buchtel, H.A. (1977) Hemispheric superiority in reaction time to faces: a sex difference. *Cortex*, **13**, 300–305.

Roberts, L. (1969) Aphasia, apraxia and agnosia in abnormal states of cerebral dominance. In: Vinken, P.J. and Bruyn, G.W. (eds) *Handbook of Clinical Neurology, Vol. 4.* Amsterdam: North Holland, pp. 312–326

Roberts, L.D. and Gregory, A.H. (1973) Ear differences and delayed auditory feedback: effect on a simple verbal repetition task and a non-verbal tapping test. *J. Exp. Psychol.*, **101**, 269–272.

Robin, D.A., Tranel, D. and Damasio, H. (1990) Auditory perception of temporal and spectral events in patients with focal left and right cerebral lesion. *Brain Lang.*, **39**, 539–555.

Robinson, G.M. and Solomon, D.J. (1974) Rhythm is processed by the speech hemisphere. *J. Exp. Psychol.*, **102**, 508–511.

Robinson, R.G. (1985) Lateralized behavioral and neurochemical consequences of unilateral brain injury in rats. In S.G. Glick (ed.), *Cerebral lateralization in non-human species*. Orlando:Academic Press, pp. 135–156.

Rodel, M., Cook, N.D., Regard, M. and Landis, T. (1992) Hemispheric dissociation in judging semantic relations: complementarity for close and distant associates. *Brain Lang.*, **43**, 448–459.

Roland, P.E., Larsen, B., Lassen, N.A. and Skinhoj, E. (1980) Supplementary motor area and other cortical areas in organization of voluntary movements in man. *J. Neurophysiol.*, **43**, 118–136.

Rose, D. (1983) An investigation into hemispheric differences in adaptation to contrast. *Percept. Psychophysics*, **34**, 89–95.

Ross, E.D. (1981) The aprosodias: functional-anatomic organization of the affective components of language in the right hemisphere. *Arch. Neurol.*, **38**, 561–569.

Ross, E.D. and Mesulam, M.-M. (1979) Dominant language functions of the right hemisphere. Prosody and emotional gesturing. *Arch. Neurol.*, **36**, 144–148.

Rossi, A., Stratta, P., D'Albeniz, L., Tartaro, A, Schiazza, G., di Michele, V., Bolino, F. and Casacchia, M. (1990) Reduced temporal lobe areas in schizophrenia: preliminary evidences from a controlled multiplanar magnetic resonance imaging study. *Biol. Psychiatry*, **27**, 61–68.

Roy, E.A. (1983) Manual performance asymmetries and motor control processes: subject-generated changes in response parameters. *Hum. Movement Sci.*, **2**, 271–277.

Roy, E.A. and Elliott, D. (1986a) Manual asymmetries in aimed movements. *Q. J. Exp. Psychol. A*, **41**, 501–516.

Roy, E.A. and Elliott, D. (1986b) Manual asymmetries in visually directed aiming. *Can. J. Psychol.*, **40**, 109–121.

Rubens, A.B., Mahowald, M.W. and Hutton, J.T. (1976) Asymmetry of the lateral (Sylvian) fissure in man. *Neurology*, **26**, 620–624.

Rudel, R.G., Denckla, M.B. and Spattern, E. (1974) The functional asymmetry of Braille letter learning in normal sighted children. *Neurology*, **24**, 733–738.

Rugg, M.D., Lines, C.R. and Milner, A.D. (1985) Further investigation of visual evoked potentials elicited by lateralized stimuli: effects of stimulus eccentricity and reference site. *Neuropsychologia*, **62**, 81–87.

Rutschmann, R. (1966) Perception of temporal order and relative visual latency. *Science*, **152**, 1099–1101.

Sackheim, H.A. and Gur, R.C. (1978) Lateral asymmetry in intensity of emotional expression. *Neuropsychologia*, **16**, 473–481.

Sackheim, H.A., Greenberg, M.S., Weiman, A.L., Gur, R.C., Hungerbuhler, J.P. and Geschwind, N. (1982) Hemispheric asymmetry in the expression of positive and negative emotion. Neurological evidence. *Arch. Neurol.*, **39**, 210–218.

Safer, M.A. and Leventhal, H. (1977) Ear differences in evaluating emotional tone of voice and verbal content. *J. Exp. Psychol. Hum. Percept. Perform.*, **3**, 75–82.

Saffran, E.M. and Marin, O.S.M. (1977) Reading without phonology: evidence from aphasia. *Q. J. Exp. Psychol.*, **29**, 515–523.

Sah, P. and Nicoll, R.A. (1991) Mechanisms underlying potentiation of synaptic transmission in rat anterior cingulate cortex in vitro. *J. Physiol. (Lond.)*, **433**, 615–630.

Salerno, J.A., Murphy, D.G.M., Horwitz, B., De Carli, C., Haxby, J.V., Rapoport, S.I., Schapiro, M.B. (1992) Brain atrophy in hypertension. A volumetric magnetic resonance imaging study. *Hypertension*, **20**, 340–348.

Samson, S. and Zattore, R.J. (1988) Melodic and harmonic discrimination following unilateral cerebral excision. *Brain Cogn.*, **7**, 348–360.

Samuels, I., Butters, N., Fedio, P. and Cox, C. (1980) Deficits in short-term auditory memory for verbal material following right temporal removals in humans. *Int. J. Neurosci.*, **11**, 101–108.

Saron, C.D. and Davidson, R.J. (1989) Visual evoked potential measures of interhemispheric transmission time in humans. *Behav. Neurosci.*, **103**, 1115–1138.

Satz, P. (1988) Primate handedness: a paradoxical link to humans? *Behav. Brain Sci.*, **11**, 729–731.

Sava, D., Liotti, M. and Rizzolatti, G. (1988) Right hemisphere superiority for programming oculomotion: evidence from simple reaction time experiments. *Neuropsychologia*, **26**, 201–211.

Scheibel, A.B., Paul, L.A., Fried, I., Forsythe, A.B., Tomiyasu, U., Wechsler, A., Kao, A., and Slotnick, J. (1985) Dendritic organization of the anterior speech area. *Exp. Neurol.*, **87**, 109–117.

Schlanger, B.B., Schlanger, P. and Gerstman, L.J. (1976) The perception of emotionally toned sentences by right hemisphere-damaged and aphasic subjects. *Brain Lang.*, **3**, 396–403.

Schlegel, S. and Kretzschmar, K. (1987) Computed tomography in affective disorders. Part II. Brain density. *Biol. Psychiatry*, **22**, 15–23.

Schmuller, J. and Goodman, R. (1979) Bilateral tachistoscopic perception, handedness and laterality. *Brain Lang.*, **8**, 81–91.

Schultz, W. (1986) Response of midbrain dopamine neurons to behavioural trigger stimuli in the monkey.. *J. Neurophysiol.*, **56**, 1439–1461.

Schultz, W., Apicella, P., Scarnati, E. and Ljungberg, T. (1992) Neuronal activity in monkey ventral striatum related to the expectation of reward. *J. Neurosci.*, **12**, 4595–4610.

Schultz, W., Apicella, P., and Ljungberg, T. (1993) Responses of monkey dopamine neurons to reward and conditioned stimuli during successive steps of learning and delayed response tasks. *J. Neurosci.*, **13**, 900–913.

Schwartz, G.E., Davidson, R.J. and Maer, F. (1975) Right-hemisphere lateralization for emotion in the human brain: interaction with cognition. *Science*, **190**, 286–288.

Schwartz, M., Creasey, H., Grady, C.L., DeLeo, J.M., Frederickson, H.A., Cutler, N.R. and Rapoport, S.I. (1985) Computed tomographic analysis of brain morphometrics in 30 healthy men aged 21–81 years. *Ann. Neurol.*, **17**, 146–157.

Schwartz, S. and Tallal, P. (1980) Rate of acoustic change may underlie hemispheric specialization for speech perception. *Science*, **207**, 1380–1381.

Seggie, J. and Berry, M. (1972) Ontogeny of interhemispheric evoked potentials in the rat: significance of myelination of the corpus callosum. *Exp. Neurol.*, **35**, 215–232.

Seidenwurm, D., Bird, C.R., Enzmann, D.R. and Marshall, W.H. (1985) Left-right temporal region asymmetry in infants and children. *Am. J. Neuroradiology*, **6**, 777–779.

Seldon, H.L. (1981) Structure of human auditory cortex. I. Cytoarchitectonics and dendritic distributions. *Brain Res.*, **229**, 277–294.

Selnes, O.A., Pestronk, A., Hart, J. and Gordon, B. (1991) Limb apraxia without aphasia froma left-sided lesion in a right handed patients. *J. Neurol. Neurosurg. Psychiatry*, **54**, 734–737.

Semmes, J. (1968) Hemispheric specialization: a possible clue to mechanism. *Neuropsychologia*, **6**, 11–26.

Sergent, J. (1982a) Theoretical and methodological consequences of variations in exposure duration in visual laterality studies. *Percept. Psychophysics*, **31**, 451–461.

Sergent, J. (1982b) Influence of luminance on hemispheric processing. *Bull. Psychon. Soc.*, **20**, 221–223.

Sergent, J. (1982c) The cerebral balance of power: confrontation or cooperation? *J. Exp. Psychol. Hum. Percept. Perform.*, **8**, 253–272.

Sergent (1983) The effects of sensory limitations on hemispheric asymmetries. *Can. J. Psychol.*, **37**, 345–366.

Sergent, J. (1984) Role of contrast, lettercase, and viewing conditions in a lateralized word-naming task. *Percept. Psychophysics*, **35**, 489–498.

Sergent, J. (1985) Influence of task and input factors on hemispheric involvement in face processing. *J. Exp. Psychol.*, **11**, 846–861.

Shallice, T. and Warrington, E.K. (1975) Word recognition in a phonemic dyslexic patient. *Q. J. Exp. Psychol.*, **27**, 187–199.

Shankweiler, D. (1966) Effects of temporal lobe damage on perception of dichotically presented melodies. *J. Comp. Physiol. Psychol.*, **62**, 115–119.

Shankweiler, D. and Studdert-Kennedy, M. (1975) A continuum of lateralization for speech perception. *Brain Lang.*, **2**, 212–225.

Shannon, B (1979) Lateralization effects in lexical decision tasks. *Brain Lang.*, **8**, 380–387.

Shapiro, B.E. and Danley, M. (1985) The role of the right hemisphere in the control of speech prosody in propositional and affective contexts. *Brain Lang.*, **25**, 19–36.

Shields, J. (1991) Semantic-pragmatic disorder: a right hemisphere syndrome? *Br. J. Commun. Disord.*, **26**, 383–392.

Shiwa, T. (1987) Cortico-cortical projections to the monkey temporal lobe with particular reference to the visual processing pathways. *Arch. Ital. Biol.*, **125**, 139–154.

Sidtis, J.J. (1980) On the nature of cortical function underlying right hemisphere auditory function. *Neuropsychologia*, **18**, 321–330.

Sidtis, J.J. and Volpe, B.T. (1988) Selective loss of complex pitch and speech discrimination after unilateral lesion. *Brain Lang.*, **34**, 235–245.

Simion, F., Bagnara, S., Bisiacchi, P., Roncato, S. and Umilta, C. (1980) Laterality effects, levels of processing and stimulus properties. *J. Exp. Psychol. Hum. Percept. Perform.*, **6**, 184–195.

Small, A.M.Jr., Brandt, J.F. and Cox, P.G. (1962) Loudness as a function of signal duration. *J. Acoust. Soc. Am.*, **34**, 513–514.

Smith, A. (1966) Speech and other functions after left (dominant) hemispherectomy. *J. Neurol. Neurosurg. Psychiatry*, **29**, 467–471.

Smith, M.O., Chu, J. and Edmonston, W.E. (1977) Cerebral lateralization of haptic perception: interaction of responses to Braille and music reveals a functional basis. *Science*, **197**, 689–690.

Sobotka, S.S., Davidson, R.J. and Senulis, J.A. (1992a) Anterior brain electrical asymmetries in response to reward and punishment. *Electroencephalogr. Clin. Neurophysiol.*, **83**, 236–247.

Sobotka, S., Grabowska, A., Grodzicka, J., Wasiliewski, R. and Budohoska, W. (1992b) Hemispheric asymmetry in event related potentials associated with positive and negative emotions. *Acta Neurobiol. Exp. (Warsz.)*, **52**, 251–260.

Springer, S.P. and Gazzaniga, M.S. (1975) Dichotic testing of partial and complete split brain subjects. *Neuropsychologia*, **13**, 341–346.

Steingruber, H.J. (1975) Handedness as a function of test complexity. *Percept. Mot. Skills*, **40**, 263–266.

Steinmetz, H., Rademaker, J., Jancke, L., Huang, Y., Thron, A. and Zilles, K. (1990) Total surface of temporoparietal intrasylvian cortex: diverging left/right asymmetries. *Brain Lang.*, **39**, 357–372.

Steriade, M. and Amzica, F. (1994) Dynamic coupling among neocortical neurons during evoked and spontaneous spike-wave activity. *J. Neurophysiol.*, **72**, 2051–2069.

Stevens, J.C. and Hall, J.W. (1966) Brightness and loudness as functions of stimulus duration. *Percept. Psychophysics*, **1**, 319–327.

Stevens, S.S. (1966) Duration, luminance and the brightness exponent. *Percept. Psychophysics*, **1**, 96–100.

Stoel-Gammon, C. and Otomo, K. (1986) Babbling development of hearing-impaired and normally hearing subjects. *J. Speech Hear. Disord.*, **51**, 33–41.

Stratta, P., Rossi, A., Gallucci, M., Amicarelli, I., Passariello, R. and Casacchia, M. (1989) Hemispheric asymmetries and schizophrenia: a preliminary magnetic resonance imaging study. *Biol. Psychiatry*, **25**, 275–284.

Strauss, E. and Moscovitch, M. (1981) Perception of facial expressions. *Brain Lang.*, **13**, 308–332.

Strauss, E. and Goldsmith, S.M. (1987) Lateral preferences and performance on non-verbal laterality tests in a normal population. *Cortex*, **23**, 495–503.

Strauss, E., Gaddes, W.H. and Wada, J. (1987) Performance on a free-recall verbal dichotic listening task and cerebral dominance determined by the carotid amytal test. *Neuropsychologia*, **25**, 747–753.

Strien, J.W. and Morpurgo, M. (1992) Opposite hemispheric activations as a result of emotionally threatening and non-threatening words. *Neuropsychologia*, **30**, 845–848.

Studdert-Kennedy, M. and Shankweiler, D. (1970) Hemispheric specialization for speech perception. *J. Acoust. Soc. Am.*, **48**, 579–594.

Studdert-Kennedy, M., Shankweiler, D. and Schulman, S. (1970) Opposed effects of a delayed channel on perception of dichotically and monotonically presented CV syllables. *J. Acoust. Soc. Am.*, **48**, 599–602.

Suberi, M. and McKeever, W.F. (1977) Differential right hemispheric memory storage of emotional and non-emotional faces. *Neuropsychologia*, **15**, 757–768.

Subirana, A. (1969) Handedness and cerebral dominance. In: Vinken, P.J. and Bruyn, G.W. (eds) *Handbook of Clinical Neurology Vol. 4.* Amsterdam: North Holland, pp. 248–272.

Suddath, R.L., Casanova, M.F., Goldberg, T.E., Daniel, D.G., Kelsoe, J.R. and Weinberger, D.R. (1989) Temporal lobe pathology in schizophrenia: a quantitative magnetic resonance imaging study. *Am. J. Psychiatry*, **146**, 464–472.

Suddath, R.L., Christison, G.W., Torrey, E.F., Casanova, M.F. and Weinberger, D.R. (1990) Anatomical abnormalities in the brains of monozygotic twins discordant for schizophrenia. *N. Engl. J. Med.*, **322**, 789–794.

Sussman, H.M. (1971) The laterality effect in lingual-auditory tracking. *J. Acoust. Soc. Am.*, **49**, 1874–1880.

Sussman, H.M., MacNeilage, P.F. and Lumbley, J. (1974) Sensorimotor dominance and the right ear advantage in mandibular-auditory tracking. *J. Acoust. Soc. Amer.* **56**, 214–216.

Sussman, H.M. and MacNeilage, P.F. (1975) Studies of hemispheric specialization for speech production. *Brain Lang.*, **2**, 131–151.

Sussman, H.M., Marquardt, T.P., MacNeilage, P.F. and Hutchinson, J.A. (1988) Anticipatory coarticulation in aphasia: some methodological considerations. *Brain Lang.*, **35**, 367–379.

Swadlow, H.A. (1974) Properties of antidromically activated callosal neurons and neurons responsive to callosal input in rabbit. *Exp. Neurol.*, **43**, 424–444.

Swadlow, H.A. (1982) Impulse conduction in the mammalian brain: physiological properties of individual axons, monitored for several months. *Science*, **218**, 911–913.

Swadlow, H.A. (1985) Physiological properies of individual cerebral axons studied in vivo for as long as one year. *J. Neurophysiol.*, **54**, 1346–1362.

Swadlow, H.A. (1989) Efferent neurons and suspected interneurons in S-1 vibrissa cortex of the awake rabbit: receptive fields and axonal properties. *J. Neurophysiol.*, **62**, 288–308.

Swadlow, H.A. (1990) Efferent neurons and suspected interneurons in S-1 forelimb representation of the awake rabbit: receptive field and axonal properties. *J. Neurophysiol.*, **63**, 1477–1498.

Swadlow, H.A. (1991) Efferent neurons and suspected interneurons in second somatosensory cortex of awake rabbits: receptive fields and axonal properties. *J. Neurophysiol.*, **66**, 1392–1409.

Swadlow, H.A. (1994) Efferent neurons and suspected interneurons in motor cortex of the awake rabbit: axonal properties, sensory receptive fields, and subthreshold synaptic inputs. *J. Neurophysiol.*, **71**, 437–453.

Swadlow, H.A. and Waxman, S.G. (1976) Variations of conduction velocity and excitability following single and multiple impulses of visual callosal axons in the rabbit. *Exp. Neurol.*, **53**, 128–150.

Swadlow, H.A., Rosene, D.L. and Waxman, S.G. (1978) Characteristics of interhemispheric impulse conduction between prelunate gyri of the rhesus monkey. *Exp. Brain Res.*, **33**, 455–467.

Swadlow, H.A. and Weyand, T.G. (1981) Efferent systems of the rabbit visual cortex: laminar distribution of the cells of origin, axonal conduction velocities and identification of axonal branches. *J. Comp. Neurol.*, **203**, 799–822.

Swayze, V.W., Andreasen, N.C., Alliger, R.J., Yuh, W.T.C. and Ehrhardt, J.C. (1990) Subcortical and temporal structures in affective disorder and schizophrenia: a magnetic resonance imaging study. *Biol. Psychiatry*, **31**, 221–240.

Swisher, L. and Hirsh, T.J. (1972) Brain damage and the ordering of two temporally successive stimuli. *Neuropsychologia*, **10**, 137–152.

Tallal, P. and Newcombe, F. (1978) Impairment of auditory perception and language comprehension in dysphasia. *Brain Lang.*, **5**, 13–24.

Tallal, P. and Schwartz, J. (1980) Temporal processing, speech perception and hemispheric asymmetry. *Trends Neurosci.*, **3**, 309–311.

Tank, D.W. and Hopfield, J.J. (1987) Neural computation by concentrating information in time. *Proc. Natl. Acad. Sci. USA*, **84**, 1896–1900.

Taylor, A.T. and Hellige, J.B. (1987) Effects of retinal size on visual laterality. *Bull. Psychon. Sci.*, **25**, 444–446.

Taylor, H.G. and Heilman, K.M. (1982) Monoaural recall and right-ear advantage. *Brain Lang.*, **15**, 334–339.

Tegner, R. and Levander, M. (1991) Through a looking glass: a new technique to demonstrate directional hypokinesia in unilateral neglect. *Brain*, **114**, 1943–1951.

Terzian, H. (1964) Behavioural and EEG effects of intracarotid sodium amobarbital injection. *Acta Neurochirurgia*, **12**, 230–239.

Teszner, D., Tzavaras, A., Gruner, J. and Hecaen, H. (1972) L'asymmetrie droite-gauche du planum temporale a propos de l'etude anatomique des 100 cerveaux. *Rev. Neurol. (Paris.)*, **126**, 444–449.

Teuber, H.-L. (1962) Effects of brain wounds implicating right or left hemisphere in man: Hemisphere differences and hemisphere interactions in vision, audition and somesthesis. In: V.B. Mountcastle (ed.). *Interhemispheric relations and cerebral dominance*. Baltimore: Johns Hopkins Press, pp. 131–157.

Thatcher, R.W., Krause, P.J. and Hrybyk, M. (1986) Cortico-cortical associations and EEG coherence: a two-compartment model. *Electroencephalogr. Clin. Neurophysiol.*, **64**, 123–143.

Thompson, J.K. (1985) Right brain, left brain; left face, right face: hemisphericity and expression of facial emotion. *Cortex*, **21**, 281–299.

Thompson, J.R., Engelhart, J.E., Hasso, A.N. and Hinshaw, D.B.Jr. (1985) Normal frontal lobe grey matter-white matter CT volume ratio in children. *Neuroradiology*, **27**, 108–111.

Thorndike, E.L. (1898) Animal intelligence: An experimental study of the associative processes in animals. *Psychol. Rev. Monogr. Suppl.*, 2 (No. 8) 1–109.

Thorndike, E.L. (1911) *Animal intelligence*. New York: MacMillan.

Todor, J.I. and Doane, T. (1978) Handedness and hemispheric asymmetry in the control of movement. *J. Motor Behav.*, **10**, 295–300.

Todor, J.I. and Cisneros, J. (1985) Accomodation to increased accuracy demands by the right and left hands. *J. Motor Behav.*, **17**, 355–372.

Tognola, G. and Vignolo, L.A. (1980) Brain lesions associated with oral apraxia in stroke patients: a clinico-neuroradiological investigation with the CT scan. *Neuropsychologia*, **18**, 257–272.

Tompkins, C.A. and Flowers, C.R. (1985) Perception of emotional intonation by brain-damaged adults: the influence of task processing levels. *J. Speech Hear. Res.*, **28**, 527–538.

Tompkins, C.A. and Mateer, C.A. (1985) Right hemisphere appreciation of prosodic and linguistic indications of implicit attitude. *Brain Lang.*, **24**, 185–203.

Tompkins, C.A. and Flowers, C.R. (1987) Contextual mood priming following left and right hemisphere damage. *Brain Cogn.*, **6**, 361–376.

Tramo, M.J. and Bharucha, J.J. (1991) Musical priming by the right hemisphere post-callosotomy. *Neuropsychologia*, **29**, 313–325.

Tsumoto, T. (1992) Long-term potentiation and long-term depression in the neocortex. *Prog. Neurobiol.*, **39**, 209–228.

Tsunoda, T. and Oka, M. (1971) Cerebral hemisphere dominance test and localization of speech. *J. Audit. Res.*, **11**, 177–189.

Tsunoda, T. (1975) Functional differences between right- and left-cerebral hemispheres detected by the key-tapping method. *Brain Lang.*, **2**, 152–170

Tucker, D.M., Roth, R.S., Arneson, B.A. and Buckingham, V. (1977a) Right hemisphere activation during stress. *Neuropsychologia*, **15**, 697–700.

Tucker, D.M., Watson, R.T. and Heilman, K.M. (1977b) Discrimination and evocation of affectively intoned speech in patients with right parietal disease. *Neurology*, **27**, 947–950.

Tucker, D.M., Stenslie, C.E., Roth, R.S. and Shearer, S.L. (1981) Right frontal lobe activation and right hemisphere performance. Decrement during a depressed mood. *Arch. Gen. Psychiatry*, **38**, 169–174.

Tucker, D.M., Roth, D.L. and Blair, T.B. (1986) Functional coherence among cortical regions: topography of EEG coherence. *Electroencephalogr. Clin. Neurophysiol.*, **63**, 242–256.

Tuller, B. and Kelso, J.A.S. (1990) Phase transitions in speech production and their perceptual consequences. In: Jeannerod, M. (ed.), *Attention and Performance XIII.* Hillsdale, New Jersey, L. Erlbaum and Associates, pp. 430–452.

Umilta, C., Rizzolatti, G., Marzi, C.A., Zamboni, G., Franzoni, C. Carmada, R. and Berlucchi, G. (1974) Hemispheric differences in the discrimination of line orientation. *Neuropsychologia*, **12**, 165–174.

Uyehara, J. and Cooper, W. (1980) Hemispheric differences for verbal and non-verbal stimuli in Japanese- and English-speaking subjectss assessed to Tsunoda's method. *Brain Lang.,* **10**, 405–417.

Vaadia, E. and Abeles, M. (1987) Temporal firing patterns of single units, pairs and triplets of units in the auditory cortex. *Isr. J. Med. Sci.,* **23**, 75–83.

Vaid, J., Bellugi, U. and Poizner, H. (1989) Hand dominance for signing: clues to brain lateralization of language. *Neuropsychologia*, **27**, 949–960.

Valenstein, E., Heilman, K.M., Watson, R.T. and van den Abell, T. (1982) Non-sensory neglect from parieto-temporal lesions in monkeys. *Neurology*, **32**, 1198–1201.

Valler, G. and Perani, D. (1987) The anatomy of spatial neglect in humans. In: M. Jeannerod (ed.), *Neurophysiological and neuropsychological aspects of spatial neglect.* Amsterdam: Elsevier Science Publishers, North Holland.

van Lancker, D. and Sidtis, J.J. (1992) The identification of affective-prosodic stimuli by left- and right-hemisphere-damaged subjects: all errors are not created equal. *J. Speech Hear. Res.,* **35**, 963–970.

Venkatagiri, H.S. and Hinkle, W.G. (1975) A study of hemispheric asymmetry for auditory feedback control of speech. *J. Audit. Res.,* **15**, 289–294.

Verfaellie, M. and Heilman, K.M. (1990) Hemispheric asymmetry in attentional control: implications for hand preference in sensorimotor tasks. *Brain Cogn.,* **14**, 70–80.

Villa, A.E. and Abeles, M. (1990) Evidence for spatiotemporal firing patterns within the auditory thalamus of the cat. *Brain Res.,* **509**, 325–327.

Villardita, C., Grioli, S, and Quattropani, M.C. (1988) Concreteness/abstractness of stimulus-words and semantic clustering in right brain-damaged patients. *Cortex*, **24**, 563–571.

Viviani, P. and Terzuolo, C. (1982) Trajectory determines movement dynamics. *Neuroscience*, **7**, 431–437.

Virostek, S. and Cutting, J.E. (1979) Asymmetries for Ameslan handshapes and other forms in signers and non-signers. *Percept. Psychophysics*, **26**, 505–508.

von Bonin, G. (1962). Anatomical asymmetries of the cerebral hemispheres. In: Mountcastle, V.B. (ed.) *Interhemispheric relations and cerebral dominance.* Baltimore: Johns Hopkins Press, pp. 1–6.

von Economo, C and Horn, L. (1930) Ueber Windungsrelief, Masse und Rindenarchitecktonik der Supratemporalflasche, ihre individuellen und ihre Seitenunterschieden. *Z. Neurol. Psychiat.*, **130**, 687–757.

Voronin, L.L. (1983) Long-term potentiation in the hippocampus. *Neuroscience*, **10**, 1051–1069.

Wada, J. and Rasmussen, T. (1960) Intracarotid injection of sodium amytal for the lateralization of cerebral speech dominance. *J. Neurosurg.,* **17**, 266–282.

Wada, J.A., Clarke, R. and Hamm, A. (1975) Cerebral hemispheric asymmetry in humans. Cortical speech zones in 100 adult and 100 infant brains. *Arch. Neurol.,* **32**, 239–246.

Wapner, W., Hamby, S. and Gardner, H. (1981) The role of the right hemisphere in the apprehension of complex linguistic material. *Brain Lang.,* **14**, 15–33.

Watson, N.V. and Kimura, D. (1989) Right hand superiority for throwing but not for intercepting. *Neuropsychologia*, **27**, 1399–1414.

Watson, R.T., Heilman, K.M., Cauthen, J.C. and King, F.A. (1973) Neglect after congulectomy. *Neurology*, **23**, 1003–1007.

Watson, R.T., Valenstein, E. and Heilman, K.M. (1984) The effect of corpus callosum lesions on unilateral neglect in monkey. *Neurology*, **34**, 812–815.

Waxman, S.G. and Swadlow, H.A. (1976) Ultrastructure of visual callosal axons in the rabbit. *Exp. Neurol.,* **53**, 115–127.

Weinberger, D.R., Luchin, D.L., Morihisa, J. and Wyatt, R.J. (1982) Asymmetrical volumes of the right and left frontal and occipital regions of the human brain. *Ann. Neurol.,* **11**, 97–100.

Weintraub, S., Mesulam, M.-M. and Kramer, L. (1981) Disturbance in prosody. A right hemisphere contribution to language. *Arch. Neurol.,* **38**, 742–744.

Weis, S., Haug, H., Holoubek, B. and Orun, H. (1989) The cerebral dominances: quantitative morphology of the human cerebral cortex. *Int. J. Neurosci.,* **47**, 165–168.

Welch, K. and Stuteville, P. (1958) Experimental production of unilateral neglect in monkeys. *Brain*, **81**, 341–357.

Weniger, D. (1984) Dysprosody as part of the aphasic language disorder. *Adv. Neurol.,* **42**, 41–50.

Wenkstern, D., Pfaus, J.G. and Fibiger, H.C. (1993) Dopamine transmission increases in the nucleus accumbens of male rats during their first exposure to sexually receptive female rats. *Brain Res.,* **618**, 41–46.

Wexler, B.E. and Halwes, T. (1981) Right ear bias in the perception of loudness of pure tones. *Neuropsychologia*, **19**, 147–150.

Wexler, B.E., Schwartz, G., Warrenburg, S., Servis, M. and Tarlatzis, I. (1986) Effects of emotion on perceptual asymmetry: interactions with personality. *Neuropsychologia*, **24**, 699–710.

Wey, T.V., Cook, N.D., Landis, T. and Regard, M. (1993) Lateralized lexical decisions and the effects of hemifield masks: a study of interhemispheric inhibition and release. *Int. J. Neurosci.*, **71**, 37–44.

Weylman, S.T., Brownell, H.H., Roman, M. and Gardner, H. (1989) Appreciation of indirect requests by left- and right-brain damaged patients: the effects of verbal context and conventionality of wording. *Brain Lang.*, **36**, 580–591.

Wible, C.G., Shenton, M.E., Hokama, H., Kikinis, R., Jolesz, F.C., Metcalf, D. and McCarley, R.W. (1995) Prefrontal cortex and schizophrenia; a quantitiative magnetic resonance imaging study. *Arch. Gen. Psychiatry*, **52**, 279–288.

Wickens, J.R. (1988) Electrically coupled but chemically isolated synapses: dendritic spines and calcium in a rule for synaptic modification. *Prog. Neurobiol.*, **31**, 507–528.

Wickens, J.R. (1990) Striatal dopamine in motor activation and reward-mediated learning: steps towards a unifying model. *J. Neural Trans.*, **80**, 9–31.

Wickens, J.R. (1991) *A theory of the striatum.* Ph. D. thesis, Otago University Press (also published by Pergamon, Oxford, [Studies in Neuroscience, No. 7, 1993].)

Wickens, J.R., Alexander, M.E. and Miller, R. (1991) Two dynamic modes of striatal function under dopaminergic-cholinergic control: simulation and analysis of a model. *Synapse*, **8**, 1–12.

Wilbur, R.B. and Allen, G.D. (1991) Perceptual evidence against internal structure in American sign language syllables. *Lang. Speech* **34**, 27–46.

Wilkins, A. and Stewart, A. (1974) The time course of lateral asymmetries in viusal perception of letters. *J. Exp. Psychol.*, **102**, 905–908.

Wilkins, A.J., Shallice, T. and McCarthy, R. (1987) Frontal lesions and sustained attention. *Neuropsychologia*, **25**, 359–365.

Wilkinson, J.M. and Carr, T.H. (1987) Strategic hand use preference and hemispheric specialization in tactual reading: impact of the demands of perceptual coding. *Brain Lang.*, **32**, 97–123.

Williams, S.M. (1981) Immediate free recall of unilateral spoken word lists. *Cortex*, **17**, 465–470.

Winer, E. and Gardner, H. (1977) The comprehension of metaphor in brain-damaged patients. *Brain*, **100**, 716–727.

Wise, R., Chollet, F., Hadar, U., Friston, K., Hoffner, E. and Frackowiak, R. (1991) Distribution of cortical neural networks involved in word comprehension and word retrieval. *Brain*, **114**, 1803–1817.

Witelson, S.F. (1976) Sex and the single hemisphere: specialization of the right hemisphere for spatial processing. *Science*, **193**, 425–427.

Witelson, S.F. and Pallie, W. (1976) Left hemisphere specialization for language in the newborn. *Brain*, **96**, 641–646.

Wolff, P.H. and Hurwitz, I. (1976) Sex differences in finger tapping: a developmental study. *Neuropsychologia*, **14**, 35–41.

Wolff, P.H., Hurwitz, I. and Moss, H. (1977) Serial organization of motor skills in left and right-handed adults. *Neuropsychologia*, **15**, 539–546.

Worrall, N. and Coles, P. (1978) Visual field differences in recognizing letters. *Percept. Psychophysics*, **20**, 21–24.

Wright, J.J. and Sergejew, A. (1991) Radial coherence, wave velocity and damping of electrocortical waves. *Electroencephalogr. Clin. Neurophysiol.*, **79**, 403–412.

Wurst, S.A. and Long, G.M. (1983) Laterality and visual persistence: Still a two sided issue. *Percept. Psychophysics.*, **33**, 595–598.

Wyke, M. (1967) Effect of brain lesions on the rapidity of arm movement. *Neurology*, **17**, 1113–1120.

Wyke, M. (1968) The effect of brain lesions in the performance of an arm-hand task. *Neuropsychologia*, **6**, 125–134.

Yeni-Komshian, G.H. and Gordon, J.F. (1974) The effect of memory load on the right ear advantage in dichotic listening. *Brain Lang.*, **1**, 375–381.

Yeni-Komshian, G.H. and Benson, D.A. (1976) Anatomical study of cerebral asymmetry in the temporal lobe of humans, chimpanzees and rhesus monkeys. *Science*, **192**, 387–389.

Yeo, R.A., Turkheimer, E., Raz, N. and Bigler, E.D. (1987) Volumetric asymmetries of the human brain: intellectual correlates. *Brain Cogn.*, **6**, 15–23.

Yokoyama, K., Jennings, R., Ackles, P., Hood, P. and Boller, F. (1987) Lack of heart rate changes during an attention-demanding task after right hemisphere lesions. *Neurology*, **37**, 624–630.

Yoshii, N. and Hockaday, W.J. (1958) Conditioning of frequency-characteristic repetitive electroencephalographic responses with intermittent photic stimulation. *Electroencephalogr. Clin. Neurophysiol.*, **10**, 487–502.

Young, A.W. and Ellis, H.D. (1980) Ear asymmetry for the perception of monoaurally presented words accompanied by binaural white noise. *Neuropsychologia*, **18**, 107–110.

Zaidel, D. and Sperry, R.W. (1973) Performance on the Raven's coloured progressive matrices test by subjects with cerebral commissurotomy. *Cortex*, **9**, 34–39.

Zaidel, E. (1976) Auditory vocabulary of the right hemisphere following brain bisection or hemidecortication. *Cortex*, **12**, 191–211.

Zaidel, E. (1977) Unilateral auditory language comprehension on the Token test following cerebral commissurotomy and hemispherectomy. *Neuropsychologia*, **15**, 1–18.

Zaidel, E (1978) Lexical organization in the right hemisphere. In: P.A. Buser and A. Rogeul-Buser (eds). *Cerebral correlates of conscious experience*. Amsterdam: Elsevier, pp. 177–196.

Zarzecki, P. (1989) Influence of somatosensory cortex on different classes of motor cortex output neuron. *J. Neurophysiol.*, **62**, 487–494.

Zatorre, R.J. (1988) Pitch perception of complex tones and human temporal lobe function. *J. Acoust. Soc. Am.*, **84**, 566–572.

Zatorre, R.J. (1989) Perceptual asymmetry on the dichotic fused words test and cerebral speech lateralization determined by the carotid sodium amytal test. *Neuropsychologia*, **27**, 1207–1219.

Zatorre, R.J., Evans, A.C., Meyer, E. and Gjedde, A. (1992) Lateralization of phonetic and pitch discrimination in speech processing. *Science*, **256**, 846–849.

Ziegler, W. (1989) Anticipatory coarticulation in aphasia: More methodology: a reply to Sussman *et al.* (1988) and Katz (1988). *Brain Lang.*, **37**, 172–176.

Ziegler, W. and von Cramon, D. (1985) Anticipatory coarticulation in a patients with apraxia of speech. *Brain Lang.*, **26**, 117–130

Zurif, E.B. (1974) Auditory lateralization: prosodic and syntactic factors. *Brain Lang.*, **1**, 391–404.

Zurif, E.B. and Bryden, M.P. (1969) Familial handedness and left-right differences in auditory and visual perception. *Neuropsychologia*, **7**, 179–187.

Zurif, E.B., Caramazza, A. and Myerson, R. (1972) Grammatical judgments of agrammatic aphasics. *Neuropsychologia*, **10**, 405–417.

Zurif, E.B. and Blumstein, S.E. (1978) Language and the brain. In: M. Halle, J. Bresnan and G.A. Miller (eds). *Linguistic theory and psychological reality*. Cambridge (Mass.) and London: MIT Press, pp. 229–245.

Zwislocki, J. (1960) Theory of temporal auditory summation. *J. Acoust. Soc. Am.* **32**, 1046–1060.

Zwislocki, J. (1965) Analysis of some auditory characteristics. In: R.D. Luce, R.R. Bush and E. Galanter (eds). *Handbook of Mathematical Psychology, Volume III*, New York: John Wiley and Sons.

Zwislocki, J. (1969) Temporal summation of loudness: an analysis. *J. Acoust. Soc. Am.* **46**, 431–441.

Author Index

F = Figure; N = Footnote; T = Table
Includes first and second authors mentioned in text

Subject Index